STOCHASTIC OPTIMAL
LINEAR ESTIMATION
AND CONTROL

McGRAW-HILL SERIES IN ELECTRONIC SYSTEMS

JOHN G. TRUXAL and RONALD A. ROHRER   *Consulting Editors*

STOCHASTIC OPTIMAL LINEAR ESTIMATION AND CONTROL

J. S. MEDITCH

Boeing Scientific Research Laboratories
Seattle, Washington

McGRAW-HILL BOOK COMPANY

New York   St. Louis   San Francisco   London
Sydney   Toronto   Mexico   Panama

Library of Congress Catalog Card Number 69-17148
07-041230-8

456789   KPKP   798765

To Theresa Claire
James Stephen, Jr.
and Sandra Anne

Preface

Two important problems in technology and applied science involve the estimation and control of the behavior of physical processes, subject to random disturbances and random measurement errors, such that a performance measure is optimized. The estimation problem is that of obtaining an approximation to the time history of a process's behavior from noisy measurements. The approximation is typically chosen as one which is "best" in the sense of some error criterion. The control problem is that of determining inputs to a process in order to achieve desired goals such as maximum yield or minimum expenditure of fuel in spite of the random disturbances which are present. These problems fall within the domain of stochastic optimal estimation and control theory. They are related both because the mathematical techniques utilized in approaching one are also relevant in attacking the other and because estimation is usually a first step in implementing a control input; that is, it is necessary to infer a process's behavior before effective control can be applied.

This book is devoted to the linear theory of stochastic optimal estimation and control. Specifically, attention is focused upon optimal estimation and control for processes which can be modeled as discrete-time (sampled-data) or continuous-time linear dynamic systems. The disturbances and measurement errors are modeled as gaussian stochastic processes, and the random initial conditions are described by a gaussian probability distribution. The fundamental viewpoints which are adopted here are those of conditional expectation and minimum mean square error in estimation, and regulation in the sense of minimum mean square deviation from a set of desired operating conditions for control.

The linear theory which is developed, illustrated, and discussed in this book is being used to solve practical optimization problems. Such problems involve navigation, guidance, attitude control, and postflight data analysis for aircraft and spacecraft, and the control of large-scale production and chemical processes. Other applications include the processing of seismological and biological data. In addition to its applications, the linear theory provides a base for follow-up study of stochastic optimal estimation and control for nonlinear processes.

The purpose of this book is to make the basic results of optimal linear estimation and control theory available to a reasonably wide segment of the technological and scientific community. The intended audience includes the beginning graduate student in engineering or applied science who intends to specialize in this field and the practicing engineer or applied scientist who is a nonspecialist, but who is interested in a working knowledge of the fundamentals. In the former case, the book provides an introduction to stochastic optimal estimation and control, and in the latter case, it gives the basic tools for applications and the background for additional study. For the specialist, on the other hand, the book is intended as a reference.

In order to make stochastic optimal linear estimation and control theory available to as wide an audience as is practical, the presentation in this book is initiated at a modest level of requisite mathematical background. Some familiarity with matrix analysis, ordinary differential equations, and probability theory at an introductory level is presupposed on the part of the reader. Chapters 2, 3, and 4 present the necessary fundamental notions and concepts from the areas of linear system theory, probability theory, and the theory of stochastic processes, respectively. However, many topics within these areas are omitted, not because they are unimportant or insignificant, but because of limitations of space and the fact that they play no explicit role in the sequel.

The principal goal in the above three chapters is the formulation and discussion of the system models for the estimation and control problems which are studied in Chapters 5 to 10. The more advanced

reader, who has an intermediate level grasp of the areas of linear system
theory, probability theory, and the theory of stochastic processes, can
begin his study with Sections 4.3 and 4.4, where the models are presented.

The book's presentation is in three parts. The first part comprises
Chapters 1 to 4, where the material in Chapters 2 to 4 is as discussed
above. Chapter 1 presents a qualitative description of estimation and
control, discusses some practical applications, and gives an overview
of the book in greater detail than is presented in this preface.

The second part of the book, Chapters 5 to 8, covers the theory of
optimal estimation, and the third part, Chaps. 9 and 10, treats the theory
of optimal control.

The primary concern in Chapters 5 to 10 is the development of
algorithms for optimal estimation and control. Conspicuous by their
absence in this book are results which deal with the question of stability
for these algorithms. This may seem a possible shortcoming, but it is
not without some justification. Specifically, the question is a nontrivial
one and would have required considerably more space to treat adequately
for the particular audience of concern than is practical.

Throughout the book, principal results are summarized in the form
of theorems and corollaries. This is not only a convenient way to
emphasize the main points, but it also permits easy reference to them.

The examples in this book are purposely kept simple, so that concepts may be illustrated without undue involvement in algebraic and
computational details. However, some computational problems which
one can expect to encounter in practice are discussed qualitatively.

Problems are given at the end of each chapter with the exception of
Chapter 1. They range in difficulty from the very simple, which require
only direct application of the text material, to those of moderate difficulty, wherein text results are to be extended. In the latter case, the
results of the extension are usually indicated or given explicitly in the
problem statement.

References on the subject matter in each chapter are cited at the
end of that chapter. In giving these references, every attempt has been
made to be objective and fair to the researchers upon whose work the
material is based.

Preliminary work on the manuscript for this book was initiated in
the fall of 1964 in connection with a professional development course
which was given while the author was affiliated with the Aerospace
Corporation, Los Angeles, California. Additional work was carried out
at Northwestern University, Evanston, Illinois, where much of the
material in this book was presented in an introductory graduate course on
estimation and control in the Department of Electrical Engineering
during the 1965 to 1966 and 1966 to 1967 academic years. The author

expresses his gratitude to both the Aerospace Corporation and Northwestern University for their support.

Special appreciation is extended to the Boeing Scientific Research Laboratories, Seattle, Washington, for support during preparation of the final manuscript. The patience, diligence, and perserverance of Miss Karen Harles of the Boeing Scientific Research Laboratories in typing the final manuscript is most sincerely acknowledged and appreciated.

Finally, but by no means last, the author is in the debt of his colleagues and students, whose helpful suggestions and constructive criticisms have been of immeasurable aid in the writing of this book.

J. S. Meditch

Contents

1
Introduction

1.1 PRELIMINARY REMARKS

Our work in this book will begin with a qualitative description of the nature of estimation and control and a discussion of certain aspects of the class of problems with which we shall be concerned. This we do in Sec. 1.2.

In Sec. 1.3, we motivate our work by presenting and discussing some problems which arise in technology and in which questions of estimation and control are of fundamental importance. Here also, our treatment is qualitative.

With this preliminary background, we then proceed to give an outline of the book in Sec. 1.4 to indicate the particular path which our work will follow.

Many of the terms which are introduced in this chapter in a qualitative way will be made precise in later chapters. For the present, the intent is that of developing a general, intuitive understanding of the class of problems of interest.

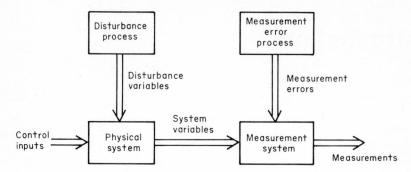

Fig. 1.1   Block diagram for description of general problem of estimation and control.

1.2 ESTIMATION AND CONTROL

Our concern in this book is with a class of problems within the framework of a general problem of estimation and control which involves dynamic systems.   The general problem can be readily described with the aid of the block diagram in Fig. 1.1.

In the figure, a physical system is shown as subject to two sets of input variables†: a control input which can usually be manipulated as desired and a disturbance input which reflects the presence of internal and/or external phenomena which cannot be manipulated, i.e., those phenomena which are inherent in the system and its environment, such as noise in electronic circuits, signal interference due to stray radiation, and turbulence in aircraft flight which is caused by random wind gusts.

The system's behavior or response‡ is then observed with the aid of some suitable collection of sensors termed the *measurement system*.   The sensors are, of course, subject to random and systematic instrument and phenomenon errors.   For example, a horizon scanner is used on some satellites to determine the direction of the local vertical by bisecting the angle formed by "limbs" from the sensor to the horizon.   However, the instrument gives erroneous results not only because of internal electronic noise but also because irregularities in the earth's atmosphere do not permit a sharp definition of the horizon.   The situation is depicted in the simplified two-dimensional model in Fig. 1.2.

In general then, measurements yield only crude information about a system's behavior and, in themselves, may be unsatisfactory for assessing the system's performance.

† Double lines are used in the figure to denote that "signal and energy flow" involves, in general, more than one variable.

‡ More precisely, we are speaking here of the *system's state*.   This notion is treated in Chap. 2.   For the present, it is sufficient to think in terms of the system's "behavior" or "response."

Basically, the estimation problem consists in determining an approximation to the time history of the system's response variables from the erroneous measurements. If a *performance measure* is introduced to assess the quality of the approximation or *estimate*, and the estimate is to be chosen so that this measure is either minimized or maximized (whichever is desired), the problem is termed one of *optimal estimation*. Implicit here is the development of an algorithm for processing the measurements. The problem is, of course, far from completely formulated at this point, but the general idea should be clear.

On the other hand, the control problem is that of specifying a manner in which the control input should be manipulated to force the system to behave in some desired fashion. If, as in the estimation problem, a performance measure is introduced to evaluate the quality of the system's behavior, and the control input is to be specified to minimize or maximize this measure, the problem is one of *optimal control*. Also, as in the estimation problem, there is implicit here the derivation of an algorithm (for control).

In approaching this latter problem, one is motivated to separate it into two problems: first, that of estimating the system's response, and second, that of specifying a control algorithm which utilizes the estimates. This separation is intuitively appealing, since one feels that the system's behavior must be determined before a control input which is to modify it can be specified. However, it has not been shown that this separation is mathematically valid in all cases.

If, in the optimal estimation and control problems, the disturbance and measurement error processes are modeled as random or stochastic phenomena, the adjective stochastic is used in the problem description; e.g., we might use the term *stochastic optimal control problem*. Further, if the models for the physical system and its associated measurement system are linear, then we use the term *stochastic optimal linear estimation and control problems*, which describes the class of problems of interest in this book.

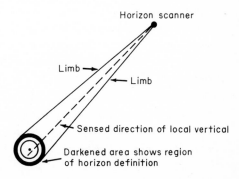

Horizon scanner

Limb →

← Limb

← Sensed direction of local vertical

Darkened area shows region
← of horizon definition

Fig. 1.2   Horizon scanner geometry.

We conclude this section by outlining the procedure which we shall follow in pursuing the above class of problems.

1. Development of models.   This, of course, involves specification of the models for (a) the physical system, (b) the measurement system, (c) the disturbance process, and (d) the measurement error process. As in any formulation of mathematical models for physical processes, the task is one of developing models which are sufficiently complete to represent the physical phenomena of interest but, on the other hand, not so complex as to be intractable for purposes of analysis and computation.
2. Specification of performance measures.   This step really amounts to a definition of goals.   As in the step above, the performance measures should be chosen to be realistic in terms of the physical problem of concern, yet tractable mathematically.
3. Problem formulation.   Here, the information in the above two steps is combined, along with a statement of any constraints that are to be imposed, to define the problem.
4. Development of estimation and control algorithms.   The task in this step is clear.   However, one does not stop at this point.   Indeed, it is essential that the results which are obtained be examined in practical applications.   Of primary concern is the computational complexity of the algorithms.

1.3 APPLICATION AREAS

COMMUNICATION SYSTEMS

An important problem in communication theory is that of processing a received signal to extract the "message."   The general scheme is shown in Fig. 1.3.

The transmitted signal is composed not only of the message but also of errors in coding and transmission.   This signal is further corrupted by disturbances in the communication channel such as atmospheric noise. At the receiving end then, the task is that of estimating what message is being or was sent.

NAVIGATION

The navigation problem is that of determining position and velocity of a vehicle in some suitable coordinate system by utilizing data from "navigational fixes" such as range, range rate, and angular measurements. The importance of navigation in the flight of aircraft and spacecraft, as well as in the movement of surface ships and submarines, is obvious.

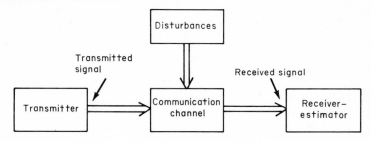

Fig. 1.3 Schematic for communication system.

Since navigation fixes are subject to inherent instrument and phenomena errors, it is usually necessary to have available an estimation algorithm to process these measurements to obtain useful information.

Going one step further, the position and velocity estimates may be utilized to effect a change in course to achieve a specified goal, e.g., an orbit transfer to achieve a rendezvous in space.

POSTEXPERIMENTAL DATA ANALYSIS

Following the completion of an experiment, it is typically desired to reduce the data which were taken during the experiment to assess the experiment.   The problem is one of estimation, since the measurements are subject to errors.

For example, tracking and telemetry data taken during a space vehicle launch, orbit injection, and subsequent orbital flight are processed to obtain estimates of the actual flight path, the guidance system error parameters (gyro drift, accelerometer bias and scale factor errors, onboard guidance computer errors, etc.), and the accuracy of orbit injection. Such information is useful to evaluate the mission relative to the desired goals, to detect weaknesses in the system, to determine estimates of the parameters in the various subsystem models, and to provide guidelines in the consideration of future missions.   In this example, estimation is termed postflight reconstruction.

In a similar vein, measurements taken during the operation of a chemical process can be processed after shutdown to determine such parameters as reaction rates and initial concentrations of the reacting substances.

PROCESS CONTROL

Successful operation of large-scale processes or systems typically requires regulation or control for purposes of efficiency of operation, quality of output products, and/or the attainment of other specified goals.   This is typically accomplished by processing sensed data during operation to

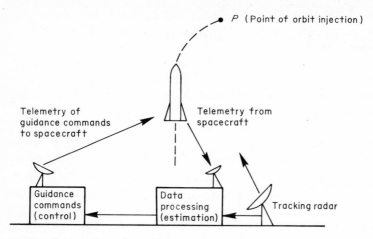

Fig. 1.4 Scheme for radio guidance.

estimate the process's current "state" and supplying this information to a control algorithm which then effects a control "policy" to offset internal and external disturbances and to achieve the desired operating conditions. The task is obviously that of *real-time* estimation and control.

Specific examples include machine tool control, aircraft and space-craft flight control, and multistage chemical processes such as catalytic cracking.

A scheme for radio guidance for a spacecraft launch where the goal is to inject the satellite onto a specified orbit at a given point P is shown in Fig. 1.4. Both telemetry data from the spacecraft and radar tracking data (e.g., range, range rate, and elevation and azimuth angle measurements) are processed to estimate the spacecraft's "state." This estimate is then supplied to an algorithm which generates guidance commands (e.g., steering signals and engine cutoff signals), which are, in turn, telemetered to the spacecraft, where they are executed. The feedback nature of the scheme is apparent, as is the need for considerable computation.

Relevant performance measures here are (1) estimation of the vehicle's "state" in such a way as to minimize the effects of errors in telemetry and tracking data and (2) specification of a control algorithm to minimize some function of the position and velocity errors at injection.

1.4 OUTLINE

Chapters 2, 3, and 4 provide background material from the areas of linear system theory, probability theory, and the theory of stochastic processes, respectively. The principal result here is the development of the two classes of system models in Secs. 4.3 and 4.4, for which estimation and control problems are formulated and solved in the remainder of the book. As remarked in the preface, the reader who has an intermediate level grasp of the above three areas can readily begin with these two sections after a perusal of the background material to gain some acquaintance with the notation and terminology.

Chapter 5 begins with the formulation of a specific optimal estimation problem for discrete-time systems and the development of three basic theorems and a corollary for the problem. In this formulation, estimates are classified into three categories in terms of the time point at which the estimate is desired relative to the time span over which measurements are available (see Table 1.1).

Attention is then focused on the development of the algorithms for optimal prediction and filtering (Kalman filter) for the first of the two classes of system models, the so-called Gauss-Markov sequence model (discrete-time).

The question of optimal smoothing is examined in Chap. 6 for the same class of system models. It is found convenient there to further subdivide the smoothing category into three cases and to develop the algorithm for optimal smoothing for each case separately.

In Chap. 7, the estimation problem for the second of the two classes of system models, the Gauss-Markov process model (continuous-time), is treated. The algorithms for optimal prediction, filtering, and the three smoothing cases are developed by application of a formal limiting procedure to the results of Chap. 6.

An alternate approach to the estimation problem for the Gauss-Markov process model is given in Chap. 8. The approach supplements the one in Chap. 7 by formulating and solving the problem entirely in the continuous-time domain. The central result is the time-varying Wiener-

Table 1.1 Classification of estimates

Classification	Measurement data interval	Estimate time
Prediction	$t_0 \leq \tau \leq t_1$	Any or all $t > t_1$
Filtering	$t_0 \leq \tau \leq t_1$	$t = t_1$
Smoothing (interpolation)	$t_0 \leq \tau \leq t_1$	Any or all t for which $t_0 \leq t < t_1$

t_0 is a fixed time point.
t, τ denote variable time.
t_1 is either a fixed or variable time point.

Hopf integral equation, which is developed using the classical calculus of variations. Since the algorithms for optimal prediction, filtering, and smoothing are already known at this point from Chap. 7, the Wiener-Hopf integral equation is solved only for the optimal filtering problem (Kalman-Bucy filter) and one of the optimal smoothing cases to illustrate the procedure.

Chapters 9 and 10 present the formulation and solution of the so-called discrete and continuous stochastic linear regulator problems, respectively. Dynamic programming is introduced and applied to develop a separation principle whereby the complete control system is specified as the "cascade" of the optimal filter with a time-varying feedback gain, each of which is determined by independent algorithms.

References upon which the material in each chapter is based are given at the end of each chapter. A list of some general references in the area of stochastic estimation and control theory is given below.

GENERAL REFERENCES

1. Aoki, M., "Optimization of Stochastic Systems: Topics in Discrete-Time Systems," Academic Press Inc., New York, 1967.
2. Bryson, A. E., Jr., and Y. C. Ho, "Optimization, Estimation, and Control," Blaisdell Publishing Company, Inc., Waltham, Mass., 1968.
3. Bucy, R. S., and P. D. Joseph, "Filtering for Stochastic Processes and Applications to Guidance," Interscience Publishers, Inc. (John Wiley & Sons, Inc.), New York, 1968.
4. Deutsch, R., "Estimation Theory," Prentice-Hall, Inc., Englewood Cliffs, N.J., 1965.
5. Kushner, H. J., "Stochastic Stability and Control," Academic Press Inc., New York, 1967.
6. Lee, R. C. K., "Optimal Estimation, Identification, and Control," M.I.T. Press, Cambridge, Mass., 1964.
7. Liebelt, P. B., "An Introduction to Optimal Estimation," Addison-Wesley Publishing Company, Inc., Reading, Mass., 1967.
8. Sage, A. P., "Optimal Systems Control," Prentice-Hall, Inc., Englewood Cliffs, N.J., 1968.
9. Tou, J. T., "Optimal Design of Digital Control Systems," Academic Press Inc., New York, 1963.
10. Wonham, W. M., "Lecture Notes on Stochastic Control—Part I," Lecture Notes 67-2, Division of Applied Mathematics, Brown University, Providence, R.I., February 1967.

2
Elements of Linear
System Theory

2.1 INTRODUCTION

The purpose of this chapter is twofold: (1) to provide a general introduction to the two classes of systems models with which we shall be concerned in this book, and (2) to examine and illustrate certain fundamental properties of these models.†

To achieve the present purpose, we begin with a review of certain selected results from matrix analysis in Sec. 2.2. This also serves to introduce much of the notation which is employed throughout the book. We then proceed in Sec. 2.3 to present the first of the two classes of system models and to illustrate its importance via some physically motivated examples. In this section, we also examine certain characteristics of the class of models by introducing and utilizing certain concepts from the theory of ordinary linear differential equations.

† We defer to Chap. 4 the detailed formulation of the models, since this requires some material from probability and stochastic processes which is given in Chap. 3 and the first part of Chap. 4.

The second class of system models is then introduced and examined in Sec. 2.4.

In Sec. 2.5, we present a qualitative discussion of the concepts of observability and controllability which are motivated by questions of estimation and control, respectively.

These two concepts are then examined in some detail in Secs. 2.6 and 2.7, respectively, for the present two classes of system models. Necessary and sufficient conditions for observability and controllability are developed to give the reader some appreciation of the underlying nature of estimation and control, and conditions under which it is meaningful to pursue estimation and control problems are described.

We conclude this chapter in Sec. 2.8 with an examination of situations in which the system models of Secs. 2.3 and 2.4 can be used to model the behavior of nonlinear systems about assumed nominal conditions.

The material in this chapter constitutes part of what is generally called linear system theory. A detailed treatment of this subject is given, for example, in DeRusso et al. [1]. A more advanced exposition may be found in Zadeh and Desoer [2].

For present purposes, the reader who wishes supplementary material on matrix analysis and ordinary linear differential equations will find helpful, for example, the appendix to chapter 6 of Kaplan [3], chapters 2, 3, 4, 6, and 10 of Bellman [4], and chapter 1 of Hildebrand [5]. An advanced treatment of the theory of ordinary differential equations is given in chapters 1 to 3 of Coddington and Levinson [6].

2.2 NOTATION AND MATHEMATICAL PRELIMINARIES

We remark initially that all variables with which we deal in this book are real unless specifically stated otherwise. Similarly, all functions are real-valued, except as noted.

MATRICES AND VECTORS

In general, we shall utilize capital Roman and Greek letters such as A, B, Γ, Θ, etc., to denote matrices and their lowercase equivalents, with suitable subscript indices, to denote the matrices' elements. In referring to a matrix, we shall usually indicate the number of rows and columns (in that order) which are involved. Thus, when we speak of the $m \times n$ matrix A, we mean

$$A = \begin{bmatrix} a_{11} & a_{12} & \cdots & a_{1n} \\ a_{21} & a_{22} & \cdots & a_{2n} \\ \cdots & \cdots & \cdots & \cdots \\ a_{m1} & a_{m2} & \cdots & a_{mn} \end{bmatrix}$$

This notation is sometimes abbreviated to $A = [a_{ij}]$. The indices i and j refer to the row and column, respectively, in which the element is located.

A matrix is *rectangular* if $m \neq n$, and *square* if $m = n$.

Two matrices A and B are said to be *equal* if and only if both are $m \times n$ and $a_{ij} = b_{ij}$ for all i and j.

Any $m \times n$ matrix all of whose elements are zero is called a *null matrix* and is usually denoted by 0.

If all the elements of a square matrix except those along the *principal diagonal* are zero, i.e.,

$$A = \begin{bmatrix} a_{11} & 0 & \cdots & 0 \\ 0 & a_{22} & \cdots & 0 \\ \cdots & \cdots & \cdots & \cdots \\ 0 & 0 & \cdots & a_{nn} \end{bmatrix}$$

the matrix is *diagonal*. If, further, $a_{ii} = 1$ for $i = 1, \ldots, n$, the matrix is called the *identity matrix*, and is denoted by I. This matrix is sometimes written as $I = [\delta_{ij}]$, where δ_{ij} is the *Kronecker delta*,

$$\delta_{ij} = \begin{cases} 0 & i \neq j \\ 1 & i = j \end{cases}$$

We shall use lowercase Roman and Greek letters x, y, z, ζ, ξ, etc., to denote $m \times 1$ (i.e., single-column) matrices having elements x_i, y_i, etc. Such matrices are called *vectors*. When the number of elements is to be specified, we use the terms m *vector*, p *vector*, etc., or m-dimensional vector, p-dimensional vector, etc. For example, the p vector ξ is

$$\xi = \begin{bmatrix} \xi_1 \\ \cdot \\ \cdot \\ \cdot \\ \xi_p \end{bmatrix}$$

When speaking of a single-row matrix, i.e., a $1 \times m$ matrix, we shall specifically use the term *row vector*. Thus, by the term n-*dimensional row vector* x, we mean

$$x = [x_1 \cdots x_n]$$

In either case, it is often convenient to think of the elements of a vector (column or row) as defining a point x in an n-dimensional euclidean space, and, subsequently, in terms of a directed line segment joining the origin of the space with the point, i.e., a vector. This geometric picture then leads to the names *coordinates* and *components*, respectively, for the elements of a column (row) vector.

Analogous to the definition of equality for two matrices, we say that two-column (row) n vectors x and y are *equal* if and only if $x_i = y_i$ for all $i = 1, \ldots, n$.

When some or all of the elements of a matrix or vector are functions of time, this is denoted by writing $A(t)$, $B(\tau)$, $x(t)$, $y(\tau)$, etc., where t and τ commonly denote continuous time.   Likewise, we write $A(i)$, $B(j)$, $x(k)$, $y(l)$, etc., where i, j, k, and l are integers denoting discrete-time instants. In the latter case, an integer index set such as $i = 0, 1, \ldots, N$ is used to denote an ordered discrete-time index set such as $t_0 < t_1 < \cdots < t_N$, where it is generally understood that the time points are not necessarily uniformly spaced.

If all the elements of a matrix or vector are continuous functions of (continuous) time, we say that the matrix or vector, respectively, is continuous.

We use the symbols d/dt and $\cdot$ (dot) to indicate the operation of the time derivative.   If each element of a matrix is differentiable with respect to time, the time derivative of $A(t)$ is defined as

$$\dot{A}(t) = [\dot{a}_{ij}(t)]$$

Similarly, for a vector,

$$\frac{dx}{dt} = \begin{bmatrix} \dfrac{dx_1}{dt} \\ \cdot \\ \cdot \\ \cdot \\ \dfrac{dx_n}{dt} \end{bmatrix}$$

In such cases, the matrix or vector is said to be differentiable with respect to time, or more simply, differentiable.

Time integrals of matrices and vectors are defined in the same way. Also, the definitions are trivially carried over to cases where the independent variable is something other than time.

Vector-matrix notation is often employed to express scalar-, vector-, and matrix-valued functions of many variables.   For example, the set of functions

$$y_1 = f_1(x_1, \ldots, x_n; u_1, \ldots, u_r; t)$$
$$\cdot \cdot \cdot \cdot \cdot \cdot \cdot \cdot \cdot \cdot \cdot \cdot \cdot \cdot \cdot \cdot \cdot \cdot \cdot \cdot \cdot \cdot \cdot \cdot \cdot \cdot \cdot \cdot$$
$$y_m = f_m(x_1, \ldots, x_n; u_1, \ldots, u_r; t)$$

can be written more compactly

$$y = f(x, u, t)$$

where y is an m vector, x an n vector, u an r vector, t a scalar (time), and f an m-dimensional vector-valued function of the $n + r + 1$ variables $x_1, \ldots, x_n; u_1, \ldots, u_r$ and t. The simplicity in this notation is evident.

This notation can obviously be used to denote systems of ordinary differential equations. For example,

$$\dot{x} = g(x,u,t)$$

where x, u, and t are the same as above and g is an n-dimensional vector-valued function of the indicated variables, denotes the system

$$\dot{x}_1 = g_1(x_1, \ldots, x_n; u_1, \ldots, u_r; t)$$
$$\cdots \cdots \cdots \cdots \cdots \cdots \cdots \cdots$$
$$\dot{x}_n = g_n(x_1, \ldots, x_n; u_1, \ldots, u_r; t)$$

TRANSPOSE AND SYMMETRY

The operation of interchanging the role of rows and columns in matrices is termed that of taking the *transpose* and is denoted by [ ]′, the prime. Thus, if $A = [a_{ij}]$ is an $m \times n$ matrix and x is an n vector, then

$$A' = [a_{ji}]$$

is an $n \times m$ matrix, where $j = 1, \ldots, n$ and $i = 1, \ldots, m$, while

$$x' = [x_1 \cdots x_n]$$

is an n-dimensional row vector. In the sequel, row vectors will always be denoted with a prime.

A square matrix A for which $A' = A$ is said to be *symmetric* and is characterized by the property $a_{ij} = a_{ji}, i, j = 1, \ldots, n$. A diagonal matrix is obviously symmetric.

It is clear that the operations of differentiation and transpose commute; e.g.,

$$\left(\frac{dx}{dt}\right)' = \frac{d}{dt}(x')$$

Similarly, for integration,

$$\left[\int_{t_0}^{t_1} A(t)\, dt\right]' = \int_{t_0}^{t_1} A'(t)\, dt$$

SUMS AND PRODUCTS

The sum and difference of two $m \times n$ matrices is defined in a natural way by the expression

$$C = A \pm B = [a_{ij} \pm b_{ij}]$$

Similarly, for two n vectors x and y, we have

$$z = x \pm y = \begin{bmatrix} x_1 \pm y_1 \\ \cdot \\ \cdot \\ \cdot \\ x_n \pm y_n \end{bmatrix}$$

In general, we shall also use lowercase Roman and Greek letters to denote scalars. If time dependence is involved, we note this by writing $\alpha(t)$, $\beta(k)$, etc.

Multiplication of a matrix or a vector by a scalar is then defined by relations such as

$$B = \alpha A = A\alpha = [\alpha a_{ij}]$$

and

$$y(t) = \beta(t)x(t) = x(t)\beta(t) = \begin{bmatrix} \beta(t)x_1(t) \\ \cdot \\ \cdot \\ \cdot \\ \beta(t)x_n(t) \end{bmatrix}$$

respectively.

The product of an $m \times n$ matrix A and an $n \times r$ matrix B is defined to be the $m \times r$ matrix

$$C = AB = [a_{ik}][b_{kj}] = \left[\sum_{k=1}^{n} a_{ik}b_{kj} \right] = [c_{ij}]$$

where $i = 1, \ldots, m$, and $j = 1, \ldots, r$. The product is, of course, defined if and only if A has as many columns as B has rows.

The terms *premultiplication* and *postmultiplication* are often used in referring to this product. In the product AB, we say that B is premultiplied by A or that A is postmultiplied by B.

If A and B are both $n \times n$, we note that $AB \neq BA$ in general. That is, the operation is not commutative, and attention must be paid to the order in which multiplication is carried out.

The n-fold product of a square matrix A with itself is denoted by A^n.

If A is an $n \times n$ matrix, I the $n \times n$ identity matrix, and 0 the $n \times n$ null matrix, it is clear that

$$AI = IA = A$$

and

$$A0 = 0A = 0$$

Of considerable importance in matrix analysis is the product Ax, where A is an $m \times n$ matrix and x is an n vector. This product is defined to be an m vector z, whose components are given by the expression

$$z_i = \sum_{j=1}^{n} a_{ij}x_j$$

where $i = 1, \ldots, m$. The appropriate shorthand notation is simply $z = Ax$. There are two interpretations associated with this product which we shall take up later. For the present, it suffices to say that A is a *linear transformation* on x.

It is easy to show from the above definitions of sums and products that

$$C' = A' \pm B' \qquad\qquad z' = x' \pm y'$$

$$(\alpha A)' = \alpha A' = A'\alpha \qquad [\beta(t)x(t)]' = \beta(t)x'(t) = x'(t)\beta(t)$$

$$(AB)' = B'A' \qquad\qquad (Ax)' = x'A'$$

The usual chain rules for differentiation of sums and products involving matrices and vectors carry over from ordinary calculus subject to the fact that multiplications are not commutative in general. For example,

$$\frac{d}{dt}[A(t)B(t)] = A(t)\dot{B}(t) + \dot{A}(t)B(t)$$

$$\frac{d}{dt}[A(t)x(t)] = A(t)\dot{x}(t) + \dot{A}(t)x(t)$$

$$\frac{d}{dt}\{[A(t)B(t)]'\} = \frac{d}{dt}[B'(t)A'(t)]$$

$$= B'(t)\dot{A}'(t) + \dot{B}'(t)A'(t)$$

$$= \left\{\frac{d}{dt}[A(t)B(t)]\right\}'$$

We note here that the operations of transpose and derivative commute.

The *inner product*, also called the *scalar* or *dot* product, of two n vectors is defined as

$$x'y = \sum_{i=1}^{n} x_i y_i$$

This product is obviously a scalar, and it follows that

$$x'y = y'x = (x'y)' = (y'x)'$$

If $x'y = 0$, the two vectors are said to be *orthogonal*.

If $x = y$, then

$$x'x = \sum_{i=1}^{n} x_i^2$$

Clearly, $x'x \geq 0$. In addition, if we once more picture x as defining a vector in n-dimensional euclidean space, $x'x$ is the square of the length of this vector. Its square root $(x'x)^{1/2}$ is, of course, the vector's length. The latter is often denoted by $\|x\|$ and is called the *euclidean norm*.

For x and y two n vectors, it is seen that $\|x - y\|$ is the distance between the two points whose coordinates are defined by x and y, and $(x - y)'(x - y)$ is the square of this distance.

The *outer product* of an m vector x and an n vector y is defined to be the $m \times n$ matrix

$$xy' = \begin{bmatrix} x_1y_1 & x_1y_2 & \cdots & x_1y_n \\ x_2y_1 & x_2y_2 & \cdots & x_2y_n \\ \cdot & \cdot & \cdots & \cdot \\ x_my_1 & x_my_2 & \cdots & x_my_n \end{bmatrix}$$

It is obvious that xx' is a symmetric matrix.

DETERMINANT AND MATRIX INVERSE

We recall from elementary algebra that there is associated with each square $n \times n$ matrix A a unique number which is termed the *determinant* of A. It is denoted by det A or $|A|$, and is commonly evaluated using the relation

$$|A| = \sum_{j=1}^{n} a_{ij}\gamma_{ij}$$

for any fixed i, $i = 1, \ldots, n$, with

$$\gamma_{ij} = (-1)^{i+j}\mu_{ij}$$

where μ_{ij} is the determinant of the $(n - 1) \times (n - 1)$ matrix obtained by deleting the row and column in A in which a_{ij} is located. The scalars γ_{ij} and μ_{ij} are called, respectively, the *cofactor* and *minor* of a_{ij}. The $n \times n$ matrix $[\gamma_{ji}]$, that is, the transpose of the matrix $[\gamma_{ij}]$, is termed the *adjoint* of A, and is denoted adj A.

We also recall from elementary algebra the following properties of $|A|$:

1. If all the elements of any row or column of A are zero, $|A| = 0$.
2. $|A| = |A'|$
3. If the corresponding elements of two rows or two columns of A are equal or are multiples of each other, $|A| = 0$.
4. If α times the elements of any row (column) are added to the corresponding elements of any other row (column) in A, the value of the determinant is unchanged.
5. For two $n \times n$ matrices A and B, $|AB| = |A| \cdot |B|$.

Let us now reconsider the expansion for $|A|$. If we replace the elements a_{ij} by a_{kj}, $i \neq k$, the result is the same as making the two rows equal or zero. Hence, by property 1 or 3, respectively,

$$\sum_{j=1}^{n} a_{kj}\gamma_{ij} = 0$$

for any fixed i and k, $i \neq k$.

Both expansions may now be combined and written

$$\sum_{j=1}^{n} a_{kj}\gamma_{ij} = |A|\delta_{ki}$$

From this expression and the definition of adj A, it follows immediately that

$$A \text{ adj } A = |A| \cdot I$$

Dividing through by $|A|$, we obtain

$$A \frac{\text{adj } A}{|A|} = I$$

under the assumption that $|A| \neq 0$. We observe that the matrix

$$\frac{\text{adj } A}{|A|}$$

has the property that when A is postmultiplied by it, the result is the identity matrix. We call this new matrix the *inverse* of A and denote it by A^{-1}. Then, $AA^{-1} = I$. Similarly, we also have that $A^{-1}A = I$.

With the aid of the matrix inverse, we have now a natural definition for the zeroth power of a matrix. Namely, $A^{-1}A = A^0$, so that $A^0 = I$ for $|A| \neq 0$.

If $|A| = 0$, the above formulation of the matrix inverse is not meaningful, and we say that A is *singular*. If A is nonsingular, it can be shown that A^{-1} is unique.

We leave as an exercise the task of showing that if A is nonsingular, $(A')^{-1} = (A^{-1})'$, that is, that the operations of transpose and inverse commute.

Now let A and B be two nonsingular $n \times n$ matrices. Assume also that AB is nonsingular and define

$$C = AB$$

Then it is clear that $A^{-1}C = B$ and subsequently that

$$B^{-1}A^{-1}C = I$$

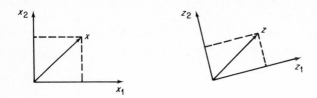

Fig. 2.1 Representation of the same vector in two different coordinate systems.

Hence, $B^{-1}A^{-1} = C^{-1}$, and we see that

$$(AB)^{-1} = B^{-1}A^{-1}$$

We remark here that a square matrix for which $A^{-1} = A'$ is called an *orthogonal matrix*. For such a matrix, it is obvious that $A'A = I$.

We present now the two interpretations associated with a linear transformation. Given an n vector x, and an $m \times n$ matrix A, we consider the relation $z = Ax$. Suppose that $m = n$. Then one possible interpretation is that z and x are one in the same vector, but the two coordinate systems in which they are expressed are different. This is illustrated in Fig. 2.1 for a simple two-dimensional case.

Since there are two distinct coordinate systems involved here, it is desirable that we be able to pass from one to the other. In one direction, we have $z = Ax$. In order to "return," we need $x = A^{-1}z$, which means that A must be nonsingular in this first interpretation.

In the second interpretation, we envision both vectors as lying in the same coordinate system. Here, z is obtained by a rotation and a scaling of x. This is also readily illustrated with the aid of a two-dimensional sketch (Fig. 2.2).

Since only one coordinate system is involved in this case, it is not necessary that A be nonsingular.

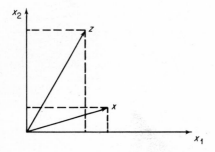

Fig. 2.2 Representation of a linear transformation as a rotation and scaling of a vector.

For those cases where $m \neq n$, we say that the transformation is a *projection*. For example, for $A = [0 \quad 1]$ and x a 2 vector,

$$z = Ax = [0 \quad 1]\begin{bmatrix} x_1 \\ x_2 \end{bmatrix} = x_2$$

A significant property of a nonsingular matrix, which is also a differentiable function of time, is that

$$\frac{d}{dt}(A^{-1}) \neq \left(\frac{dA}{dt}\right)^{-1}$$

That is, the operations of differentiation and inverse are not commutative, in general. To see this, we note that

$$A^{-1}(t)A(t) = I$$

Then

$$\left[\frac{d}{dt}(A^{-1})\right]A + A^{-1}\frac{dA}{dt} = 0$$

so that

$$\frac{d}{dt}(A^{-1}) = -A^{-1}\dot{A}A^{-1}$$

LINEAR INDEPENDENCE AND RANK

A set of n vectors, $x^1, \ldots, x^k$, is said to be *linearly dependent* if there exists a set of constants $\alpha_1, \ldots, \alpha_k$, at least one of which is nonzero, such that

$$\sum_{i=1}^{k} \alpha_i x^i = 0$$

If there does not exist such a set of constants, the vectors are *linearly independent*.

The *rank* of any $m \times n$ matrix is the order of the largest square array in the matrix, formed by deleting rows and/or columns, that is nonsingular.

By virtue of properties 3 and 4 for determinants (see page 16), it is clear that the rank of a matrix is equal to the number of linearly independent rows or columns, whichever is smaller.

CHARACTERISTIC VALUES AND CHARACTERISTIC VECTORS

These two concepts from matrix analysis play a very important role in the study of many physical problems. In linear system analysis, for example, the characteristic values define the natural frequencies of the dynamic system under consideration, and the characteristic vectors permit one to construct the so-called *Jordan canonical form* for the system equations. The results, of course, pertain only to dynamic systems which can be represented by a finite number of ordinary linear differential equations with constant coefficients.

Because we are generally interested in systems with time-varying coefficients, the role of characteristic values and characteristic vectors is minor in our work. Nevertheless, we shall review briefly the central ideas for the sake of completeness.

We consider the problem of determining those values of a scalar λ for which the homogeneous system of equations

$$(A - \lambda I)x = 0$$

has a nontrivial solution. Here, A is $n \times n$ and x is an n vector. This problem is called a *characteristic value* problem, the corresponding values of λ are the *characteristic values of A*, and the vectors for which $Ax = \lambda x$ are the *characteristic vectors* of A.

A necessary and sufficient condition for the characteristic value problem to have a nontrivial solution is that

$$|A - \lambda I| = 0$$

This nth-order algebraic equation, termed the *characteristic equation* of A, yields the n (not necessarily distinct) characteristic values of $A : \lambda_1, \ldots, \lambda_n$; this is also denoted $\lambda_i(A)$, $i = 1, \ldots, n$.

It can be shown that

$$|A| = \prod_{i=1}^{n} \lambda_i(A)$$

and

$$\operatorname{tr} A \triangleq \sum_{i=1}^{n} a_{ii} = \sum_{i=1}^{n} \lambda_i(A)$$

The latter quantity, tr A, which is termed the *trace* of A, is seen by its definition to be the sum of the terms along the principal diagonal of A. Two other properties of the trace which are readily derived are

$$\operatorname{tr} (A + B) = \operatorname{tr} A + \operatorname{tr} B$$
$$\operatorname{tr} (ABC) = \operatorname{tr} (BCA) = \operatorname{tr} (CAB)$$

If A is symmetric, it can be shown that its characteristic values are real and that its characteristic vectors are (or can be chosen to be) orthogonal.

Finally, if the characteristic values of A (not necessarily a symmetric matrix) are distinct, and M is the matrix of the characteristic vectors of A, then

$$M^{-1}AM = \begin{bmatrix} \lambda_1 & 0 & \cdots & 0 \\ 0 & \lambda_2 & \cdots & \cdots \\ \cdots & \cdots & \cdots & \cdots \\ 0 & \cdots & \cdots & \lambda_n \end{bmatrix}.$$

where the characteristic vectors of M are ordered according to $\lambda_1, \ldots, \lambda_n$. If A is symmetric in addition, then M is an orthogonal matrix and

$$M'AM = \begin{bmatrix} \lambda_1 & 0 & \cdots & 0 \\ 0 & \lambda_2 & \cdots & \cdot \\ \cdots & \cdots & \cdots & \cdots \\ 0 & \cdots & \cdots & \lambda_n \end{bmatrix}$$

QUADRATIC FORMS

For A an $n \times n$ matrix and x an n vector, it is easily shown by direct expansion that

$$x'Ax = \sum_{i=1}^{n} \sum_{j=1}^{n} a_{ij}x_ix_j$$

This product, obviously a scalar, is called a *quadratic form*.

In dealing with quadratic forms, it is sufficient to choose A symmetric. This can be seen by noting that for $i \neq j$, the terms in the expansion are of form $(a_{ij} + a_{ji})x_ix_j$. If A is not symmetric, we can replace it by the matrix $\bar{A}$, whose diagonal terms are the same as those of A and whose off-diagonal terms are

$$\bar{a}_{ij} = \frac{a_{ij} + a_{ji}}{2} \qquad i \neq j$$

without changing the value of the quadratic form; that is, $x'Ax = x'\bar{A}x$.

For example, if

$$A = \begin{bmatrix} 1 & 0 \\ -2 & 3 \end{bmatrix}$$

then

$$x'Ax = x_1^2 - 2x_1x_2 + 3x_2^2$$

On the other hand, for

$$\bar{A} = \begin{bmatrix} a_{11} & \dfrac{a_{12} + a_{21}}{2} \\ \dfrac{a_{12} + a_{21}}{2} & a_{22} \end{bmatrix} = \begin{bmatrix} 1 & -1 \\ -1 & 3 \end{bmatrix}$$

we have

$$x'\bar{A}x = x_1{}^2 - 2x_1x_2 + 3x_2{}^2$$

If $x'Ax > 0$ for all $x \neq 0$, we say that the quadratic form is *positive definite*. In this connection, it is also common practice to say that A is a positive definite matrix. If $x'Ax \geq 0$ for all $x \neq 0$, the quadratic form and A are called *positive semidefinite*, or *nonnegative definite*. Similarly, if $x'Ax < 0$ for all nontrivial x, we employ the term *negative definite*.

It can be shown that a necessary and sufficient condition for A to be positive definite is that the principal minors of A,

$$|a_{11}| \quad \begin{vmatrix} a_{11} & a_{12} \\ a_{12} & a_{22} \end{vmatrix} \quad \cdots \quad |A|$$

all be positive. As a result, we see that if A is positive definite, $|A| > 0$, which implies that A is nonsingular.

Now let B be a nonsingular $n \times n$ matrix, x and y n vectors, and define $y = Bx$. Clearly, $y'y = x'B'Bx$. Now $y'y \geq 0$, with equality holding if and only if $y = 0$. However, $y = 0$ implies $x = 0$, since B is nonsingular. Consequently, $B'B$ is positive definite.

On the other hand, suppose that A is a (symmetric) positive definite matrix. Then there exists an orthogonal matrix M for which

$$A^* = M'AM = [a_{ii}^*]$$

is a diagonal matrix with $a_{ii}^* > 0$ for all i. Now letting

$$B^* = [\sqrt{a_{ii}^*}]$$

noting that B^* is symmetric and nonsingular with $B^*B^* = A^*$, and defining $B = B^*M'$, we have

$$A = M(M'AM)M' = MA^*M' = MB^*B^*M' = B'B$$

In other words, for every positive definite matrix A, there exists a nonsingular matrix B for which $A = B'B$. It is seen that B can be viewed as the "square root" of A.

Finally, since B is nonsingular,

$$A^{-1} = B^{-1}(B')^{-1}$$

which means that if A is positive definite, so is its inverse.

We remark that when we speak in the sequel of a positive definite, positive semidefinite, or negative definite matrix, we shall mean one which is also symmetric.

GRADIENT OPERATOR

Let $f(x)$ denote a scalar-valued function of the n vector x; that is, $f(x) = f(x_1, \ldots, x_n)$. We define the *gradient operator* ∇_x to be the row vector operator

$$\nabla_x = \left[\frac{\partial}{\partial x_1} \quad \cdots \quad \frac{\partial}{\partial x_n} \right]$$

Thus, $\nabla_x f(x)$ is the row n vector

$$\nabla_x f(x) = \left[\frac{\partial f}{\partial x_1} \quad \cdots \quad \frac{\partial f}{\partial x_n} \right]$$

under the assumption that the indicated partial derivatives exist.

If $f(x)$ is the *bilinear form* $f(x) = y'Bx$, where B is $n \times n$ but not necessarily symmetric and x and y are n vectors, then

$$\nabla_x f(x) = \nabla_x \left(\sum_{i=1}^{n} \sum_{j=1}^{n} b_{ij} x_j y_i \right)$$

$$= \left(\sum_{i=1}^{n} b_{i1} y_i \quad \cdots \quad \sum_{i=1}^{n} b_{in} y_i \right)$$

— multiply together ?

$$= y'B$$

Similarly, if $f(x) = x'By$,

$$\nabla_x f(x) = y'B'$$

Finally, if $f(x) = x'Ax$, taking A not necessarily symmetric for the moment,

$$\nabla_x f(x) = \nabla_x \left(\sum_{i=1}^{n} \sum_{j=1}^{n} a_{ij} x_i x_j \right)$$

$$= \left(\sum_{i=1}^{n} a_{i1} x_i \quad \cdots \quad \sum_{i=1}^{n} a_{in} x_i \right)$$

$$\quad + \left(\sum_{j=1}^{n} a_{1j} x_j \quad \cdots \quad \sum_{j=1}^{n} a_{nj} x_j \right)$$

$$= x'A + x'A'$$

If A is symmetric, then obviously

$$\nabla_x (x'Ax) = 2x'A$$

This completes our review of matrix analysis, and we are now ready to introduce the system models with which we shall be concerned throughout this book. Additional concepts which we shall require from matrix analysis and the theory of ordinary differential equations will be presented as the need arises.

2.3 CONTINUOUS LINEAR SYSTEMS

SYSTEM EQUATIONS

We consider first physical systems whose dynamic behavior can be modeled by the system of first-order ordinary linear differential equations

$$\dot{x} = F(t)x + G(t)w(t) + C(t)u(t) \tag{2.1}$$

for $t \geq t_0$ where $x = x(t)$. In Eq. (2.1), x is an n vector called the *state* of the system, w is a p vector called the *disturbance*, u is an r vector called the *control*, and t denotes time. Also $F(t)$ is an $n \times n$ matrix, $G(t)$ an $n \times p$ matrix, and $C(t)$ an $n \times r$ matrix. These three matrices, commonly called the *system matrices*, are assumed to be continuous in t.

It is assumed that the initial time t_0 is fixed and that the initial state $x(t_0)$ is known.

The elements of x are termed the *state variables*, and those of w and u the *disturbance* and *control variables*, respectively. The three vectors x, w, and u are usually referred to as the *state vector*, *disturbance vector*, and *control vector*, respectively.

For the present, we assume that $w(t)$ and $u(t)$ are arbitrary Riemann integrable functions for all $t \geq t_0$.

We assume that the collection of sensors which are attached to the system, which is modeled by Eq. (2.1), to monitor its behavior has an output (collection of measured variables) which can be modeled by the relation

$$z(t) = H(t)x(t) + v(t) \tag{2.2}$$

In Eq. (2.2), z is an m vector termed the *measurement vector*, v is also an m vector called the *measurement error vector*, and $H(t)$ is a continuous $m \times n$ matrix which relates the state to the measurement. It is commonly called the *measurement matrix*. Also, z is sometimes called the *system output* (vector).

The elements of z are called the *measurement* or *measured variables*, and those of v the *measurement error* or *measurement noise variables*. For the present, $v(t)$ is assumed arbitrary for all $t \geq t_0$.

The system description which is defined by Eqs. (2.1) and (2.2) is called a *continuous linear system*. The adjective continuous is used in reference to the time variable t. A block diagram for this description is shown in Fig. 2.3, in which double lines are used to emphasize that signal flow involves vector variables rather than scalars.

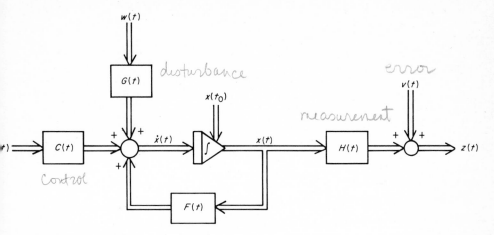

Fig. 2.3 Block diagram for continuous linear system description.

In dealing with such systems, we assume that we have access only to the measurement vector $z(t)$ and the control vector $u(t)$. By processing the first, we hope to determine "what the system is doing" and by manipulating the second, we hope to force the system "to behave in some desired fashion." In this connection, $w(t)$ and $v(t)$ are undesirable signals for which we must compensate if possible.

Let us now give some examples to show how the above system description arises in practice.

Example 2.1 We examine first a highly simplified model for the problem of yaw control of an aircraft in level flight. We assume that wind gusts force the aircraft to yaw from its desired heading, and we wish to correct yaw errors by suitable rudder deflections.

We consider only planar motion, idealize the aircraft to a rigid beam, and assume small angle deflections. Geometrically, the model is as shown below where θ = yaw error and ϕ = rudder deflection. The yaw axis, located at the center of gravity, is normal to the plane of the paper, and we assume that the aircraft has a moment of inertia J about this axis (Fig. 2.4). We assume further that the restoring torque is proportional to the rudder deflection, that rotational viscous drag and roll-yaw coupling are negligible.

Then, from Newton's second law,

$$J\ddot{\theta} = -k_1\phi(t) + w_\theta(t)$$

where k_1 = restoring torque-rudder deflection proportionality constant > 0 and $w_\theta(t)$ = torque caused by wind gusts.

Dividing through by J, and defining $b = -k_1/J$, $u(t) = \phi(t)$, and $w(t) = 1/Jw_\theta(t)$, we have

$$\ddot{\theta} = bu(t) + w(t)$$

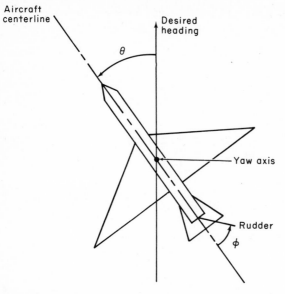

Fig. 2.4   Schematic for aircraft model.

The system is obviously second-order, and, in addition to $u(t)$ and $w(t)$, requires the two constants of integration $\theta(t_0) =$ initial yaw error and $\dot{\theta}(t_0) =$ initial yaw-rate error in order to obtain its solution for $t \geq t_0$.

Defining $x_1 = \theta$ and $x_2 = \dot{\theta} = \dot{x}_1$, we see that the system equation can be written as

$$\dot{x}_1 = x_2$$
$$\dot{x}_2 = bu(t) + w(t)$$

or equivalently as

$$\dot{x} = \begin{bmatrix} 0 & 1 \\ 0 & 0 \end{bmatrix} x + \begin{bmatrix} 0 \\ 1 \end{bmatrix} w(t) + \begin{bmatrix} 0 \\ b \end{bmatrix} u(t)$$

where x is the state vector

$$x = \begin{bmatrix} x_1 \\ x_2 \end{bmatrix}$$

which is obviously composed of the two physical variables, yaw error and yaw-rate error.

Comparing this result with Eq. (2.1), we see that

$$F(t) = \begin{bmatrix} 0 & 1 \\ 0 & 0 \end{bmatrix} \qquad G(t) = \begin{bmatrix} 0 \\ 1 \end{bmatrix} \qquad C(t) = \begin{bmatrix} 0 \\ b \end{bmatrix}$$

and that the disturbance $w(t)$ and the control $u(t)$ are scalars.

Under the assumption that both yaw error and yaw-rate error can be sensed onboard the aircraft with the aid of an inertial reference system, the measurement equation for the system is

$$z(t) = \begin{bmatrix} 1 & 0 \\ 0 & 1 \end{bmatrix} x(t) + v(t)$$

where

$$z(t) = \begin{bmatrix} z_1(t) \\ z_2(t) \end{bmatrix}$$

is the measurement vector and

$$v(t) = \begin{bmatrix} v_1(t) \\ v_2(t) \end{bmatrix}$$

is the measurement error vector; that is, $v_1(t)$ = error in measuring yaw and $v_2(t)$ = error in measuring yaw rate. Comparing this result with Eq. (2.2), we see that

$$H(t) = \begin{bmatrix} 1 & 0 \\ 0 & 1 \end{bmatrix}$$

Example 2.2   We examine next the linearized model for a fixed-field-current dc motor with an inertia and viscous friction load.   The problem involves control of the angular position θ and the angular velocity $\dot{\theta}$ of the output shaft by manipulation of the armature voltage e_a.   The schematic diagram is indicated in Fig. 2.5,

where J = inertia load
B = viscous friction coefficient
R = armature circuit resistance
L = armature circuit inductance
i_a = armature current
e_a = input voltage
e_b = armature back emf
i_f = field current = constant

We assume that the output torque produced by the motor is proportional to the armature current with the constant of proportionality being K_T.   Also, we assume that the back emf is proportional to the angular velocity, viz., $e_b = K_g \dot{\theta}$, where K_g = constant > 0.

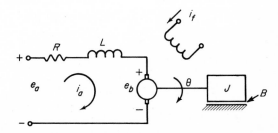

Fig. 2.5   Schematic for armature-controlled dc motor.

For the mechanical part of the system, we have from Newton's second law that

$$J\ddot{\theta} = -B\dot{\theta} + K_T i_a \tag{2.3}$$

From Kirchhoff's voltage law, we have

$$L\frac{di_a}{dt} + Ri_a + e_b = e_a \tag{2.4}$$

for the armature circuit, where $e_b = K_g \dot{\theta}$.

Since the first equation is second-order and the second is first-order, we pick θ, $\dot{\theta}$, and i_a as the state variables.

Proceeding as in Example 2.1, we write

$$\ddot{\theta} = -\frac{B}{J}\dot{\theta} + \frac{K_T}{J} i_a \tag{2.5}$$

and

$$\frac{di_a}{dt} = -\frac{R}{L} i_a - \frac{K_g}{L}\dot{\theta} + \frac{1}{L} e_a \tag{2.6}$$

We define $f_1 = -B/J$, $f_2 = K_T/J$, $f_3 = -K_g/L$, $f_4 = -R/L$, $c = 1/L$, $x_1 = \theta$, $x_2 = \dot{\theta}$, $x_3 = i_a$; and since e_a is the control variable input, we set $e_a = u$. We see now that the system in Eqs. (2.5) and (2.6) can be put in the form

$$\dot{x}_1 = x_2$$
$$\dot{x}_2 = f_1 x_2 + f_2 x_3$$
$$\dot{x}_3 = f_3 x_2 + f_4 x_3 + cu(t)$$

or with

$$x = \begin{bmatrix} x_1 \\ x_2 \\ x_3 \end{bmatrix} \triangleq \begin{bmatrix} \theta \\ \dot{\theta} \\ i_a \end{bmatrix}$$

in the form

$$\dot{x} = \begin{bmatrix} 0 & 1 & 0 \\ 0 & f_1 & f_2 \\ 0 & f_3 & f_4 \end{bmatrix} x + \begin{bmatrix} 0 \\ 0 \\ c \end{bmatrix} u(t)$$

Assuming next that measurements are made on the system by inserting an ammeter in the armature circuit and a position (angle) indicator on the output shaft, the measurement is a 2 vector, and we write

$$z(t) = \begin{bmatrix} 1 & 0 & 0 \\ 0 & 0 & 1 \end{bmatrix} x(t) + v(t)$$

In this case,

$$v(t) = \begin{bmatrix} v_1(t) \\ v_2(t) \end{bmatrix}$$

where $v_1(t)$ is the error in measuring $\theta(t)$ and $v_2(t)$ is the error in measuring $i_a(t)$.

Generally, the number of state variables is determined by the order of the system—third-order in this example, second-order in Example 2.1. However, the choice of state variables is not unique.

We illustrate this point in connection with Example 2.2. Solving Eq. (2.3) for i_a and substituting the result into Eq. (2.4) along with the relation $e_b = K_g \dot{\theta}$, we obtain

$$L \frac{d}{dt} \left(\frac{J}{K_T} \ddot{\theta} + \frac{B}{K_T} \dot{\theta} \right) + R \left(\frac{J}{K_T} \ddot{\theta} + \frac{B}{K_T} \dot{\theta} \right) + K_g \dot{\theta} = e_a$$

Grouping terms, we have

$$\frac{LJ}{K_T} \ddot{\theta} + \frac{LB + RJ}{K_T} \ddot{\theta} + \frac{RB + K_T K_g}{K_T} \dot{\theta} = e_a$$

which we choose to rewrite in the form

$$\ddot{\theta} = -\frac{LB + RJ}{LJ} \ddot{\theta} - \frac{RB + K_T K_g}{LJ} \dot{\theta} + \frac{K_T}{LJ} e_a \qquad (2.7)$$

We now let $a = -(LB + RJ)/LJ$, $b = -(RB + K_T K_g)/LJ$, $c = K_T/LJ$, and choose $x_1 = \theta$, $x_2 = \dot{\theta}$, and $x_3 = \ddot{\theta}$ as the state variables, with $u = e_a$ as the control input. Then, clearly,

$$\dot{x}_1 = x_2$$
$$\dot{x}_2 = x_3$$
$$\dot{x}_3 = bx_2 + ax_3 + cu(t)$$

or in vector matrix form

$$\dot{x} = \begin{bmatrix} 0 & 1 & 0 \\ 0 & 0 & 1 \\ 0 & b & a \end{bmatrix} x + \begin{bmatrix} 0 \\ 0 \\ c \end{bmatrix} u(t)$$

where x is the 3 vector

$$x = \begin{bmatrix} x_1 \\ x_2 \\ x_3 \end{bmatrix} \triangleq \begin{bmatrix} \theta \\ \dot{\theta} \\ \ddot{\theta} \end{bmatrix}$$

The armature current in the first formulation has evidently been replaced as a state variable in this second formulation by the angular acceleration $\ddot{\theta}$. In either case, it is clear that the procedure for putting the system equations into this so-called *state variable form* consists of first writing each system equation with the highest-order derivative on the left-hand side.

Because of the change in state variables in this second formulation, the first measurement scheme no longer pertains. In this case, we may, for example, assume that the measurement is simply the angular position of the output shaft. We then have

$$z(t) = [1 \quad 0 \quad 0] \, x(t) + v(t)$$

That is, $z(t) = \theta(t) +$ error in measuring $\theta(t)$.

NOTION OF STATE

At this point, it is worth our while to reflect briefly on the notion of state which we have employed above.

Let us consider a dynamic system whose behavior we choose to model by a system of ordinary differential equations

$$\dot{x} = f(x, \xi, t) \tag{2.8}$$

where x is a k vector, ξ is a q vector, and f is a k-dimensional vector-valued function of the indicated variables. Equation (2.1) is obviously a special case of Eq. (2.8).

We say that x is a state vector for the dynamic system under consideration if and only if $x(t_1)$ can be determined unambiguously from a knowledge of $x(t_0)$, $t_1 > t_0$ and $\xi(t)$, $t_0 \leq t \leq t_1$.

Basically, this means that k must be at least equal to the order of the dynamic system of interest. Thus, if we were to choose

$$x = \begin{bmatrix} \theta \\ \dot{\theta} \end{bmatrix}$$

in Example 2.2, we would *not* have a state vector, since the dynamic system of interest is third-order. In other words, knowledge of $\theta(0)$ and $\dot{\theta}(0)$, along with $e_a(t)$, $0 \leq t \leq t_1$, is not sufficient to determine $\theta(t_1)$ and $\dot{\theta}(t_1)$. This is clear from Eq. (2.7).

On the other hand, there is nothing to preclude our choosing

$$x = \begin{bmatrix} \theta \\ \dot{\theta} \\ \ddot{\theta} \\ \dot{i}_a \end{bmatrix}$$

as the state vector in Example 2.2. The extra equation however is redundant, and we are creating unnecessary work for ourselves.

In general then, one picks a state vector which has as many elements as the order of the system of differential equations which describe the dynamics of the phenomenon of interest.

All of the state variables in a given description may not be of physical interest. Referring to Example 2.2 again, the only variable of interest in a position control system would be $\theta(t)$.   Still, $\dot{\theta}(t)$ and $\ddot{\theta}(t)$ or $i_a(t)$ are required if the description is to reflect the physics of the system.

STATE TRANSITION MATRIX

For many theoretical developments and in certain computational problems, it is desirable to have an explicit expression for the solution of Eq. (2.1).   This expression is easily represented in terms of the so-called *state transition matrix*.

We assume that t_0, $x(t_0)$, $w(t)$, and $u(t)$ are given with the latter two piecewise continuous and bounded for all $t \geq t_0$.

Since the system in Eq. (2.1) is linear, we know that its complete solution consists of a linear combination of its *homogeneous* and *particular* solutions.   The former is sometimes called the *force-free* solution, the latter the *forced* solution.

We obtain first the homogeneous solution by treating the system of equations

$$\dot{x} = F(t)x \tag{2.9}$$

for $t \geq t_0$ with $x(t_0)$ arbitrary.

Substituting a trial solution of the form $x(t) = X(t)x(t_0)$, where $X(t)$ is an unknown $n \times n$ matrix, into Eq. (2.9), we have that

$$[\dot{X} - F(t)X]x(t_0) = 0$$

must hold for all $t \geq t_0$.   Since $x(t_0)$ is arbitrary, this relation is satisfied if and only if $X(t)$ satisfies the system of $n \times n$ matrix differential equations

$$\dot{X} = F(t)X \tag{2.10}$$

for all $t \geq t_0$.   Moreover, at $t = t_0$, we must have $x(t_0) = X(t_0)x(t_0)$ or equivalently,

$$[I - X(t_0)]x(t_0) = 0$$

which implies that

$$X(t_0) = I \tag{2.11}$$

is the initial condition for Eq. (2.10).

The homogeneous solution of Eq. (2.1) can then be written

$$x(t) = X(t)x(t_0) \tag{2.12}$$

where $X(t)$ is subject to Eqs. (2.10) and (2.11).

We obtain next the particular solution for Eq. (2.1) by using the Lagrange variation of parameters technique.

We assume a solution of the form

$$x(t) = X(t)y(t) \tag{2.13}$$

where $X(t)$ is as above and $y(t)$ is an unknown n vector. Substituting this result into Eq. (2.1), we have

$$\dot{X}(t)y(t) + X(t)\dot{y}(t) = F(t)X(t)y(t) + G(t)w(t) + C(t)u(t)$$

However, since $\dot{X}(t) = F(t)X(t)$, this expression reduces to

$$X(t)\dot{y}(t) = G(t)w(t) + C(t)u(t)$$

or, equivalently, to

$$\dot{y}(t) = X^{-1}(t)[G(t)w(t) + C(t)u(t)] \tag{2.14}$$

under the assumption that $X(t)$ is nonsingular for all $t \geq t_0$. (We verify this latter fact below.)

Integrating in Eq. (2.14), we have

$$y(t) = \int_{t_0}^{t} X^{-1}(\tau)[G(\tau)w(\tau) + C(\tau)u(\tau)] \, d\tau$$

which means that the particular solution of Eq. (2.1) is

$$x(t) = X(t) \int_{t_0}^{t} X^{-1}(\tau)[G(\tau)w(\tau) + C(\tau)u(\tau)] \, d\tau \tag{2.15}$$

Combining the results in Eqs. (2.12) and (2.15), we obtain the complete solution

$$x(t) = X(t)x(t_0) + X(t) \int_{t_0}^{t} X^{-1}(\tau)[G(\tau)w(\tau) + C(\tau)u(\tau)] \, d\tau \tag{2.16}$$

Let us now show that $X(t)$ is nonsingular for all $t \geq t_0$. From Eq. (2.10),

$$\dot{x}_{ij} = \sum_{k=1}^{n} f_{ik}(t)x_{kj} \tag{2.17}$$

for $i, j = 1, \ldots, n$, where $X(t) = [x_{ij}(t)]$. Then,

$$\frac{d}{dt}|X(t)| = \begin{vmatrix} \dot{x}_{11} & \dot{x}_{12} & \cdots & \dot{x}_{1n} \\ x_{21} & x_{22} & \cdots & x_{2n} \\ \cdots & \cdots & \cdots & \cdots \\ x_{n1} & x_{n2} & \cdots & x_{nn} \end{vmatrix} + \begin{vmatrix} x_{11} & x_{12} & \cdots & x_{1n} \\ \dot{x}_{21} & \dot{x}_{22} & \cdots & \dot{x}_{2n} \\ \cdots & \cdots & \cdots & \cdots \\ x_{n1} & x_{n2} & \cdots & x_{nn} \end{vmatrix}$$

$$+ \cdots + \begin{vmatrix} x_{11} & x_{12} & \cdots & x_{1n} \\ x_{21} & x_{22} & \cdots & x_{2n} \\ \cdots & \cdots & \cdots & \cdots \\ \dot{x}_{n1} & \dot{x}_{n2} & \cdots & \dot{x}_{nn} \end{vmatrix}$$

Substituting Eq. (2.17) into the first determinant on the right-hand side, we have

$$
\begin{vmatrix}
\sum\limits_{k=1}^{n} f_{1k}x_{k1} & \sum\limits_{k=1}^{n} f_{1k}x_{k2} & \cdots & \sum\limits_{k=1}^{n} f_{1k}x_{kn} \\
x_{21} & x_{22} & \cdots & x_{2n} \\
\cdots & \cdots & \cdots & \cdots \\
x_{n1} & x_{n2} & \cdots & x_{nn}
\end{vmatrix}
$$

The value of this determinant is unchanged if we subtract from the first row f_{12} times the second row plus f_{13} times the third row up to f_{1n} times the nth row. This permits us to write the above determinant as

$$
\begin{vmatrix}
f_{11}x_{11} & f_{11}x_{12} & \cdots & f_{11}x_{1n} \\
x_{21} & x_{22} & \cdots & x_{2n} \\
\cdots & \cdots & \cdots & \cdots \\
x_{n1} & x_{n2} & \cdots & x_{nn}
\end{vmatrix}
$$

whose value is obviously $f_{11}(t)|X(t)|$.

Repeating this procedure for the other determinants, we get

$$
\frac{d}{dt}|X(t)| = f_{11}(t)|X(t)| + f_{22}(t)|X(t)| + \cdots + f_{nn}(t)|X(t)|
$$
$$
= [\operatorname{tr} F(t)]|X(t)|
$$

Since $|X(t_0)| = |I| = 1$, it follows that the solution of this scalar differential equation is

$$
|X(t)| = \exp \int_{t_0}^{t} [\operatorname{tr} F(\tau)]\, d\tau
$$

If $F(\tau)$ is continuous for all $\tau \geq t_0$, then it is clear that $|X(t)| \neq 0$ for all $t \geq t_0$ and $X(t)$ is therefore nonsingular.

The matrix $X(t)$ is called the *fundamental matrix* of the system in Eq. (2.1). We note that it depends only on the system matrix $F(t)$.

Returning now to the result in Eq. (2.16), we define

$$
\Phi(t,\tau) = X(t)X^{-1}(\tau) \tag{2.18}
$$

and call the $n \times n$ matrix $\Phi(t,\tau)$ the *state transition matrix* for the system shown in Eq. (2.1). We note that $\Phi(t,t_0) = X(t)X^{-1}(t_0) = X(t)$, since $X^{-1}(t_0) = I^{-1} = I$.

We next write Eq. (2.16) as

$$
x(t) = \Phi(t,t_0)x(t_0) + \int_{t_0}^{t} \Phi(t,\tau)[G(\tau)w(\tau) + C(\tau)u(\tau)]\, d\tau \tag{2.19}
$$

where $t \geq t_0$. We shall have considerable occasion to utilize this form in many of our theoretical developments in the sequel.

Differentiating in Eq. (2.18) with respect to t and utilizing Eq. (2.10), we see that

$$\frac{\partial \Phi(t,\tau)}{\partial t} = \dot{X}(t)X^{-1}(\tau) = F(t)X(t)X^{-1}(\tau) = F(t)\Phi(t,\tau)$$

When there is no ambiguity regarding the variable with which the differentiation is to be carried out, we write this expression as

$$\dot{\Phi}(t,\tau) = F(t)\Phi(t,\tau) \tag{2.20}$$

From Eq. (2.18), we note that

$$\Phi(t,t) = X(t)X^{-1}(t) = I \tag{2.21}$$

for all $t \geq t_0$. Clearly, $\Phi(t,\tau)$ is specified by Eqs. (2.20) and (2.21).

In solving for $\Phi(t,\tau)$, it is sufficient to solve the differential equation $\dot{\Phi}(t,t_0) = F(t)\Phi(t,t_0)$, subject to the initial condition $\Phi(t_0,t_0) = I$, and then replace t_0 by τ to obtain $\Phi(t,\tau)$.

Example 2.3   For

$$F(t) = \begin{bmatrix} 0 & \dfrac{1}{(t+1)^2} \\ 0 & 0 \end{bmatrix}$$

we have

$$\begin{bmatrix} \dot{\phi}_{11} & \dot{\phi}_{12} \\ \dot{\phi}_{21} & \dot{\phi}_{22} \end{bmatrix} = \begin{bmatrix} 0 & \dfrac{1}{(t+1)^2} \\ 0 & 0 \end{bmatrix} \begin{bmatrix} \phi_{11} & \phi_{12} \\ \phi_{21} & \phi_{22} \end{bmatrix}$$

with $\phi_{11}(t_0,t_0) = \phi_{22}(t_0,t_0) = 1$ and $\phi_{12}(t_0,t_0) = \phi_{21}(t_0,t_0) = 0$.
The system of equations of interest is seen to be

$$\dot{\phi}_{11}(t,t_0) = \frac{1}{(t+1)^2}\,\phi_{21}(t,t_0)$$

$$\dot{\phi}_{12}(t,t_0) = \frac{1}{(t+1)^2}\,\phi_{22}(t,t_0)$$

$$\dot{\phi}_{21}(t,t_0) = 0$$

$$\dot{\phi}_{22}(t,t_0) = 0$$

We see immediately from the last two equations and their corresponding initial conditions that

$$\phi_{21}(t,t_0) = 0 \qquad \text{and} \qquad \phi_{22}(t,t_0) = 1$$

It then follows that $\dot{\phi}_{11}(t,t_0) = 0$, so that $\phi_{11}(t,t_0) = 1$. Also,

$$\dot{\phi}_{12}(t,t_0) = \frac{1}{(t+1)^2}$$

This gives

$$\phi_{12}(t,t_0) = -\frac{1}{t+1} + \alpha$$

where α is the constant of integration which is to be chosen such that $\phi_{12}(t_0,t_0) = 0$. Clearly,

$$\alpha = \frac{1}{t_0+1}$$

Hence,

$$\phi_{12}(t,t_0) = \frac{t-t_0}{(t_0+1)(t+1)}$$

or

$$\phi_{12}(t,\tau) = \frac{t-\tau}{(\tau+1)(t+1)}$$

and we have

$$\Phi(t,\tau) = \begin{bmatrix} 1 & \dfrac{t-\tau}{(t+1)(\tau+1)} \\ 0 & 1 \end{bmatrix}$$

The state transition matrix possesses two important properties which we now develop.

1. $\Phi(t_2,t_1) = \Phi(t_2,\tau)\Phi(\tau,t_1)$ for all $t_1,\, t_2,\, \tau \geq t_0$ (2.22)

This follows immediately from Eq. (2.18), viz.,

$$\begin{aligned} \Phi(t_2,\tau)\Phi(\tau,t_1) &= [X(t_2)X^{-1}(\tau)][X(\tau)X^{-1}(t_1)] \\ &= X(t_2)X^{-1}(t_1) \\ &= \Phi(t_2,t_1) \end{aligned}$$

We remark that the ordering on t_1, t_2, and τ is immaterial as long as each is $\geq t_0$.

2. $\Phi(t_2,t_1) = \Phi^{-1}(t_1,t_2)$ for all $t_1,\, t_2 \geq t_0$ (2.23)

Again, from the definition,

$$\begin{aligned} \Phi^{-1}(t_1,t_2) &= [X(t_1)X^{-1}(t_2)]^{-1} \\ &= X(t_2)X^{-1}(t_1) \\ &= \Phi(t_2,t_1) \end{aligned}$$

CONSTANT COEFFICIENT SYSTEMS

If the system matrices $F(t)$, $G(t)$, $C(t)$, and $H(t)$ in Eqs. (2.1) and (2.2) are constant, as in Examples 2.1 and 2.2, the system is said to be a *constant coefficient* or a *stationary* one.   The system equations are

$$\dot{x} = Fx + Gw(t) + Cu(t) \tag{2.24}$$

$$z(t) = Hx(t) + v(t) \tag{2.25}$$

and the initial time t_0 is usually taken as zero for convenience.

For a system such as in Eq. (2.24), the state transition matrix assumes a relatively simple form.   To see this, we introduce the matrix exponential

$$e^{Ft} = I + Ft + \cdots + \frac{F^n t^n}{n!} + \cdots = \sum_{k=0}^{\infty} F^k \frac{t^k}{k!} \tag{2.26}$$

From this definition, we see that

$$e^{Ft} e^{F\tau} = \left(\sum_{k=0}^{\infty} F^k \frac{t^k}{k!} \right) \left(\sum_{l=0}^{\infty} F^l \frac{\tau^l}{l!} \right) = \sum_{k=0}^{\infty} \sum_{l=0}^{\infty} F^{k+l} \frac{t^k \tau^l}{k! l!}$$

Letting $n = k + l$ so that $l = n - k$, we have

$$e^{Ft} e^{F\tau} = \sum_{k=0}^{\infty} \sum_{n=k}^{\infty} F^n \frac{t^k \tau^{n-k}}{k!(n-k)!}$$

$$= \sum_{n=0}^{\infty} \sum_{k=0}^{\infty} F^n \frac{t^k \tau^{n-k}}{k!(n-k)!}$$

$$= \sum_{n=0}^{\infty} \frac{F^n}{n!} \sum_{k=0}^{\infty} \frac{n!}{k!(n-k)!} t^k \tau^{n-k}$$

$$= \sum_{n=0}^{\infty} \frac{F^n}{n!} (t + \tau)^n$$

$$= e^{F(t+\tau)}$$

Replacing τ by $-t$, we get

$$e^{Ft}(e^{-Ft}) = e^{F0} = I$$

Hence,

$$(e^{Ft})^{-1} = e^{-Ft}$$

which always exists.

From Eq. (2.26),

$$\frac{d}{dt}(e^{Ft}) = \sum_{k=1}^{\infty} F^k \frac{t^{k-1}}{(k-1)!} = F \sum_{j=0}^{\infty} F^j \frac{t^j}{j!} = F e^{Ft}$$

and it is clear that

$$X(t) = e^{Ft}$$

is the solution of the differential equation

$$\dot{X} = FX$$

subject to the initial condition $X(0) = I$.   Since

$$X^{-1}(\tau) = e^{-F\tau}$$

it follows from Eq. (2.18) that

$$\Phi(t,\tau) = e^{F(t-\tau)}$$

is the state transition matrix for the system in Eq. (2.24).
    The solution for this latter equation is then obviously

$$x(t) = e^{Ft}x(0) + \int_0^t e^{F(t-\tau)}[Gw(\tau) + Cu(\tau)]\, d\tau$$

INPUT-OUTPUT RELATION

Viewing the system in Eqs. (2.1) and (2.2) as having inputs $w(t)$ and $u(t)$ and an output $z(t)$, we write

$$z(t) = H(t)\left[\Phi(t,t_0)x(t_0) + \int_{t_0}^t \Phi(t,\tau)[G(\tau)w(\tau) + C(\tau)u(\tau)]\, d\tau\right] + v(t)$$

$$= H(t)\Phi(t,t_0)x(t_0) + \int_{t_0}^t H(t)\Phi(t,\tau)G(\tau)w(\tau)\, d\tau$$

$$+ \int_{t_0}^t H(t)\Phi(t,\tau)C(\tau)u(\tau)\, d\tau + v(t)$$

Defining

$$A(t,\tau) = H(t)\Phi(t,\tau)G(\tau) \tag{2.27}$$

and

$$B(t,\tau) = H(t)\Phi(t,\tau)C(\tau) \tag{2.28}$$

we have

$$z(t) = H(t)\Phi(t,t_0)x(t_0) + \int_{t_0}^t A(t,\tau)w(\tau)\, d\tau + \int_{t_0}^t B(t,\tau)u(\tau)\, d\tau + v(t) \tag{2.29}$$

In Eq. (2.29), the $m \times p$ and $m \times r$ matrices $A(t,\tau)$ and $B(t,\tau)$, respectively, are called *system weighting* or *impulse response* matrices. The latter name derives from the fact that each element $a_{ij}(t,\tau)$ or $b_{ij}(t,\tau)$ is the response of the ith component of $z(t)$ for a unit impulse input in the jth component of w or u, respectively, applied at time τ.
    The use of system impulse response matrices is convenient when one wishes only an input-output relation and is not concerned with the system's state variables.   As a simple example, suppose that $x(t_0)$,

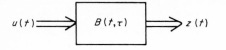

Fig. 2.6 Block diagram for system input-output relation.

$w(t)$, and $v(t)$ are zero. That is, the system is initally at rest; there is no system disturbance and no measurement error. Then, from Eq. (2.29),

$$z(t) = \int_{t_0}^{t} B(t,\tau)u(\tau) \, d\tau$$

and the system block diagram can be indicated as in Fig. 2.6.

If the system is constant coefficient, we see that Eqs. (2.27) and (2.28) can be written

$$A(t,\tau) = A(t - \tau) = H\Phi(t - \tau)G$$

and

$$B(t,\tau) = B(t - \tau) = H\Phi(t - \tau)C$$

where

$$\Phi(t - \tau) = e^{F(t-\tau)}$$

2.4 DISCRETE LINEAR SYSTEMS

SYSTEM EQUATIONS

We turn now to a consideration of physical systems whose behavior can be modeled by the system of first-order linear difference equations

$$x(k + 1) = \Phi(k + 1, k)x(k) + \Gamma(k + 1, k)w(k)$$
$$+ \Psi(k + 1, k)u(k) \quad (2.30)$$

for $k = 0, 1, \ldots$. In Eq. (2.30), x, w, and u have the same number of elements, respectively, as in the preceding section. They are also, respectively, the *state*, *disturbance*, and *control* vectors for the system. Here, however, they are expressed only at discrete instants in time.

We call the $n \times n$ matrix $\Phi(k + 1, k)$ the *state transition matrix*, the $n \times p$ matrix $\Gamma(k + 1, k)$ the *disturbance transition matrix*, and the $n \times r$ matrix $\Psi(k + 1, k)$ the *control transition matrix*. We assume that $k = 0$ specifies a fixed initial time t_0 and that $x(0)$ is given.

The sequence $\{w(0), w(1), \ldots\}$ is termed the *disturbance sequence*, and $\{u(0), u(1), \ldots\}$ the *control sequence*. Given these two sequences along with $x(0)$, we see that the *state vector sequence* $\{x(1), x(2), \ldots\}$ can be computed recursively using Eq. (2.30).

For the present, we assume that the disturbance and control sequences are arbitrary.

We assume here that the measurement system with which we are dealing can also be modeled by Eq. (2.2) but that the measurements occur only at the same discrete time points at which the state x is expressed in Eq. (2.30). Thus, we take as our measurement model

$$z(k + 1) = H(k + 1)x(k + 1) + v(k + 1) \qquad (2.31)$$

for $k = 0, 1, \ldots$, where z and v are again m vectors and H is an $m \times n$ matrix. We remark that we have arbitrarily taken $(k + 1)$ as the time argument in Eq. (2.13). We could equally as well have chosen k.

The sequence $\{z(1), z(2), \ldots\}$ is called the *measurement* or *system output sequence*, and $\{v(1), v(2), \ldots\}$ is termed the *measurement error sequence*.

The system description which is given in Eqs. (2.30) and (2.31) is designated a *discrete linear system*. The model's block diagram is given in Fig. 2.7 where the block marked "delay" refers to the act of storing or retaining the value of x from the present computational cycle for use in the succeeding one. That is, after $x(k)$ is computed, it must be stored until time $k + 1$ for use in determining $x(k + 1)$.

Most commonly the above system description arises when sampling is introduced into the continuous linear system formulation of Eqs. (2.1) and (2.2). In other words, the system model of this section is the sampled-data version of the one in the preceding section. The present formulation is of course a natural one if, say for reasons of economy, measurements are taken only at discrete-time instants. We now show that Eqs. (2.30) and (2.31) are the discrete-time analogs of Eqs. (2.1) and (2.2), respectively.

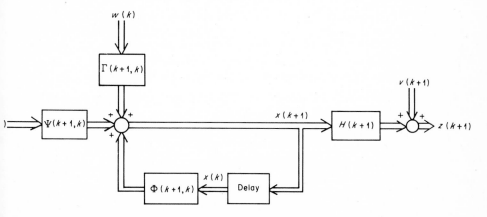

Fig. 2.7 Block diagram for discrete linear system description.

SAMPLED-DATA MODEL

Let us assume that the disturbance vector and the control vector of the system in Eq. (2.1) are piecewise constant functions of time, both of which change values at the same time points. We assume further that measurements are made only at these same time points.

We indicate representative components of $w(t)$ and $u(t)$ in Fig. 2.8 where we emphasize that the sampling times are not necessarily uniformly spaced.

We consider the time interval $t_k \leq t \leq t_{k+1}$ for some $k = 0, 1, \ldots$. We assume that $x(t_k)$ is given and that $w(t) = w(k) = $ constant and $u(t) = u(k) = $ constant, both for $t_k \leq t < t_{k+1}$. It now follows from Eq. (2.19) that

$$x(t_{k+1}) = \Phi(t_{k+1}, t_k)x(t_k) + \left[\int_{t_k}^{t_{k+1}} \Phi(t_{k+1}, \tau)G(\tau)\, d\tau \right] w(k)$$
$$+ \left[\int_{t_k}^{t_{k+1}} \Phi(t_{k+1}, \tau)C(\tau)\, d\tau \right] u(k) \quad (2.32)$$

Defining

$$x(t_{k+1}) = x(k + 1)$$
$$x(t_k) = x(k)$$
$$\Phi(t_{k+1}, t_k) = \Phi(k + 1, k) \qquad\qquad (2.33)$$
$$\int_{t_k}^{t_{k+1}} \Phi(t_{k+1}, \tau)G(\tau)\, d\tau = \Gamma(k + 1, k)$$

and

$$\int_{t_k}^{t_{k+1}} \Phi(t_{k+1}, \tau)C(\tau)\, d\tau = \Psi(k + 1, k)$$

we can write Eq. (2.32) as

$$x(k + 1) = \Phi(k + 1, k)x(k) + \Gamma(k + 1, k)w(k) + \Psi(k + 1, k)u(k)$$

for $k = 0, 1, \ldots$, which is identically Eq. (2.30).

At each $t = t_{k+1}$, $k = 0, 1, \ldots$, Eq. (2.2) becomes

$$z(t_{k+1}) = H(t_{k+1})x(t_{k+1}) + v(t_{k+1})$$

or, more simply, in terms of the discrete-time index k,

$$z(k + 1) = H(k + 1)x(k + 1) + v(k + 1)$$

which is obviously Eq. (2.31).

We note the need in the last two expressions in Eq. (2.33) to have the state transition matrix in the form in which its first argument is fixed,

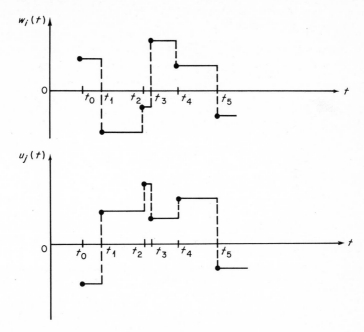

Fig. 2.8 Components of piecewise constant $w(t)$ and $u(t)$ in sampled-data model formulation.

whereas its second is variable, in order to evaluate $\Gamma(k + 1,\ k)$ and $\Psi(k + 1,\ k)$.

Example 2.4 Let us employ the above procedure to develop the discrete-time model for the second-order constant coefficient system of Example 2.1. For simplicity, let us assume that the sampling times are uniformly spaced at intervals at length T. Then $k + 1$ corresponds to the time point $(k + 1)T$ and k to the time point kT.

Since

$$F = \begin{bmatrix} 0 & 1 \\ 0 & 0 \end{bmatrix}$$

we see that $F^n = 0$ for all $n \geq 2$. Hence the state transition matrix is

$$\Phi(t,\tau) = \Phi(t - \tau)$$
$$= e^{F(t-\tau)}$$
$$= I + F(t - \tau) \quad \text{first order } \sim\text{mation}$$
$$= \begin{bmatrix} 1 & t - \tau \\ 0 & 1 \end{bmatrix}$$

This gives

$$\Phi(k + 1,\ k) = \begin{bmatrix} 1 & T \\ 0 & 1 \end{bmatrix}$$

Then, from Eq. (2.33),

$$\Gamma(k+1, k) = \int_{kT}^{(k+1)T} \begin{bmatrix} 1 & (k+1)T - \tau \\ 0 & 1 \end{bmatrix} \begin{bmatrix} 0 \\ 1 \end{bmatrix} d\tau$$

$$= \int_{kT}^{(k+1)T} \begin{bmatrix} (k+1)T - \tau \\ 1 \end{bmatrix} d\tau$$

$$= \begin{bmatrix} \frac{1}{2}T^2 \\ T \end{bmatrix}$$

Similarly

$$\Psi(k+1, k) = \int_{kT}^{(k+1)T} \begin{bmatrix} 1 & (k+1)T - \tau \\ 0 & 1 \end{bmatrix} \begin{bmatrix} 0 \\ b \end{bmatrix} d\tau$$

$$= b\Gamma(k+1, k)$$

$$= \begin{bmatrix} \frac{1}{2}bT^2 \\ bT \end{bmatrix}$$

We have, therefore, that

$$x(k+1) = \begin{bmatrix} 1 & T \\ 0 & 1 \end{bmatrix} x(k) + \begin{bmatrix} \frac{1}{2}T^2 \\ T \end{bmatrix} w(k) + \begin{bmatrix} \frac{1}{2}bT^2 \\ bT \end{bmatrix} u(k).$$

The corresponding measurement model equation is seen to be

$$z(k+1) = \begin{bmatrix} 1 & 0 \\ 0 & 1 \end{bmatrix} x(k+1) + v(k+1)$$

CONTINUOUS MODEL AS THE LIMITING CASE OF THE DISCRETE MODEL

In developing certain of our results for estimation and control for continuous linear systems, it will be convenient for us to "discretize" the system equations, develop the estimation and control algorithms for the discrete systems, and then consider the limiting behavior of these algorithms as the time interval between sampling instants goes to zero.

The technique for obtaining the discrete-time model from the continuous one has been given above. We wish now to exhibit the reverse process. In order to do this, we let the discrete times k and $k+1$ be denoted by t and $t + \Delta t$, respectively, where $\Delta t > 0$.

Then, expanding $\Phi(t + \Delta t, t)$ in a Taylor series about t, we have

$$\Phi(t + \Delta t, t) = \Phi(t,t) + \dot{\Phi}(t,t) \Delta t + 0(\Delta t^2)$$

where $0(\Delta t^2)$ is an $n \times n$ matrix all of whose terms are of order $(\Delta t)^2$ and higher.

Since $\Phi(t,t) = I$ and $\dot{\Phi}(t,t) = F(t)\Phi(t,t)$, we have

$$\Phi(t + \Delta t, t) = I + F(t) \Delta t + 0(\Delta t^2) \tag{2.34}$$

From the fourth relation in Eq. (2.33), it is clear that

$$\Gamma(t + \Delta t, t) = \int_t^{t+\Delta t} \Phi(t + \Delta t, \tau)G(\tau) \, d\tau = G(t) \Delta t + 0(\Delta t^2) \tag{2.35}$$

Similarly,

$$\Psi(t + \Delta t, t) = \int_t^{t+\Delta t} \Phi(t + \Delta t, \tau)C(\tau) \, d\tau = C(t)\Delta t + 0(\Delta t^2) \quad (2.36)$$

For k and $k + 1$ replaced by t and $t + \Delta t$, respectively, Eq. (2.30) is

$$x(t + \Delta t) = \Phi(t + \Delta t, t)x(t) + \Gamma(t + \Delta t, t)w(t) + \Psi(t + \Delta t, t)u(t)$$

Substituting into this relation from Eqs. (2.34) to (2.36) and rearranging terms, we get

$$
\begin{aligned}
x(t + \Delta t) &= [I + F(t) \Delta t + 0(\Delta t^2)]x(t) + [G(t) \Delta t + 0(\Delta t^2)]w(t) \\
&\quad + [C(t) \Delta t + 0(\Delta t^2)]u(t) \\
&= x(t) + F(t)x(t) \Delta t + G(t)w(t) \Delta t + C(t)u(t) \Delta t + 0(\Delta t^2)
\end{aligned}
$$

Transposing $x(t)$ to the left-hand side of the equation, dividing through by Δt, and taking the limit as $\Delta t \to 0$, we obtain the result

$$\dot{x} = F(t)x + G(t)w(t) + C(t)u(t)$$

for $t \geq t_0$, which is identical to Eq. (2.1).

Finally,

$$z(t + \Delta t) = H(t + \Delta t)x(t + \Delta t) + v(t + \Delta t)$$

becomes

$$z(t) = H(t)x(t) + v(t)$$

as $\Delta t \to 0$.

2.5 OBSERVABILITY AND CONTROLLABILITY

We turn now to a consideration of two fundamental concepts of linear system theory which are intimately related to the basic ideas of estimation and control. These notions, termed *observability* and *controllability*, are due originally to Kalman [7, 8], whose work formed the basis for other studies.†

The formulation and study of the concepts of observability and controllability is based, respectively, on the following two questions which are motivated by obvious physical considerations:

1. Under what conditions is it possible to establish, in a finite interval of time, the time history of the state x of a dynamic system given the time history of the measurement vector z over the same time interval?
2. Under what conditions is it possible to transfer the state of a dynamic system from a given initial state to a desired terminal state in a finite amount of time using a piecewise continuous control u?

† See, for example, References 9 to 15.

Before proceeding to a precise treatment of these questions for the two classes of linear systems which we have given in Secs. 2.3 and 2.4, let us illustrate these two notions qualitatively for a general system with the aid of some simple diagrams.

We consider a dynamic system S with state vector x, control vector u, and measurement vector z. We assume that there are no disturbances or measurement errors and that the system may be either a discrete-time or a continuous-time one.

First let us suppose that the system block diagram can be put in the form shown in Fig. 2.9 where y is a vector whose components consist of some or all of the elements $x_1, \ldots, x_k$. Because of the system's structure, there is no way that the values of $x_{k+1}, \ldots, x_n$ can be determined from an examination of z, since these variables do not affect $x_1, \ldots, x_k$, nor do they appear in z. Hence, the system is unobservable. On the other hand, if u affects all of the elements of x, the system is controllable.

In a similar way, the system in Fig. 2.10 is observable, but it is uncontrollable since u affects only the variables $x_1, \ldots, x_k$.

At this point it is clear that systems can be classified into the following four categories: (1) observable-controllable, (2) observable-uncontrollable, (3) unobservable-controllable, and (4) unobservable-uncontrollable.

2.6 OBSERVABILITY IN CONTINUOUS AND DISCRETE LINEAR SYSTEMS [7–15]

Let us examine more closely the concept of observability for the two classes of linear systems which are of concern in this book. We approach this subject (and that of controllability in Sec. 2.7) under the simplifying

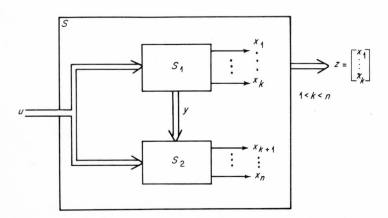

Fig. 2.9 Schematic for unobservable-controllable system.

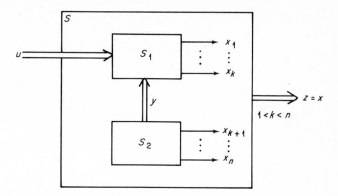

Fig. 2.10 Schematic for observable-uncontrollable system.

assumption that the system disturbances and measurement errors are zero (i.e., under ideal conditions).

CONTINUOUS LINEAR SYSTEMS

We consider first the continuous linear system

$$\dot{x} = F(t)x + C(t)u(t) \tag{2.37}$$

$$z(t) = H(t)x(t) \tag{2.38}$$

for $t \geq t_0$, where all the terms were defined in Sec. 2.3. We assume that $u(t)$ is a known function of time for all $t \geq t_0$ but that $x(t_0)$ is unknown.

Our concern is with determining $x(t)$ from an examination of $z(t)$ over some finite time interval $[t_0,t_1]$. Evidently, if $H(t)$ is $n \times n$ and non-singular for all $t \geq t_0$, then

$$x(t) = H^{-1}(t)z(t)$$

and the question of observability is resolved trivially.

Indeed, the question is also easily resolved if $H(t)$ is $n \times n$ and non-singular for only one value of $t \geq t_0$, say t_α. To see this, we note first that

$$x(t_\alpha) = H^{-1}(t_\alpha)z(t_\alpha)$$

We recall from Eq. (2.19) that

$$x(t) = \Phi(t,t_0)x(t_0) + \int_{t_0}^{t} \Phi(t,\tau)C(\tau)u(\tau) \, d\tau \tag{2.39}$$

For $t = t_\alpha$,

$$H^{-1}(t_\alpha)z(t_\alpha) = \Phi(t_\alpha,t_0)x(t_0) + \int_{t_0}^{t_\alpha} \Phi(t_\alpha,\tau)C(\tau)u(\tau) \, d\tau$$

from which it follows that

$$x(t_0) = \Phi^{-1}(t_\alpha, t_0)[H^{-1}(t_\alpha)z(t_\alpha) - \int_{t_0}^{t_\alpha} \Phi(t_\alpha, \tau)C(\tau)u(\tau) \, d\tau]$$

$$= \Phi(t_0, t_\alpha)H^{-1}(t_\alpha)z(t_\alpha) - \int_{t_0}^{t_\alpha} \Phi(t_0, \tau)C(\tau)u(\tau) \, d\tau \qquad (2.40)$$

Substitution of the value of $x(t_0)$ as obtained from Eq. (2.40) into Eq. (2.39) then permits us to determine $x(t)$ for all $t \geq t_0$, since $u(\tau)$, $\tau \geq t_0$, is known.

If, on the other hand, $H(t)$ is $n \times n$ and singular for all $t \geq t_0$ or $H(t)$ is $m \times n$ with $m \neq n$, it is not at all clear how $x(t)$ can be determined from $z(\tau)$, $t_0 \leq \tau \leq t_1$, for some finite t_1. This is the real question that we wish to pursue.

We note from Eq. (2.39) that $x(t)$ is known for all $t \geq t_0$ if the initial condition $x(t_0)$ can be determined from the measurements. With this in mind, we define observability in the following way.

DEFINITION

The continuous linear system of Eqs. (2.37) and (2.38) is observable if $x(t_0)$ can be determined from $z(t)$, $t_0 \leq t \leq t_1$, for some finite t_1. If this is true for any t_0, the system is completely observable.

A necessary and sufficient condition for complete observability is given in the following theorem.

Theorem 2.1   *The continuous linear system of Eqs. (2.37) and (2.38) is completely observable if and only if the symmetric $n \times n$ matrix*

$$M_c(t_0, t_1) = \int_{t_0}^{t_1} \Phi'(t, t_0)H'(t)H(t)\Phi(t, t_0) \, dt \qquad (2.41)$$

is positive definite for some finite $t_1 > t_0$.

Proof   Since $u(t)$ is assumed known for all $t \geq t_0$, its contribution to $x(t)$, which is

$$\int_{t_0}^{t} \Phi(t, \tau)C(\tau)u(\tau) \, d\tau$$

is easily determined. Hence, it is sufficient to consider only the system

$$\dot{x} = F(t)x \qquad (2.42)$$

$$z(t) = H(t)x(t) \qquad (2.43)$$

where $t \geq t_0$.

We prove sufficiency first. Substituting the solution

$$x(t) = \Phi(t,t_0)x(t_0)$$

of Eq. (2.42) into Eq. (2.43), we get

$$z(t) = H(t)\Phi(t,t_0)x(t_0) \tag{2.44}$$

Premultiplying this equation by $\Phi'(t,t_0)H'(t)$ and integrating the result between t_0 and t_1, we obtain

$$\left[\int_{t_0}^{t_1} \Phi'(t,t_0)H'(t)H(t)\Phi(t,t_0) \, dt \right] x(t_0) = \int_{t_0}^{t_1} \Phi'(t,t_0)H'(t)z(t) \, dt \tag{2.45}$$

Defining

$$M_c(t_0,t_1) = \int_{t_0}^{t_1} \Phi'(t,t_0)H'(t)H(t)\Phi(t,t_0) \, dt$$

it then follows from Eq. (2.45) that

$$x(t_0) = M_c^{-1}(t_0,t_1) \int_{t_0}^{t_1} \Phi'(t,t_0)H'(t)z(t) \, dt \tag{2.46}$$

and the system is obviously completely observable if $M_c(t_0,t_1)$ is positive definite for some finite $t_1 > t_0$.

To prove necessity, we assume that the system is completely observable but that $M_c(t_1,t_0)$ is not positive definite; that is,

$$x'(t_0)M_c(t_0,t_1)x(t_0) = 0$$

for some $x(t_0) \neq 0$.†

From Eq. (2.45), this means that

$$x'(t_0) \int_{t_0}^{t_1} \Phi'(t,t_0)H'(t)z(t) \, dt = 0$$

Substituting into this expression from Eq. (2.44), we get

$$\int_{t_0}^{t_1} z'(t)z(t) \, dt = 0$$

which implies that $z(t) = 0$ for $t_0 \le t \le t_1$.

Hence, we are forced to conclude that there exist nonzero $x(t_0)$ which cannot be determined from $z(t)$, $t_0 \le t \le t_1$, which contradicts the hypothesis that the system is observable. This completes the proof.

In general, the condition in Eq. (2.41) is difficult to apply in practice because of the large amount of computation which may be required. However, if the system is constant coefficient, we have the following corollary, which we state without proof, that is very easy to apply.

† The case $x'(t_0)M_c(t_0,t_1)x(t_0) < 0$ is excluded by virtue of the definition of $M_c(t_0,t_1)$.

Corollary 2.1 *The constant coefficient continuous linear system* $\dot{x} = Fx$, $z(t) = Hx(t)$, *where* $t \geq t_0$, *is completely observable if and only if the* $n \times mn$ *matrix*

$$[H', F'H', \ldots, (F')^{n-1}H'] \tag{2.47}$$

has rank n.

Example 2.5 For the system in Example 2.1, where

$$F = \begin{bmatrix} 0 & 1 \\ 0 & 0 \end{bmatrix} \quad \text{and} \quad H = \begin{bmatrix} 1 & 0 \\ 0 & 1 \end{bmatrix}$$

we have

$$H' = \begin{bmatrix} 1 & 0 \\ 0 & 1 \end{bmatrix} \quad \text{and} \quad F'H' = \begin{bmatrix} 0 & 0 \\ 1 & 0 \end{bmatrix}$$

Then

$$\text{Rank } [H', F'H'] = \text{rank } \begin{bmatrix} 1 & 0 & 0 & 0 \\ 0 & 1 & 1 & 0 \end{bmatrix} = 2$$

and the system is completely observable.

On the other hand, we see that if the measurement were only the yaw-rate error, so that $H = [0 \quad 1]$, we would have

$$H' = \begin{bmatrix} 0 \\ 1 \end{bmatrix} \quad F'H' = \begin{bmatrix} 0 \\ 0 \end{bmatrix}$$

and

$$\text{Rank } [H', F'H'] = \text{rank } \begin{bmatrix} 0 & 0 \\ 1 & 0 \end{bmatrix} = 1$$

in which case the system is not observable.

DISCRETE LINEAR SYSTEMS

Proceeding in a fashion analogous to that above, we examine next the question of observability for discrete linear systems.

We have the system

$$x(k+1) = \Phi(k+1, k)x(k) + \Psi(k+1, k)u(k) \tag{2.48}$$

$$z(k+1) = H(k+1)x(k+1) \tag{2.49}$$

where $k = 0, 1, \ldots$, $x(0)$ is unknown, $\{u(0), u(1), \ldots\}$ is assumed given, and all the other terms were defined in Sec. 2.4.

DEFINITION

The discrete linear system of Eqs. (2.48) and (2.49) is observable if $x(0)$ *can be determined from the set of measurements* $\{z(1), \ldots, z(N)\}$ *for some finite* N. *If this is true for any initial time* ($k = 0$ *corresponds to* t_0), *the system is completely observable.*

Analogous to Theorem 2.1, we have the following result.

Theorem 2.2 *The discrete linear system of Eqs. (2.48) and (2.49) is completely observable if and only if the symmetric $n \times n$ matrix*

$$M_d(0,N) = \sum_{i=1}^{N} \Phi'(i,0)H'(i)H(i)\Phi(i,0) \qquad (2.50)$$

is positive definite for some $N > 0$, where $\Phi(i,0) = \Phi(i, i-1) \cdots \Phi(1,0)$, $i = 1, \ldots, N$.

Proof As for the continuous-time case, it is sufficient to consider only the system

$$x(k+1) = \Phi(k+1, k)x(k) \qquad (2.51)$$

$$z(k+1) = H(k+1)x(k+1) \qquad (2.52)$$

$k = 0, 1, \ldots$, since $u(k)$ is assumed known for all k.

SUFFICIENCY Consider the sequence of measurements $\{z(1), \ldots, z(N)\}$. From Eqs. (2.51) and (2.52), we have

$$z(1) = H(1)x(1) = H(1)\Phi(1,0)x(0)$$
$$z(2) = H(2)x(2) = H(2)\Phi(2,1)x(1) = H(2)\Phi(2,1)\Phi(1,0)x(0)$$

$$\cdots\cdots\cdots\cdots\cdots\cdots\cdots\cdots\cdots\cdots\cdots\cdots$$

$$z(N) = H(N)x(N) = H(N)\Phi(N, N-1)x(N-1)$$
$$= H(N)\Phi(N, N-1) \cdots \Phi(1,0)x(0)$$

Now define

$$z_N = \begin{bmatrix} z(1) \\ \cdot \\ \cdot \\ \cdot \\ z(N) \end{bmatrix}$$

and

$$\Phi(i,0) = \Phi(i, i-1) \cdots \Phi(1,0)$$

for $i = 1, \ldots, N$. It is clear that z_N is an mN vector.
Letting

$$H_N = \begin{bmatrix} H(1)\Phi(1,0) \\ \cdot \\ \cdot \\ \cdot \\ H(N)\Phi(N,0) \end{bmatrix} \qquad (2.53)$$

which is obviously an $mN \times n$ matrix, we obtain the relation

$$z_N = H_N x(0) \tag{2.54}$$

Premultiplying Eq. (2.54) by H'_N, we get

$$[H'_N H_N] x(0) = H'_N z_N \tag{2.55}$$

From the definition of H_N in Eq. (2.53), we see that

$$H'_N H_N = \sum_{i=1}^{N} \Phi'(i,0) H'(i) H(i) \Phi(i,0)$$

We denote this symmetric $n \times n$ matrix by $M_d(0,N)$.

It now follows from Eq. (2.55) that

$$x(0) = M_d^{-1}(0,N) H'_N z_N \tag{2.56}$$

which shows that the system is completely observable if $M_d(0,N)$ is positive definite for some $N > 0$.

NECESSITY Now assume that the system is observable, but suppose that for some $x(0) \neq 0$, $x'(0) M_d(0,N) x(0) = 0$. Then, from Eq. (2.55), we see that

$$x'(0) H'_N z_N = 0$$

which, in view of Eq. (2.54), means that

$$z'_N z_N = 0 \tag{2.57}$$

This implies that $z(i) = 0$ for all i, $i = 1, \ldots, N$, and, as in the continuous-time case, we are left with the conclusion that there exist nonzero $x(0)$ which cannot be determined from $\{z(1), \ldots, z(N)\}$. This, however, contradicts the hypothesis of observability, and the proof is complete.

For the constant coefficient case, we have

Corollary 2.2 *The constant coefficient discrete linear system $x(k + 1) = \Phi x(k)$, $z(k + 1) = H x(k + 1)$, $k = 0, 1, \ldots$, is completely observable if and only if the $n \times mn$ matrix*

$$[H', \Phi'H', \ldots, (\Phi')^{n-1}H']$$

has rank n.

This corollary can be proved by essentially paralleling the steps in the proof of Theorem 2.2. The details are left as an exercise.

Discussion In our proofs of the above two theorems on observability, we showed that since $u(t)$ and $u(k)$ were assumed known, we needed only to consider the homogeneous or force-free system equations. The assumption that the control vector is known is reasonable on physical grounds, since we are generally free to specify it.

The relation between observability and estimation should be clear at this point. Indeed, in the proofs of the sufficiency in both Theorems 2.1 and 2.2, we have presented algorithms for determining $x(t_0)$ and $x(0)$ from available measurements [cf. Eqs. (2.46) and (2.56), respectively]. These results, along with the two expressions

$$x(t) = \Phi(t,t_0)x(t_0) + \int_{t_0}^{t} \Phi(t,\tau)C(\tau)u(\tau)\,d\tau$$

and

$$x(k + 1) = \Phi(k + 1, k)x(k) + \Psi(k + 1, k)u(k)$$

$k = 0, 1, \ldots$ permit us to determine the states' time histories exactly. In this sense, we have solved the estimation problem under ideal conditions—that is, with no disturbances and no measurement errors.

When we reintroduce these disturbances and measurement errors, we have, of course, a much more difficult problem. However, if the estimation problem cannot be solved under ideal circumstances, we would be ill-advised to pursue the problem when disturbances and measurement errors are present. For this reason, we shall assume that the systems with which we deal in the sequel in our study of estimation are completely observable.

2.7 CONTROLLABILITY IN CONTINUOUS AND DISCRETE LINEAR SYSTEMS [7–15]

CONTINUOUS LINEAR SYSTEMS

We consider the continuous linear system

$$\dot{x} = F(t)x + C(t)u(t) \tag{2.58}$$

for $t \geq t_0$, where $x(t_0)$ is known but $u(t)$ is not specified. We concern ourselves here with the problem of transferring the state of the system of Eq. (2.58) from $x(t_0)$ to some desired terminal state $x(t_1) = x^1$, where t_1 is finite. By introducing the change of coordinates $y(t) = x(t) - x^1$, the problem becomes that of a transfer from some $x(t_0)$ to the origin in a finite time.

We now introduce the following definition.

DEFINITION

The continuous linear system of Eq. (2.58) is controllable at time t_0 if there exists a piecewise continuous control function $u(t)$ depending on $x(t_0)$ and *defined over some finite interval $t_0 \leq t \leq t_1$ for which $x(t_1) = 0$. If this is true for all $x(t_0)$ and t_0, the system is completely controllable.*

A necessary and sufficient condition for controllability is given in the following theorem.

Theorem 2.3 *The continuous linear system of Eq. (2.58) is completely controllable if and only if the symmetric $n \times n$ matrix*

$$W_c(t_0, t_1) = \int_{t_0}^{t_1} \Phi(t_0, t) C(t) C'(t) \Phi'(t_0, t) \, dt \tag{2.59}$$

is positive definite for some $t_1 > t_0$.

Proof

SUFFICIENCY From Eq. (2.39) for $t = t_1$,

$$x(t_1) = \Phi(t_1, t_0) x(t_0) + \int_{t_0}^{t_1} \Phi(t_1, \tau) C(\tau) u(\tau) \, d\tau \tag{2.60}$$

Now let

$$u(\tau) = -C'(\tau) \Phi'(t_0, \tau) W_c^{-1}(t_0, t_1) x(t_0) \tag{2.61}$$

for $t_0 \leq \tau \leq t_1$. Then, substituting Eq. (2.61) into Eq. (2.60), we have

$$x(t_1) = \Phi(t_1, t_0) x(t_0) \\ - \left[\int_{t_0}^{t_1} \Phi(t_1, \tau) C(\tau) C'(\tau) \Phi'(t_0, \tau) \, d\tau \right] W_c^{-1}(t_0, t_1) x(t_0) \tag{2.62}$$

But, from Eq. (2.22),

$$\Phi(t_1, \tau) = \Phi(t_1, t_0) \Phi(t_0, \tau)$$

for all t_0. Therefore, the integral in Eq. (2.62) becomes

$$\Phi(t_1, t_0) \int_{t_0}^{t_1} \Phi(t_0, \tau) C(\tau) C'(\tau) \Phi'(t_0, \tau) \, d\tau = \Phi(t_1, t_0) W_c(t_0, t_1)$$

from which it follows that

$$x(t_1) = \Phi(t_1, t_0) x(t_0) - \Phi(t_1, t_0) W_c(t_0, t_1) W_c^{-1}(t_0, t_1) x(t_0)$$

giving

$$x(t_1) = 0$$

Hence, the system is completely controllable if $W_c(t_0, t_1)$ is positive definite for some finite $t_1 > t_0$.

NECESSITY Suppose now that the system is controllable, but that for some $x(t_0) \neq 0$,

$$x'(t_0) W_c(t_0,t_1) x(t_0) = 0 \qquad (2.63)$$

Let a control $u^*(t)$ be defined by

$$u^*(t) = -C'(t)\Phi'(t_0,t)x(t_0) \qquad (2.64)$$

for $t_0 \leq t \leq t_1$. Then,

$$u^{*\prime}(t)u^*(t) = x'(t_0)\Phi(t_0,t)C(t)C'(t)\Phi'(t_0,t)x(t_0)$$

Integrating this result between the limits of t_0 and t_1 and utilizing the definition of $W_c(t_0,t_1)$ and Eq. (2.63), we obtain

$$\int_{t_0}^{t_1} u^{*\prime}(t)u^*(t) \, dt = x'(t_0) \left[\int_{t_0}^{t_1} \Phi(t_0,t)C(t)C'(t)\Phi'(t_0,t) \, dt \right] x(t_0)$$
$$= x'(t_0) W_c(t_0,t_1) x(t_0) = 0 \qquad (2.65)$$

But $C(t)$ and $\Phi(t_0,\tau)$ in Eq. (2.64) are continuous functions of time, and, therefore, $u^*(t)$ is also. Hence, it follows from Eq. (2.65) that $u^*(t) = 0$ for all t, where $t_0 \leq t \leq t_1$.

Since the system is completely controllable, there must exist a control function $u(t)$ for which $x(t_1) = 0$. Then Eq. (2.60) becomes

$$\Phi(t_1,t_0)x(t_0) = - \int_{t_0}^{t_1} \Phi(t_1,\tau)C(\tau)u(\tau) \, d\tau$$

which can also be expressed as

$$x(t_0) = - \int_{t_0}^{t_1} \Phi(t_0,\tau)C(\tau)u(\tau) \, d\tau \qquad (2.66)$$

Now from Eqs. (2.66) and (2.64), we obtain the result

$$x'(t_0)x(t_0) = -x'(t_0) \int_{t_0}^{t_1} \Phi(t_0,\tau)C(\tau)u(\tau) \, d\tau$$
$$= - \int_{t_0}^{t_1} x'(t_0)\Phi(t_0,\tau)C(\tau)u(\tau) \, d\tau$$
$$= \int_{t_0}^{t_1} u^{*\prime}(\tau)u(\tau) \, d\tau$$

which means that

$$x'(t_0)x(t_0) = 0$$

since $u^*(t) = 0$ for $t_0 \leq t \leq t_1$. Hence, we are led to the conclusion that $x(t_0) = 0$, which contradicts the hypothesis that $x(t_0) \neq 0$.

For constant coefficient continuous linear systems, we have the following corollary which we state without proof.

Corollary 2.3 *The constant coefficient continuous linear system $\dot{x} = Fx + Cu$, $t \geq 0$, is completely controllable if and only if the $n \times nr$ matrix*

$$[C, FC, \ldots, F^{n-1}C]$$

has rank n.

Example 2.6 For the system in Example 2.2, in which we expressed the system equations in terms of two different sets of state variables, we had

$$F = \begin{bmatrix} 0 & 1 & 0 \\ 0 & f_1 & f_2 \\ 0 & f_3 & f_4 \end{bmatrix} \quad \text{and} \quad C = \begin{bmatrix} 0 \\ 0 \\ c \end{bmatrix}$$

in the first formulation and

$$F = \begin{bmatrix} 0 & 1 & 0 \\ 0 & 0 & 1 \\ 0 & b & a \end{bmatrix} \quad \text{and} \quad C = \begin{bmatrix} 0 \\ 0 \\ c \end{bmatrix}$$

in the second.

For the first formulation, we get

$$[C, FC, F^2C] = \begin{bmatrix} 0 & 0 & f_2 c \\ 0 & f_2 c & (f_1 f_2 + f_2 f_4)c \\ c & f_4 c & (f_2 f_3 + f_4{}^2)c \end{bmatrix}$$

which has rank three if $f_2 c = K_T/JL \neq 0$. In the second formulation,

$$[C, FC, F^2C] = \begin{bmatrix} 0 & 0 & c \\ 0 & c & ac \\ c & ac & bc + a^2 c \end{bmatrix}$$

which has rank three as long as $c = K_T/JL \neq 0$. (We recall from Example 2.2 that in the first formulation $c = 1/L$, whereas in the second formulation, $c = K_T/JL$.)

We observe in this example that controllability is a property of the system and not of the particular coordinate system (choice of state variables) in which the system is represented. It can be shown that this is true in general for both controllability and observability.

DISCRETE LINEAR SYSTEMS

Finally, we treat the question of controllability for the discrete linear system

$$x(k + 1) = \Phi(k + 1, k)x(k) + \Psi(k + 1, k)u(k) \tag{2.67}$$

where $k = 0, 1, \ldots$, $x(0)$ is assumed known and all the other terms have been defined previously.

Analogous to the definition of controllability for continuous linear systems, we have the following definition.

DEFINITION

The discrete linear system of Eq. (2.67) is controllable at time $k = 0$ (corresponding to an initial time t_0) if there exists a control sequence $\{u(0),$ $u(1), \ldots, u(N-1)\}$ depending on $x(0)$ and the initial time for which $x(N) = 0$ where N is finite. If this is true for all $x(0)$ and initial times, the system is completely controllable.

We are led now to the following theorem.

Theorem 2.4   *The discrete linear system of Eq. (2.67) is completely controllable if and only if the symmetric $n \times n$ matrix*

$$W_d(0,N) = \sum_{i=1}^{N} \Phi(0,i)\Psi(i, i-1)\Psi'(i, i-1)\Phi'(0,i) \tag{2.68}$$

is positive definite for some finite $N > 0$ where

$$\Phi(0,i) = \Phi(0,1) \cdots \Phi(i-1, i), \, i = 1, 2, \ldots, N.$$

Proof

SUFFICIENCY   It can be shown by recursive application of Eq. (2.67) that

$$x(N) = \Phi(N,0)x(0) + \sum_{i=1}^{N} \Phi(N,i)\Psi(i, i-1)u(i-1) \tag{2.69}$$

where $\Phi(N,i) = \Phi(N, N-1) \cdots \Phi(i+1, i), \, i = N-1, \ldots,$ 1, 0.   Now let

$$u(i-1) = -\Psi'(i, i-1)\Phi'(0,i)W_d^{-1}(0,N)x(0) \tag{2.70}$$

for $i = 1, \ldots, N$.   Substituting Eq. (2.70) into Eq. (2.69), we have

$$x(N) = \Phi(N,0)x(0)$$

$$- \sum_{i=1}^{N} \Phi(N,i)\Psi(i, i-1)\Psi'(i, i-1)\Phi'(0,i)W_d^{-1}(0,N)x(0)$$

But $\Phi(N,i) = \Phi(N,0)\Phi(0,i)$ for all initial times 0.   Hence,

$$x(N) = \Phi(N,0)x(0) - \Phi(N,0) \sum_{i=1}^{N}$$

$$\Phi(0,i)\Psi(i, i-1)\Psi'(i, i-1)\Phi'(0,i)[W_d^{-1}(0,N)x(0)]$$

$$= \Phi(N,0)x(0) - \Phi(N,0)W_d(0,N)W_d^{-1}(0,N)x(0)$$

from which $x(N) = 0$.   Therefore, the system is completely controllable if $W_d(0,N)$ is positive definite for some finite $N > 0$.

NECESSITY The proof of necessity essentially parallels that for continuous linear systems with certain modifications, and is left as an exercise.

For the constant coefficient case, we have the following result.

Corollary 2.4 *The constant coefficient discrete linear system* $x(k + 1) = \Phi x(k) + \Psi u(k)$, $k = 0, 1, \ldots$, *is completely controllable if and only if the* $n \times nr$ *matrix*

$$[\Psi, \Phi\Psi, \ldots, \Phi^{(n-1)}\Psi]$$

has rank n.

Example 2.7 For the second-order constant coefficient discrete linear system of Example 2.4, we recall that

$$\Phi = \begin{bmatrix} 1 & T \\ 0 & 1 \end{bmatrix} \quad \text{and} \quad \Psi = \begin{bmatrix} \frac{1}{2}bT^2 \\ bT \end{bmatrix}$$

for which we see that

$$[\Psi, \Phi\Psi] = \begin{bmatrix} \frac{1}{2}bT^2 & \frac{3}{2}bT^2 \\ bT & bT \end{bmatrix}$$

This matrix obviously has rank two for all $T > 0$ and nonzero b.

Discussion Similarities between Theorems 2.1 and 2.2 for observability and Theorems 2.3 and 2.4 for controllability, respectively, are quite evident. For example, Theorem 2.3 can be obtained from Theorem 2.1 by making the following changes in the latter:

Observable $\rightarrow$ Controllable
$M_c(t_0,t_1) \quad \rightarrow W_c(t_0,t_1)$
$\Phi(t,t_0) \quad \rightarrow \Phi'(t_0,t)$
$H(t) \quad \rightarrow C'(t)$

The situation is similar for Theorems 2.2 and 2.4, as well as for the corresponding corollaries.

This property was first observed by Kalman [7], who termed it *duality*. Thus, observability and controllability are dual properties of linear systems. The exploration of the many ramifications of duality is beyond the scope of this book, and no attempt will be made to pursue this topic further. However, much of the discussion which was presented earlier on observability is pertinent here for controllability in this sense of duality.

We note for example that Eq. (2.61) in the proof of Theorem 2.3 and Eq. (2.70) in the proof of Theorem 2.4 are both algorithms for effecting a transfer of a given initial state to the origin in finite time. This makes evident the connection between control and controllability. In fact, it can be shown [12] that the control input which is given by Eq. (2.61) minimizes the control "energy"

$$\int_{t_0}^{t_1} u'(t)u(t) \, dt$$

while transferring $x(t_0)$ to the origin during the fixed time interval $[t_0, t_1]$. In other words, we have here the solution of a particular optimal control problem. The same is true of Eq. (2.70) for the discrete-time case, where the control energy is

$$\sum_{i=1}^{N} u'(i-1)u(i-1)$$

With this introduction to observability and controllability, the reader should have a good understanding of the nature of estimation and control. Basically, our work in the sequel is concerned with developing and applying techniques to resolve questions regarding observability and controllability when system disturbances and measurement errors of a particular type are present.

2.8 NONLINEAR SYSTEMS

In many cases of practical interest, physical systems cannot be modeled by ordinary linear differential equations or linear difference equations, but must be represented by nonlinear systems of equations.

In such problems, it is often convenient to linearize the system equations about some assumed set of nominal conditions and develop algorithms for estimation and control about these nominal conditions.

We choose to conclude this chapter with a development and illustration of the techniques for obtaining the linearized system equations for a given nonlinear system.

We consider systems which can be modeled by the relations

$$\dot{x} = f[x, w(t), u(t), t] \tag{2.71}$$

and

$$z(t) = h[x(t), v(t), t] \tag{2.72}$$

where $t \geq t_0$ and x, w, u, and v are the same as before. We assume that f is an n-dimensional vector-valued function of the indicated variables which is continuous and continuously differentiable with respect to all the elements of x, w, and u. Also, we take h to be an m-dimensional vector-valued function of the indicated variables which is continuous and continuously differentiable with respect to the elements of x and v.

For a given $x(t_0)$, and given piecewise continuous $w(t) = \bar{w}(t)$ and $u(t) = \bar{u}(t)$, we know from the theory of ordinary differential equations that Eq. (2.71) can be solved to obtain some $x(t) = \bar{x}(t)$. Then, for a given $v(t) = \bar{v}(t)$, piecewise continuous, we can obtain $z(t) = \bar{z}(t)$ by substituting $\bar{x}(t)$ and $\bar{v}(t)$ into Eq. (2.72).

We take $\bar{x}(t)$, $\bar{w}(t)$, $\bar{u}(t)$, $\bar{v}(t)$, and $\bar{z}(t)$, where $t \geq t_0$, as the nominal values about which we wish to linearize the system in Eqs. (2.71) and (2.72). To carry out this linearization, we define

$$x(t) = \bar{x}(t) + \Delta x(t) \qquad w(t) = \bar{w}(t) + \Delta w(t)$$

$$u(t) = \bar{u}(t) + \Delta u(t) \qquad v(t) = \bar{v}(t) + \Delta v(t)$$

$$z(t) = \bar{z}(t) + \Delta z(t)$$

and then perform a Taylor-series expansion of the two systems of equations.

Let us consider first the ith member of the set of equations in Eq. (2.71),

$$\dot{x}_i = f_i[x, w(t), u(t), t]$$

for some $i = 1, \ldots, n$. Expanding in a Taylor series, we get

$$\Delta\dot{x}_i = \left(\frac{\partial f_i}{\partial x_1}\right)_0 \Delta x_1 + \cdots + \left(\frac{\partial f_i}{\partial x_n}\right)_0 \Delta x_n + \left(\frac{\partial f_i}{\partial w_1}\right)_0 \Delta w_1 + \cdots$$

$$+ \left(\frac{\partial f_i}{\partial w_p}\right)_0 \Delta w_p + \left(\frac{\partial f_i}{\partial u_1}\right)_0 \Delta u_1 + \cdots + \left(\frac{\partial f_i}{\partial u_r}\right)_0 \Delta u_r$$

$$+ \text{ second-order terms} \quad (2.73)$$

where the subscript 0 denotes that the indicated partial derivatives are to be evaluated at $x = \bar{x}(t)$, $w = \bar{w}(t)$, and $u = \bar{u}(t)$.

Assuming that the second- and higher-order terms in Eq. (2.73) are negligible, and utilizing vector-matrix notation, we can write

$$\Delta\dot{x}_i = \left[\frac{\partial f_i}{\partial x_1} \cdots \frac{\partial f_i}{\partial x_n}\right]_0 \Delta x + \left[\frac{\partial f_i}{\partial w_1} \cdots \frac{\partial f_i}{\partial w_p}\right]_0 \Delta w(t)$$

$$+ \left[\frac{\partial f_i}{\partial u_1} \cdots \frac{\partial f_i}{\partial u_r}\right]_0 \Delta u(t) \quad (2.74)$$

for each $i = 1, \ldots, n$, where

$$\Delta x = \begin{bmatrix} \Delta x_1 \\ \cdot \\ \cdot \\ \cdot \\ \Delta x_n \end{bmatrix} \qquad \Delta w = \begin{bmatrix} \Delta w_1 \\ \cdot \\ \cdot \\ \cdot \\ \Delta w_p \end{bmatrix} \qquad \Delta u = \begin{bmatrix} \Delta u_1 \\ \cdot \\ \cdot \\ \cdot \\ \Delta u_r \end{bmatrix}.$$

Then, the entire system of equations in Eq. (2.74) can be put into the more compact form

$$\Delta \dot{x} = F(t) \, \Delta x + G(t) \, \Delta w(t) + C(t) \, \Delta u(t) \qquad (2.75)$$

for $t \geq t_0$, where $F(t)$ is the $n \times n$ matrix

$$F(t) = [f_{ij}(t)]$$

$$f_{ij}(t) = \left(\frac{\partial f_i}{\partial x_j}\right)_0$$

where $i, j = 1, \ldots, n$; $G(t)$ is the $n \times p$ matrix

$$G(t) = [g_{ij}(t)]$$

$$g_{ij}(t) = \left(\frac{\partial f_i}{\partial w_j}\right)_0$$

where $i = 1, \ldots, n, j = 1, \ldots, p$; and $C(t)$ is the $n \times r$ matrix

$$C(t) = [c_{ij}(t)]$$

$$c_{ij}(t) = \left(\frac{\partial f_i}{\partial u_j}\right)_0$$

where $i = 1, \ldots, n, j = 1, \ldots, r$.

Equation (2.75) is obviously of the same form as Eq. (2.1).

In a similar way, we obtain

$$\Delta z(t) = H(t) \, \Delta x(t) + A(t) \, \Delta v(t)$$

where $H(t)$ is the $m \times n$ matrix

$$H(t) = [h_{ij}(t)]$$

$$h_{ij}(t) = \left(\frac{\partial h_i}{\partial x_j}\right)_0$$

where $i = 1, \ldots, m, j = 1, \ldots, n$; and $A(t)$ is the $m \times m$ matrix

$$A(t) = [a_{ij}(t)]$$

$$a_{ij}(t) = \left(\frac{\partial h_i}{\partial v_j}\right)_0$$

where $i = 1, \ldots, m, j = 1, \ldots, m$. Defining $\Delta v^*(t) = A(t) \, \Delta v(t)$, we get

$$\Delta z(t) = H(t) \, \Delta x(t) + \Delta v^*(t) \qquad (2.76)$$

which is of the same form as Eq. (2.2).

It must be borne in mind here that the linearized system of equations is only valid for sufficiently small perturbations about the assumed nominal values.

Example 2.8 To illustrate use of the above procedure, we consider the motion of a satellite of mass m about a spherical planet of mass M in which the motion is considered in a planet-centered inertial coordinate system (cylindrical coordinates) as shown. Assuming that the planet's force field obeys an inverse square law, that the only other forces present are the two thrust forces $u_r(t)$ and $u_\theta(t)$, and that the satellite's initial position and velocity vectors lie in the plane, we know from elementary particle mechanics that the satellite's motion is confined to the plane and is governed by the two equations

$$\ddot{r} = r\dot{\theta}^2 - \frac{\gamma}{r^2} + \frac{1}{m} u_r(t)$$

and

$$\ddot{\theta} = -\frac{2\dot{r}\dot{\theta}}{r} + \frac{1}{m} u_\theta(t)$$

where $\gamma = GM$ and G is the universal gravitational constant.

Defining $x_1 = r$, $x_2 = \dot{r}$, $x_3 = \theta$, $x_4 = \dot{\theta}$, $u_1 = u_r$, and $u_2 = u_\theta$, we have

$$\dot{x}_1 = x_2$$

$$\dot{x}_2 = x_1 x_4^2 - \frac{\gamma}{x_1^2} + \frac{1}{m} u_1(t)$$

$$\dot{x}_3 = x_4 \qquad\qquad\qquad\qquad (2.77)$$

$$\dot{x}_4 = -\frac{2x_2 x_4}{x_1} + \frac{1}{m} u_2(t)$$

which is of the form in Eq. (2.71).

Letting the bar denote nominal values again, we obtain the relations

$$\Delta\dot{x}_1 = \Delta x_2$$

$$\Delta\dot{x}_2 = \left[\bar{x}_4^2(t) + \frac{2\gamma}{\bar{x}_1^3(t)}\right]\Delta x_1 + [2\bar{x}_1(t)\bar{x}_4(t)]\,\Delta x_4 + \frac{1}{m}\Delta u_1(t)$$

$$\Delta\dot{x}_3 = \Delta x_4$$

$$\Delta\dot{x}_4 = \left[\frac{2\bar{x}_2(t)\bar{x}_4(t)}{\bar{x}_1^2(t)}\right]\Delta x_1 - \left[\frac{2\bar{x}_4(t)}{\bar{x}_1(t)}\right]\Delta x_2 - \left[\frac{2\bar{x}_2(t)}{\bar{x}_1(t)}\right]\Delta x_4 + \frac{1}{m}\Delta u_2(t)$$

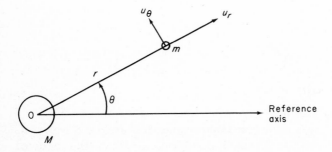

Fig. 2.11 Schematic for satellite-planet system.

It is now clear in this example that

$$F(t) = \begin{bmatrix} 0 & 1 & 0 & 0 \\ \bar{x}_4{}^2(t) + \dfrac{2\gamma}{\bar{x}_1{}^3(t)} & 0 & 0 & 2\bar{x}_1(t)\bar{x}_4(t) \\ 0 & 0 & 0 & 1 \\ \dfrac{2\bar{x}_2(t)\bar{x}_4(t)}{\bar{x}_1{}^2(t)} & \dfrac{-2\bar{x}_4(t)}{\bar{x}_1(t)} & 0 & \dfrac{-2\bar{x}_2(t)}{\bar{x}_1(t)} \end{bmatrix}$$

and that

$$C(t) = \begin{bmatrix} 0 & 0 \\ \dfrac{1}{m} & 0 \\ 0 & 0 \\ 0 & \dfrac{1}{m} \end{bmatrix}$$

The nominal values $\bar{x}_1(t)$, $\bar{x}_2(t)$, $\bar{x}_3(t)$, and $\bar{x}_4(t)$ are obtained by solving the system in Eq. (2.77) for given $u_1(t) = \bar{u}_1(t)$ and $u_2(t) = \bar{u}_2(t)$, along with the initial conditions $x_1(t_0) = r(t_0)$, $x_2(t_0) = \dot{r}(t_0)$, $x_3(t_0) = \theta(t_0)$, and $x_4(t_0) = \dot{\theta}(t_0)$.

Assuming finally that the measurement made on the satellite during its motion is simply its distance from the surface of the planet, we have the scalar measurement equation

$$\begin{aligned} z(t) &= r(t) - r_0 + v(t) \\ &= x_1(t) - r_0 + v(t) \end{aligned}$$

where r_0 is the planet's radius. Then obviously

$$\Delta z(t) = \Delta x_1(t) + \Delta v(t)$$

or

$$\Delta z(t) = [1 \quad 0 \quad 0 \quad 0] \, \Delta x(t) + \Delta v(t)$$

where Δx is the four vector with elements Δx_1, Δx_2, Δx_3, and Δx_4. Hence, we have

$$H(t) = [1 \quad 0 \quad 0 \quad 0]$$

in this example.

The discrete-time analog of the system in Eqs. (2.71) and (2.72) is

$$x(k + 1) = f[x(k), w(k), u(k), k + 1] \tag{2.78}$$

and

$$z(k + 1) = h[x(k + 1), v(k + 1), k + 1] \tag{2.79}$$

respectively.

The procedure for obtaining the linearized system equations for this system is the same as above except that partial derivatives are now evaluated only at the discrete time points $k = 0, 1, \ldots$. For example, $\Phi(k + 1, k)$ is the $n \times n$ matrix

$$\Phi(k + 1, k) = [\phi_{ij}(k + 1, k)]$$

where

$$\phi_{ij}(k + 1, k) = \left(\frac{\partial f_i}{\partial x_j}\right)_{\substack{x = \bar{x}(k) \\ w = \bar{w}(k) \\ u = \bar{u}(k)}}$$

and the bar denotes the assumed nominal values.

PROBLEMS

2.1. For the matrices

$$A = \begin{bmatrix} 2 & 0 \\ -5 & 3 \end{bmatrix} \qquad B = \begin{bmatrix} 1 & 1 \\ 0 & 2 \end{bmatrix} \qquad C = \begin{bmatrix} 2 & 1 & 1 \\ 0 & 3 & -2 \\ 0 & 3 & 1 \end{bmatrix}$$

determine

(a) AB, BA, and $A^2 - B$
(b) C, $|C|$, $|C'|$, adj C, and C^{-1}
(c) $|AB|$ and $(AB)^{-1}$
(d) Rank AB and rank C
(c) The characteristic values and characteristic vectors of C
(d) $M^{-1}CM$, where M is a 3×3 matrix whose columns are the characteristic vectors of C. (Is M an orthogonal matrix?)
(e) tr AB^2
(f) A 3×3 symmetric matrix C^* for which $x'Cx = x'C^*x$, where x is a 3 vector
(g) Whether or not C is positive definite

2.2. Show that for a square matrix A,

(a) $(A')^{-1} = (A^{-1})'$ if A is nonsingular

(b) $|A^{-1}| = |A|^{-1} = \dfrac{1}{|A|}$ if $|A| \neq 0$

(c) $\nabla_x[(y - Ax)'B(y - Ax)] = 2(x'A'BA - y'BA) = -2(y - Ax)'BA$ where B is symmetric.

(d) tr $(Axx') = x'Ax$

(e) Rank $(AB) \leq \min$ [rank A, rank B]

2.3. Under what conditions is it true that

(a) $(A - B)(A + B) = A^2 - B^2$?
(b) $A = B$ if $x'Ax = x'Bx$?
(c) $B = C$ if $AB = AC$ where A is square?

2.4. Determine $\dot{A}(t)$, $\int A(t) \, dt$, and $(d/dt)[A^{-1}(t)]$ for

$$A(t) = \begin{bmatrix} t + 1 & \sin^2 t \\ 1 & 3 \end{bmatrix}$$

2.5. Consider the given electrical circuit (Fig. P 2.5) where $e(t)$ is the input (control) variable and the output $z(t)$ is the voltage measured across the resistor R_2.

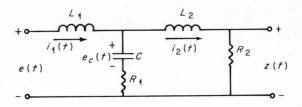

Fig. P 2.5

(a) Let $x_1 = i_1$, $x_2 = i_2$, and $x_3 = e_c$ be the system state variables and express the system equations in state variable form. Assume that the measurement error in $z(t)$ is negligible.

(b) Formulate (but do not solve) the equations from which the state transition matrix for the system can be determined. Set $t_0 = 0$ for this part.

(c) Is the system completely observable? Completely controllable?

2.6. Consider the heat exchanger depicted in Fig. P 2.6 in which a gas enters at a temperature θ_1 and is heated and expelled at a temperature θ_2. For simplicity, assume that there is uniform mixing within the chamber so that the temperature can be taken as θ_2 throughout the volume of the exchanger.

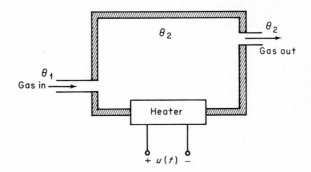

Fig. P 2.6

Heat is supplied to the exchanger at a rate $q(t)$ (Btu/sec) which is controlled by varying a supply voltage $u(t)$. Assume that the heater does not respond instantaneously to changes in $u(t)$, but that its response is adequately modeled by a first-order lag:

$$\dot{q} = -\alpha q + \beta u(t)$$

for $t \geq 0$, where α, β, $q(0) > 0$, and $u(t) \geq 0$.

Assume further that the exchanger is perfectly insulated, and that the only heat lost from it is that which is carried away by the expelled gas. The corresponding heat rate is $q_2(t) = k\theta_2$, where $k > 0$, Btu/(sec)(°F), is the conductance of the gas. Similarly, assume that the rate at which heat is supplied at the inlet is $q_1(t) = k\theta_1$.

Finally, the rate at which heat is absorbed by the gas in the exchanger can be taken as $q_a(t) = c\theta_2$, where $c > 0$, Btu/°F, is the specific heat of the gas.

(a) Express the equations for this system in state variable form taking θ_2 and q as the state variables, θ_1 as a disturbance variable, and u as the control variable. Also, assume that the measurement is $\theta_2(t)$.

(b) Determine the state transition matrix for the system and use it to express the solution for the system's state in the form of Eq. (2.19).

(c) Set $t_0 = 0$, $q(0) = 0$, $\theta_2(0) = 0$, $\theta_1(t) = 0$ and assume that there is no measurement error. Express the system input-output in the form

$$\theta_2(t) = \int_0^t B(t,\tau)u(\tau)\,d\tau$$

(d) Determine the equations for the discrete-time analog of the system for a sampling interval of $T = 1$.

(e) Is the continuous-time system completely observable? Completely controllable?

(f) Is the discrete-time system completely observable? Completely controllable?

2.7. In Eq. (2.19), the variable of integration on the right-hand side is τ. Hence, to evaluate the integral for given values of t and t_0, it is necessary to have the state transition matrix $\Phi(t,\tau)$ as a function of its second argument.

(a) Show that as a function of τ, $\Phi(t,\tau)$ must satisfy the partial differential equation

$$\frac{\partial \Phi(t,\tau)}{\partial \tau} = -\Phi(t,\tau)F(\tau)$$

with $\Phi(t,t) = I$ and $t_0 \leq \tau \leq t$.

(b) Let $\Theta(t,\tau)$ denote the state transition matrix for the system $\dot{y} = -F'(t)y$, $t \geq t_0$, where y is an n vector. What is the relation between $\Theta(t,\tau)$ and $\Phi(t,\tau)$? [The system $\dot{y} = -F'(t)y$ is called the *adjoint* of the system $\dot{x} = F(t)x$.]

2.8. If the system in Eqs. (2.37) and (2.38) is expressed in a different coordinate system by introducing the change of variables $y(t) = Ax(t)$ where A is $n \times n$ and nonsingular, show that the result in Theorem 2.1 is unchanged. In other words, show that observability is a property of the system which is independent of the particular coordinate frame in which the system equations are expressed. The same is then obviously also true of controllability.

2.9. Prove Corollary 2.2 by paralleling the steps in the proof of Theorem 2.2 for a constant coefficient discrete linear system.

2.10. Is the system in Example 2.4 completely observable?

2.11. Develop Eq. (2.69) by the recursive application of Eq. (2.67).

2.12. If the continuous linear system

$$\dot{x}_1 = x_2$$
$$\dot{x}_2 = -x_1 + u(t)$$
$$z(t) = x_1(t)$$

where $t \geq 0$, is "discretized" for a sampling interval T, is the resulting discrete linear system completely observable and completely controllable for all $T > 0$?

2.13. If in Example 2.8, $\bar{u}_1(t) = \bar{u}_r(t) = 0$, $\bar{u}_2(t) = \bar{u}_\theta(t) = 0$, and the satellite's nominal orbit is circular with radius r_0, show that the system is constant coefficient with

$$
F = \begin{bmatrix}
0 & 1 & 0 & 0 \\
3\omega_0{}^2 & 0 & 0 & 2r\omega_0 \\
0 & 0 & 0 & 1 \\
0 & -\dfrac{2\omega_0}{r_0} & 0 & 0
\end{bmatrix}
$$

where $\omega_0 = \sqrt{\gamma/r_0{}^3}$ is the orbital rate. Is this system completely observable for the measurement scheme in Example 2.8 where $H = [1 \quad 0 \quad 0 \quad 0]$? Is it completely controllable? What is the physical significance of these last two results?

2.14. In a second-order chemical process, two reacting substances A and B combine to form a compound C according to the chemical equation A + B → C. The rate at which C is produced is governed by the scalar differential equation

$$\dot{c} = ka(t)b(t) \qquad \text{for } t \geq 0$$

where $c(t) =$ amount of C present
$a(t) =$ amount of A present
$b(t) =$ amount of B present
$k =$ reaction constant > 0

If the initial amount of A is a_0 and that of B is b_0, and if αc units of A combine with βc units of B to form $(\alpha + \beta)c = c$ units of C, where $0 < \alpha < 1, 0 < \beta < 1, \alpha + \beta = 1$, then $a(t) = a_0 - \alpha c$ and $b(t) = b_0 - \beta c$. In this case, the reaction is governed by the equation

$$\dot{c} = k(a_0 - \alpha c)(b_0 - \beta c)$$

Physically, it should be noted that the equation makes sense if and only if c, $a_0 - \alpha c$, and $b_0 - \beta c$ are nonnegative.

(a) What is the range on $c(0)$ for which the system equation is physically meaningful?

(b) In a particular estimation problem, it is desired to determine the reaction constant k by measuring and processing $c(t)$ for $t \geq 0$. Assuming that $c(0)$, a_0, b_0, α, and β are known quite accurately, it is clear that a nominal $\bar{c}(t)$ can be computed for a nominal $\bar{k}$. Take $c(t) = \bar{c}(t) + \Delta c(t)$ and $k = \bar{k} + \Delta k$ and determine the linearized system equations where the state vector consists of $\Delta c(t)$ and Δk. Assume that the measurement equation is $z(t) = c(t) + v(t)$, where $v(t)$ is the measurement error.

(c) Repeat part b for the case in which k is known with sufficient accuracy, but a_0 is not, that is, where the roles of k and a_0 are interchanged.

2.15. Consider a rocket of mass m in vertical flight as shown in the figure. Assume that the rocket is subject to a drag force $d = m\rho(\dot{h})^2 e^{-h}$, where $\rho > 0$. (The model here, Fig. P 2.15, is that of an "exponential atmosphere" wherein the air density, and therefore the drag, decreases exponentially with altitude.) Assume further that the rocket is acted upon also by a gravitational force

$$g = \frac{\mu m}{(r_0 + h)^2}$$

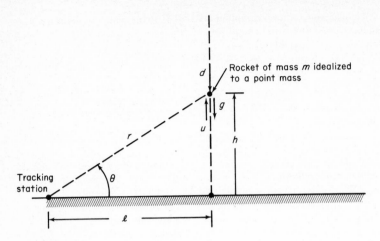

Fig. P 2.15

where μ and r_0 are positive constants, and by a thrust force u (upward). For simplicity, assume also that the mass of the rocket may be considered constant over the region of flight which is of interest.

(a) Determine the linearized equations of motion for a nominal $\bar{h}(t)$, $\dot{\bar{h}}(t)$, and $\bar{u}(t)$ all ≥ 0.

(b) Assuming that the rocket is "tracked" during its vertical flight by measuring its range r and its elevation angle θ at a tracking station located as shown ($l > 0$ is a known distance), determine the linearized measurement equations. Assume that measurement errors are additive, i.e.,

$$z(t) = \begin{bmatrix} r(t) + v_r(t) \\ \theta(t) + v_\theta(t) \end{bmatrix}$$

where $v_r(t)$ = error in range measurement and $v_\theta(t)$ = error in elevation angle measurement.

REFERENCES

1. DeRusso, P. M., R. J. Roy, and C. M. Close, "State Variables for Engineers," John Wiley & Sons, Inc., New York, 1965.
2. Zadeh, L. A., and C. A. Desoer, "Linear System Theory," McGraw-Hill Book Company, New York, 1963.
3. Kaplan, W., "Ordinary Differential Equations," Addison-Wesley Publishing Company, Inc., Reading, Mass., 1960.
4. Bellman, R. E., "Introduction to Matrix Analysis," McGraw-Hill Book Company, New York, 1960.
5. Hildebrand, F. B., "Methods of Applied Mathematics," 2d ed., Prentice-Hall, Inc., Englewood Cliffs, N.J., 1965.
6. Coddington, E. A., and N. Levinson, "Theory of Ordinary Differential Equations," McGraw-Hill Book Company, New York, 1955.
7. Kalman, R. E., On the General Theory of Control Systems, *Proc. 1st Intern. Congr. Autom. Control, London*, vol. 1, p. 481, 1961.

8. ———, Contributions to the Theory of Optimal Control, *Proc. Mexico City Conf. Ordinary Differential Equations*, Mexico City, 1959. Also, *Bol. Soc. Mat. Mexico*, Ser. 2, vol. 5, p. 102, 1960.

9. ———, New Methods in Wiener Filtering Theory, "Proceedings of the First Symposium on Engineering Applications of Random Function Theory and Probability," John Wiley & Sons, Inc., New York, 1963, pp. 270–388. Also, New Methods and Results in Linear Prediction and Filtering Theory, *Tech. Rept.* 61-1, Research Institute for Advanced Studies (RIAS), Martin Company, Baltimore, 1961.

10. ———, Canonical Structure of Linear Dynamical Systems, *Proc. Natl. Acad. Sci.*, vol. 48, p. 596, 1962.

11. ———, Mathematical Description of Linear Dynamical Systems, *J. SIAM*, Ser. A, vol. 1, p. 152, 1963.

12. Kalman, R. E., Y. C. Ho, and K. S. Narendra, Controllability of Linear Dynamical Systems, in "Contributions to Differential Equations," vol. 1, John Wiley & Sons, Inc., New York, 1962, pp. 189–213.

13. Kalman, R. E., and R. S. Bucy, New Results in Linear Filtering and Prediction Theory, *J. Basic Eng.*, vol. 83, p. 95, 1961.

14. Gilbert, E. G., Controllability and Observability in Multivariable Control Systems, *J. SIAM*, Ser. A, vol. 1, p. 128, 1963.

15. Kreindler, E., and P. E. Sarachik, On the Concepts of Controllability and Observability of Linear Systems, *IEEE Trans. Autom. Control*, vol. AC-9, p. 129, 1964.

3
Elements of Probability Theory

3.1 INTRODUCTION

We recall that the disturbance and measurement error vectors as well as the initial condition vector for both discrete and continuous linear systems were assumed to be arbitrary in Chap. 2.   In the sequel, we shall be concerned with a probabilistic description for these quantities.· Our motivation for such a description stems from the fact that, in many physical systems, we expect initial conditions, disturbances, and measurement errors to be random rather than deterministic in character.   The immediate goal which we have in mind, then, is to study those aspects of probability theory that we can utilize in the sequel to obtain a meaningful and useful probabilistic description of linear system behavior.

In brief, we shall examine the basic concepts of probability theory and conclude with a consideration of the multidimensional gaussian or normal distribution which plays a central role in the sequel.   Additionally, we shall pay special attention to the notion of conditional expectation because of its importance in both estimation and control.

We assume that the reader has had some previous exposure to probability theory and elementary set theory.

A more than adequate background for present purposes may be found, for example, in Feller [1, 2], Gnedenko [3], Papoulis [4], or Parzen [5].

3.2 DEFINITION OF PROBABILITY AND RANDOM VARIABLE

Let us consider an experiment in which chance or uncertainty is involved, e.g., the toss of a die or the measurement of the noise voltage present in the output of an electronic circuit at some specified time. The set of all possible outcomes of such an experiment is called the *sample space* of the experiment. We denote this set by Ω and its elements by ω. If Ω has a finite or denumerably infinite number of elements, it is called a *discrete* sample space. An example in point is the sample space for the game of roulette.

On the other hand, the sample space may consist of a nondenumerable number of elements such as the set of all real-valued continuous functions on the interval [0,1]. In such cases, Ω is a *continuous* sample space.

An *event* is defined as some specific class of outcomes of the experiment. In this connection, an event A is said to occur if and only if the observed outcome of the experiment is an element of A.

Now suppose that the experiment is performed N times and that in these N trials, an event A occurs $N(A)$ times. Then we say that the *probability* of the event A, denoted by $P(A)$, is defined by the relation

$$P(A) = \lim_{N \to \infty} \frac{N(A)}{N} \tag{3.1}$$

assuming that the indicated limit exists.

Since $0 \leq N(A) \leq N$, we have that $0 \leq P(A) \leq 1$. Also, we note that $P(\Omega) = 1$ and $P(\phi) = 0$, where ϕ is the empty set; i.e., there is no outcome.

This is the definition of probability in terms of the physically intuitive notion of "relative frequency of occurrence." However, an intuitive base is hardly adequate for a rigorous development of a theory of probability. Hence, the modern theory proceeds from a base in which probability is defined by three fundamental axioms. We now present this axiomatic definition.

Let Ω be a space of elements ω and $\mathfrak{F}$ a Borel field of subsets of Ω, that is, a class of subsets $A_1, A_2, \ldots, A_n, \ldots$ for which

  i. $\Omega \in \mathcal{F}$.

  ii. If $A \in \mathcal{F}$, then $A^* \in \mathcal{F}$, where A^* is the complement of A.

iii. If $A_1, A_2, \ldots, A_n, \ldots \in \mathcal{F}$, then

$$\bigcup_{i=1}^{\infty} A_i \in \mathcal{F}$$

It follows from (i) and (ii) that

$$\phi \in \mathcal{F}$$

since $\phi = \Omega^*$, and from (ii) and (iii) that

$$\bigcap_{i=1}^{\infty} A_i \in \mathcal{F}$$

since

$$[\bigcup_{i=1}^{\infty} A_1^*]^* = \bigcap_{i=1}^{\infty} A_i$$

In addition, for $A, B \in \mathcal{F}$, it is clear that

$$A - B = A \cap B^* \in \mathcal{F}$$

Now let $P(\cdot)$ be a real scalar-valued function on the family of events $\mathcal{F}$. We say that $P(\cdot)$ is a *probability function* which defines the *probability of the events* $A \in \mathcal{F}$ if and only if it satisfies the following three axioms:

(1) $P(A) \geq 0$ for all A

(2) $P(\Omega) = 1$

(3) For every collection of events $A_1, A_2, \ldots, A_n, \ldots$ in $\mathcal{F}$ for which $A_i \cap A_j = \phi$, $i \neq j$,

$$P(\bigcup_{i=1}^{\infty} A_i) = \sum_{i=1}^{\infty} P(A_i)$$

In axiom 3, the events are said to be *mutually exclusive*.

That these axioms agree with our intuitive notions of probability is evident from their statement. We have also the following two results which agree with our intuition.

First, we assert that $A \subset B$ implies that $P(A) \leq P(B)$. To see this, we note that $B = A \cup (A^* \cap B)$ and that $A \cap (A^* \cap B) = \phi$. Hence, the events A and $(A^* \cap B)$ are mutually exclusive, and from axiom 3 it follows that

$$P(B) = P(A) + P(A^* \cap B)$$

However, $P(A^* \cap B) \geq 0$ as a consequence of axiom 1, and the result follows immediately.

Now letting $B = \Omega$ in the above result and utilizing axiom 2, we see that $P(A) \leq P(\Omega) = 1$. In conjunction with axiom 1, this means that $0 \leq P(A) \leq 1$, which is the second result.

The triple $(\Omega, \mathcal{F}, P)$ is said to define a *probability space*. Obviously, $\mathcal{F}$ must be chosen such that the three axioms are satisfied, and it is not clear that any choice of $\mathcal{F}$ will suffice for this purpose. Indeed, it can be shown by more advanced methods that there exist cases where no probability function can be found which satisfies axiom 3 when $\mathcal{F}$ is taken as the family of all subsets of Ω. In such a case, $\mathcal{F}$ is too large. It is then natural to select $\mathcal{F}$ as a smaller family of subsets in Ω which includes the events of interest and satisfies conditions (i) to (iii) and for which a probability function which satisfies the three axioms can be found.

For our work, it suffices for us to consider Ω as the set of points in n-dimensional euclidean space and to let $\mathcal{F}$ be the collection of subsets in Ω of the form

$$\{\omega: \omega \leq a, \, \omega \in \Omega\}$$

where ω is an n vector and a is an n vector of specified value. By taking countable unions and intersections of such subsets, it is clear that subsets such as $\{\omega: a \leq \omega \leq b\}$, $\{\omega: a < \omega \leq b\}$, and $\{\omega: a < \omega < b\}$ are elements in $\mathcal{F}$, as are individual points such as $\omega = a$, where a and b are given n vectors.

The motivation for this is that in physical systems involving random phenomena, one is concerned with probabilities associated with quantities such as voltages, forces, pressures, concentrations of reacting substances, velocities, etc. For example, one might be concerned with the probability that the noise voltage on each of n channels in a complex communication system lies within prescribed limits at some specified time. Letting $x(T)$ denote the n vector of noise voltages at time T, the quantity of interest is

$$P(A) = P(a \leq x(T) \leq b)$$

where a and b are given n vectors and A is obviously the event

$$a \leq x(T) \leq b$$

The sample space is clearly continuous in this formulation.

In a more general context involving an arbitrary probability space $(\Omega, \mathcal{F}, P)$, it is desirable to introduce a mapping from the sample space to the real numbers in order to facilitate quantitative analysis. This is done via the notion of *random variable*.

DEFINITION

A random variable is a real-valued function $X(\omega)$ on Ω such that every set $\{\omega\colon X(\omega) \leq x\}$, x real, is an element of $\mathfrak{F}$, where $\omega \in \Omega$.

The function $X(\cdot)$ may be either scalar or vector-valued. In the latter case, it is called a *vector-valued random variable* or more simply a *random vector*. In the scalar case, the terms *scalar random variable* or simply *random variable* are then employed.

For our work, where Ω is n-dimensional euclidean space and $\mathfrak{F}$ is the collection of subsets of the form $\{\omega\colon \omega \leq a,\ \omega \in \Omega\}$, we take $X(\cdot)$ to be the identity mapping, thereby establishing a simple and useful one-to-one correspondence between euclidean n space and itself. The random vector $X(\cdot)$ is obviously continuous in this case; i.e., each of its components can take on any value in the range $(-\infty, \infty)$.

Since only continuous random vectors are of consequence in our work, we shall pay primary attention to them in our treatments of probability in this chapter and stochastic processes in the next chapter.

3.3 PROBABILITY FUNCTIONS

PROBABILITY DISTRIBUTION AND DENSITY FUNCTIONS

The usual description of a random vector is in terms of probability distribution, which is defined in the following way. Let X be a random n vector and x an arbitrary n vector. The scalar-valued function that specifies the probability that $X \leq x$ is called the *probability distribution function* of X and is denoted by

$$F_X(x) = P(X \leq x) = P(X_1 \leq x_1,\ \ldots,\ X_n \leq x_n) \tag{3.2}$$

Letting A denote the event $X \leq x$, we see that $F_X(x) = P(A)$, and we are actually talking about the function $P(\cdot)$, which is defined by the three axioms.

The function $F_X(x)$ is also called the *joint probability distribution* function of the random variables $X_1,\ \ldots,\ X_n$. However, we shall reserve the use of the adjective joint for cases involving two or more random vectors. Also, for simplicity, we shall write $F(x)$ instead of $F_X(x)$, where there is no ambiguity. The same simplification will also be used in other probability functions which we introduce.

Since Ω corresponds to the event $X \leq +\infty$, while ϕ corresponds to the event $X \leq -\infty$, it is clear that $F(+\infty) = 1$ and $F(-\infty) = 0$. Hence, $0 \leq F(x) \leq 1$.

Now let a and b be two n vectors for which $a < b$. Suppose that A is the event $X \leq a$ and B the event $X \leq b$. Clearly, $A \subset B$, so that $P(A) \leq P(B)$, or equivalently,

$$F(a) \leq F(b)$$

This means that $F(x)$ is a monotone nondecreasing function of x.

If there exists a function $f(\cdot)$ for which

$$F(x) = \int_{-\infty}^{x_1} \cdots \int_{-\infty}^{x_n} f(\xi_1, \ldots, \xi_n) \, d\xi_1 \cdots d\xi_n$$

holds for all x, this function is called the *probability density function* of X. The above expression is sometimes written more simply by using $f(\xi)$ to denote $f(\xi_1, \ldots, \xi_n)$ and/or $d\xi$ to denote $d\xi_1 \cdots d\xi_n$.

It is clear that

$$f(x) = \frac{\partial^n F}{\partial x_1 \cdots \partial x_n}$$

and that $f(x) \geq 0$. These two results follow from the fundamental theorem of calculus and the monotone nondecreasing nature of $F(\cdot)$, respectively.

Let us now consider a situation where X is a 2 vector and examine the event $\{x_1 < X_1 \leq x_1 + \Delta x_1, x_2 < X_2 \leq x_2 + \Delta x_2\}$, where Δx_1 and $\Delta x_2 > 0$ under the assumption that

$$f(x_1, x_2) = \frac{\partial^2 F(x_1, x_2)}{\partial x_1 \, \partial x_2}$$

exists and is continuous. From a consideration of events in the $X_1 X_2$ plane (see Fig. 3.1), it is relatively easy to show that

$$P(x_1 \leq X_1 \leq x_1 + \Delta x_1, x_2 \leq X_2 \leq x_2 + \Delta x_2)$$
$$= F(x_1 + \Delta x_1, x_2 + \Delta x_2) - F(x_1, x_2 + \Delta x_2) - F(x_1 + \Delta x_1, x_2)$$
$$+ F(x_1, x_2) \quad (3.3)$$

From calculus, we recall the definition

$$\frac{\partial^2 F(x_1, x_2)}{\partial x_1 \, \partial x_2} = \lim_{\substack{\Delta x_1 \to 0 \\ \Delta x_2 \to 0}} \frac{F(x_1 + \Delta x_1, x_2 + \Delta x_2) - F(x_1, x_2 + \Delta x_2) - F(x_1 + \Delta x_1, x_2) + F(x_1, x_2)}{\Delta x_1 \, \Delta x_2}$$

Hence,

$$\lim_{\substack{\Delta x_1 \to 0 \\ \Delta x_2 \to 0}} \frac{P(x_1 \leq X_1 \leq x_1 + \Delta x_1, x_2 \leq X_2 \leq x_2 + \Delta x_2)}{\Delta x_1 \, \Delta x_2} = f(x_1, x_2)$$

and we can write

$$P(x_1 \leq X_1 \leq x_1 + dx_1, x_2 \leq X \leq x_2 + dx_2) = f(x_1, x_2) \, dx_1 \, dx_2.$$

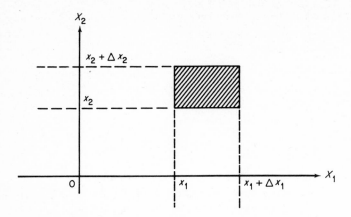

Fig. 3.1 Two-dimensional representation of the event $\{x_1 \leq X_1 \leq x_1 + \Delta x_1,\ x_2 \leq X_2 \leq x_2 + \Delta x_2\}$.

Then, for an arbitrary closed region R in the plane,

$$P(X \in R) = \iint_R f(\xi_1, \xi_2)\, d\xi_1\, d\xi_2$$

where X is a 2 vector.   In particular, for $R = \{\xi : a \leq \xi \leq b\}$, where

$$\xi = \begin{bmatrix} \xi_1 \\ \xi_2 \end{bmatrix} \qquad a = \begin{bmatrix} a_1 \\ a_2 \end{bmatrix} \qquad \text{and} \qquad b = \begin{bmatrix} b_1 \\ b_2 \end{bmatrix}$$

we have

$$P(a \leq X \leq b) = \int_{a_1}^{b_1} \int_{a_2}^{b_2} f(\xi_1, \xi_2)\, d\xi_2\, d\xi_1$$

The above argument can be extended to n-dimensional euclidean space to obtain the results

$$P(X \in R) = \int \cdots \int_R f(\xi)\, d\xi_1 \cdots d\xi_n$$

and

$$P(a \leq X \leq b) = \int_{a_1}^{b_1} \cdots \int_{a_n}^{b_n} f(\xi)\, d\xi_1 \cdots d\xi_n \qquad (3.4)$$

where a, b, X, and ξ are now n vectors and R is a closed region in euclidean n space.

Returning to the two-dimensional case, we consider the event $\{X_1 \leq x_1,\ X_2 \leq \infty\}$.   We call the function

$$F(x_1, \infty) = P(X_1 \leq x_1,\ X_2 \leq \infty)$$

the *marginal probability distribution function of* X_1.   Since $\{X_2 \leq \infty\}$ is the certain event, we see that

$$\{X_1 \leq x_1, X_2 \leq \infty\} = \{X_1 \leq x_1\}$$

Hence, $F(x_1, \infty) = F_{X_1}(x_1)$; that is, it is the probability distribution function of X_1 alone. Similarly, $F(\infty, x_2) = F_{X_2}(x_2)$.

If $f(x_1, x_2)$ exists, we see that

$$F(x_1, \infty) = F_{X_1}(x_1) = \int_{-\infty}^{x_1} \int_{-\infty}^{\infty} f(\xi_1, \xi_2)\, d\xi_2\, d\xi_1$$

and it is clear that

$$f_{X_1}(\xi_1) = \int_{-\infty}^{\infty} f(\xi_1, \xi_2)\, d\xi_2$$

which is termed the *marginal probability density function of X_1*, is the density function for X_1 alone. Alternately, we observe that

$$f_{X_1}(x_1) \triangleq \frac{\partial F_{X_1}(x_1)}{\partial x_1} = \int_{-\infty}^{\infty} f(x_1, \xi_2)\, d\xi_2$$

Similar results follow if the roles of X_1 and X_2 are interchanged.

The notions of marginal distribution and density functions are readily extended to n dimensions. For example,

$$F(x_1, x_2, \infty, \ldots, \infty)$$
$$= P(X_1 \leq x_1, X_2 \leq x_2, X_3 \leq \infty, \ldots, X_m \leq \infty) = F_{X_1 X_2}(x_1, x_2)$$

and

$$f_{X_1 X_2}(x_1, x_2) = \int_{-\infty}^{\infty} \cdots \int_{-\infty}^{\infty} f(x_1, x_2, \xi_3, \ldots, \xi_n)\, d\xi_n \cdots d\xi_3$$

Let us take an example to illustrate the above concepts.

Example 3.1 A random variable X_1 is said to be uniformly distributed on the interval $[a_1, b_1]$, $a_1 < b_1$, if its distribution function is of the form shown in Fig. 3.2. We see that

$$F(x_1) = \begin{cases} 0 & \text{if } x_1 < a_1 \\ \dfrac{x_1 - a_1}{b_1 - a_1} & \text{if } a_1 \leq x_1 \leq b_1 \\ 1 & \text{if } b_1 < x_1 \end{cases}$$

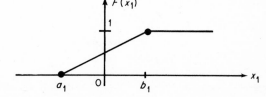

Fig. 3.2 Distribution function for a uniformly distributed random variable.

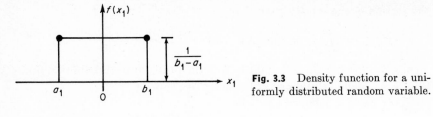

Fig. 3.3 Density function for a uniformly distributed random variable.

Its density function is obviously

$$f(x_1) = \begin{cases} 0 & \text{if } x_1 < a_1 \text{ or } x_1 > b_1 \\ \dfrac{1}{b_1 - a_1} & \text{if } a_1 \leq x_1 \leq b_1 \end{cases}$$

which is indicated in Fig. 3.3.

For the n-dimensional case, we have

$$F(x) = \begin{cases} 0 & \text{if } x_i < a_i \text{ for at least one } i \\ \displaystyle\prod_{i=1}^{n} \dfrac{\zeta_i - a_i}{b_i - a_i} & \text{where } \zeta_i = x_i \text{ if } a_i \leq x_i \leq b_i \text{ and} \\ & \qquad \zeta_i = b_i \text{ if } x_i > b_i \end{cases}$$

and

$$f(x) = \begin{cases} 0 & \text{if } x_i < a_i \text{ for at least one } i \\ \displaystyle\prod_{i=1}^{n} \dfrac{1}{b_i - a_i} & \text{if } a_i \leq x_i \leq b_i \text{ for all } i \\ 0 & \text{if } x_i > b_i \text{ for all } i \end{cases}$$

Determination of the marginal distribution and density functions then follows easily from these two expressions.

Now let X be a random n vector, x an arbitrary n vector, Y a random m vector, and y an arbitrary m vector. Then, by a trivial extension of Eq. (3.2),

$$F(x,y) = P(X_1 \leq x_1, \ldots, X_n \leq x_n; Y_1 \leq y_1, \ldots, Y_m \leq y_m)$$
$$= P(X \leq x, Y \leq y)$$

defines the *joint probability distribution function* of X and Y.

The *joint probability density function* is seen to be

$$f(x,y) = \frac{\partial^{n+m} F}{\partial x_1 \cdots \partial x_n \, \partial y_1 \cdots \partial y_m} \tag{3.5}$$

when it exists.

The following properties are easily established:

$$F(x,y) = \int_{-\infty}^{x_1} \cdots \int_{-\infty}^{x_n} \int_{-\infty}^{y_1} \cdots \int_{-\infty}^{y_m} f(\xi,\zeta) \, d\xi_1 \cdots$$
$$d\xi_n \, d\zeta_1 \cdots d\zeta_m$$

$$\int_{-\infty}^{\infty} \cdots \int_{-\infty}^{\infty} f(\xi,\zeta) \, d\xi_1 \cdots d\xi_n \, d\zeta_1 \cdots d\zeta_m = 1$$

and

$$P(X \leq x; a \leq Y \leq b) = \int_{-\infty}^{x_1} \cdots \int_{-\infty}^{x_n} \int_{a_1}^{b_1} \cdots$$
$$\int_{a_m}^{b_m} f(\xi,\zeta) \, d\xi_1 \cdots d\xi_n \, d\zeta_1 \cdots d\zeta_m \quad (3.6)$$

where ξ is an n vector, ζ an m vector, and a and b are arbitrary m vectors.

Marginal distribution and density functions also follow in a simple way. For example,

$$P(X \leq x) = P(X \leq x; Y \leq \infty)$$
$$= F(x, \infty)$$
$$\triangleq F_X(x)$$

Assuming that F is differentiable with respect to all x_i, $i = 1, \ldots, n$, we have

$$f_X(x) = \frac{\partial^n F(x, \infty)}{\partial x_1 \cdots \partial x_n}$$

$$= \frac{\partial^n}{\partial x_1 \cdots \partial x_n} \int_{-\infty}^{x_1} \cdots \int_{-\infty}^{x_n} \int_{-\infty}^{\infty} \cdots$$
$$\int_{-\infty}^{\infty} f(\xi,\zeta) \, d\xi_1 \cdots d\xi_n \, d\zeta_1 \cdots d\zeta_m$$

$$= \int_{-\infty}^{\infty} \cdots \int_{-\infty}^{\infty} f(x,\zeta) \, d\zeta_1 \cdots d\zeta_m \quad (3.7)$$

In identically the same way,

$$f_Y(y) = \int_{-\infty}^{\infty} \cdots \int_{-\infty}^{\infty} f(\xi,y) \, d\xi_1 \cdots d\xi_n \quad (3.8)$$

CONDITIONAL PROBABILITY DISTRIBUTION AND DENSITY FUNCTIONS

Of considerable importance in probability theory, and also in our work in estimation and control, is the notion of conditional probability, wherein the probability function for an event is "conditioned" or dependent upon the prior occurrence of some other event.

To define this notion, we consider two events A and B in a sample space Ω. The probability of the joint occurrence of A and B is $P(A \cap B)$ and that of B alone is $P(B)$. We denote the conditional probability of A

occurring given that B has already occurred by $P(A|B)$ and define it by the relation

$$P(A|B) = \frac{P(A \cap B)}{P(B)} \tag{3.9}$$

under the assumption that $P(B) \neq 0$.

From the relative frequency of occurrence point of view, this definition arises in the following way. Suppose that the experiment is repeated a large number of times N. Let $N(B)$ be the number of times B has occurred and $N(A \cap B)$ the number of times $A \cap B$, assumed to be nonempty, has occurred in these N trials. Then, for N large, we intuitively expect $P(A|B)$ to behave as the ratio $N(A \cap B)/N(B)$. Noting that

$$\frac{N(A \cap B)}{N(B)} = \frac{N(A \cap B)/N}{N(B)/N}$$

we are led to the definition in Eq. (3.9).

Now let $f(x,y)$ denote the joint density function of the random vectors X and Y, and let $f_Y(y)$ denote the marginal density function of Y [see Eq. (3.8)]. Assume that both functions are continuous. In addition, let A denote the event $X \leq x$ and B the event $y \leq Y \leq y + \Delta y$ where x is an n vector and y and Δy are m vectors with $\Delta y_i > 0$ for $i = 1, \ldots, m$. Then from Eq. (3.9),

$$P(X \leq x | y \leq Y \leq y + \Delta y) = \frac{P(X \leq x; y \leq Y \leq y + \Delta y)}{P(y \leq Y \leq y + \Delta y)}$$

With the aid of Eqs. (3.4) and (3.6), we have

$$P(X \leq x | y \leq Y \leq y + \Delta y)$$

$$= \frac{\int_{-\infty}^{x_1} \cdots \int_{-\infty}^{x_n} \int_{y_1}^{y_1+\Delta y_1} \cdots \int_{y_m}^{y_m+\Delta y_m} f(\xi,\zeta) \, d\xi_1 \cdots d\xi_n \, d\zeta_1 \cdots d\zeta_m}{\int_{y_1}^{y_1+\Delta y_1} \cdots \int_{y_m}^{y_m+\Delta y_m} f_Y(\zeta) \, d\zeta_1 \cdots d\zeta_m}$$

Utilizing the mean value theorem and letting $\Delta y_i \to 0$, $i = 1, \ldots, m$, we see that

$$P(X \leq x | y = Y) = \frac{\int_{-\infty}^{x_1} \cdots \int_{-\infty}^{x_n} f(\xi,y) \, d\xi_1 \cdots d\xi_n}{f_Y(y)} \tag{3.10}$$

if $f_Y(y) \neq 0$.

The function

$$F(x|Y = y) \triangleq P(X \leq x | Y = y)$$

is called the *conditional probability distribution function of X* given Y. We then define the function

$$f(x|y) = \frac{\partial^n F(x|Y = y)}{\partial x_1 \cdots \partial x_n}$$

to be the *conditional probability density function of X* given Y. From its definition and Eq. (3.10), it is clear that

$$f(x|y) = \frac{f(x,y)}{f(y)} \tag{3.11}$$

where the subscript Y has been dropped from $f_Y(y)$ for simplicity. Similarly,

$$f(y|x) = \frac{f(x,y)}{f(x)}$$

where $f(x) = f_X(x)$. We shall have occasion in the sequel to make use of this result, which is termed *Bayes' rule*.

We note that if $f(x,y)$ is given, $f(x)$ and $f(y)$ can be computed using Eqs. (3.7) and (3.8), respectively. Then $f(x|y)$ and $f(y|x)$ follow immediately from Bayes' rule. We illustrate this with the following example.

Example 3.2   Suppose that x and y are scalars whose joint density function is cylindrical, as shown in Fig. 3.4.

$$f(x,y) = \begin{cases} \dfrac{1}{\pi} & \text{for } x^2 + y^2 \leq 1 \\ 0 & \text{elsewhere} \end{cases}$$

Then it is clear that

$$f(y) = \int_{-\sqrt{1-y^2}}^{\sqrt{1-y^2}} f(\xi,y) \, d\xi = \frac{1}{\pi} \int_{-\sqrt{1-y^2}}^{\sqrt{1-y^2}} d\xi = \frac{2}{\pi} \sqrt{1-y^2}$$

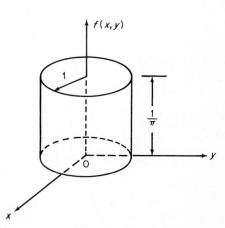

Fig. 3.4   Density function for Example 3.2.

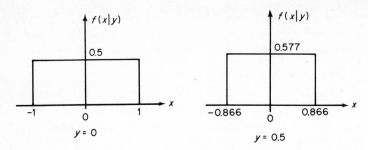

Fig. 3.5 Conditional density functions for $y = 0$ and $y = 0.5$.

for $-1 \leq y \leq 1$ and $f(y) = 0$ zero elsewhere. Bayes' rule then leads to the result that

$$f(x|y) = \frac{1}{2\sqrt{1 - y^2}}$$

for $-\sqrt{1 - y^2} \leq x \leq \sqrt{1 - y^2}$ with $-1 \leq y \leq 1$, and $f(x|y) = 0$ elsewhere. This conditional density function is sketched in Fig. 3.5 for the two values of $y = 0$ and $y = 0.5$.

3.4 EXPECTED VALUE AND CHARACTERISTIC FUNCTION

EXPECTATION, MEAN, AND COVARIANCE

The *expectation* or *expected value* of a scalar or vector-valued function $g(\cdot)$ of a random n vector X is defined as

$$E[g(X)] = \int_{-\infty}^{\infty} \cdots \int_{-\infty}^{\infty} g(x)f(x)\, dx_1 \cdots dx_n \qquad (3.12)$$

Similarly, we write

$$E[g(X,Y)] = \int_{-\infty}^{\infty} \cdots \int_{-\infty}^{\infty} g(x,y)f(x,y)\, dx_1 \cdots dx_n\, dy_1 \cdots dy_m$$

for the expected value of the function $g(\cdot,\cdot)$ of the two random vectors X and Y. In the above expressions, x is an n vector and y an m vector. When $g(X) = X$, we have

$$E(X) = \int_{-\infty}^{\infty} \cdots \int_{-\infty}^{\infty} xf(x)\, dx_1 \cdots dx_n \qquad (3.13)$$

which is called the *mean value* or simply the *mean* of X. It is commonly denoted by $\bar{X}$ or $\bar{x}$, and is obviously an n vector.

The name *mean* is used to denote that $\bar{x}$ is the *average* value of X. To see this, we consider an experiment from the relative frequency of occurrence point of view. We assume that the experiment is repeated a large number of times N, and we let x^i denote the value that the random n vector X takes on for each trial. Then, for some n vector Δx^i sufficiently small, the average value of X for N trials of the experiment is

$$\sum_i \frac{x^i N(x^i \leq X \leq x^i + \Delta x^i)}{N}$$

where $N(x^i \leq X \leq x^i + \Delta x)$ denotes the number of times X lies in the indicated range. Since

$$\frac{N(x^i \leq X \leq x^i + \Delta x^i)}{N}$$

approximates $P(x^i \leq X \leq x^i + \Delta x^i)$, which in turn, is $f_X(x^i)\, \Delta x^i$ to within first-order terms in Δx^i, we have

$$\bar{X}_N \approx \sum_i x^i f(x^i)\, \Delta x^i$$

where the subscript N denotes the fact that the average is computed from N trials. In the limit as $\Delta x^i \to 0$, we obtain the expression in Eq. (3.13).

Alternately, if we let $f(x)$ denote the density with which the unit mass

$$\int_{-\infty}^{\infty} \cdots \int_{-\infty}^{\infty} f(x)\, dx_1 \cdots dx_n = 1$$

is distributed over n-dimensional euclidean space, $\bar{X}$ is the location of the *center of gravity* of the mass distribution. In other words, $E(X)$ is the *first moment* of the probability density function.

The above interpretations of $E(X)$ help us to obtain a useful mental picture of this parameter.

Continuing, if $g(X) = (X - \bar{x})(X - \bar{x})'$, we have

$$E[(X - \bar{x})(X - \bar{x})'] = \int_{-\infty}^{\infty} \cdots \int_{-\infty}^{\infty} (x - \bar{x})(x - \bar{x})' f(x)\, dx_1 \\ \cdots dx_n \quad (3.14)$$

which is obviously a symmetric $n \times n$ matrix. This particular expected value is called the *covariance matrix* of X, and is commonly denoted by P_{XX}, P_{xx}, or P, the latter being used when there is no ambiguity as to which random vector is involved.

In terms of $f(x)$ denoting once more a mass density function in n-dimensional euclidean space, P is seen to be the *second moment* of the mass distribution *about the center of gravity*. As a result, it is usually termed the *second central moment*.

The ij term of P is given by the expression

$$p_{ij} = \int_{-\infty}^{\infty} \cdots \int_{-\infty}^{\infty} (x_i - \bar{x}_i)(x_j - \bar{x}_j)f(x)\, dx_1 \cdots dx_n$$

$$= \iint_{-\infty}^{\infty} (x_i - \bar{x}_i)(x_j - \bar{x}_j)f_{X_i X_j}(x_i, x_j)\, dx_i\, dx_j$$

for $i, j = 1, \ldots, n$. When $i = j$, we have

$$p_{ii} = \iint_{-\infty}^{\infty} (x_i - \bar{x}_i)^2 f_{X_i}(x_i)\, dx_i$$

which is obviously nonnegative, since $f_{X_i}(x_i) \geq 0$ for all x_i. However, it is clear that p_{ij}, $i \neq j$, may be either positive, negative, or zero.

The term p_{ii} is called the *variance* of X_i, while its square root is the *standard deviation*. On the other hand, p_{ij}, $i \neq j$, is termed the covariance of X_i and X_j.

Just as the variance of a random variable is taken as a measure of the "spread" of the random variable about its mean value, the covariance matrix is given the same interpretation for a random vector. Of course, the diagonal terms of P are simply the variances of the components of X. However, the off-diagonal terms, which are of a different nature, give a measure of the "coupling" or correlation between the pairs of random variables X_i and X_j; $i, j = 1, \ldots, n$; $i \neq j$. We shall have more to say about correlation in Sec. 3.5.

In connection with its definition in Eq. (3.14), we note that P is positive semidefinite. This follows from the fact that in the integrand, the matrix $(x - \bar{x})(x - \bar{x})'$ is positive semidefinite and $f(x) \geq 0$ for all x.

We note also from Eq. (3.14) that an alternate expression for P is

$$P = \int_{-\infty}^{\infty} \cdots \int_{-\infty}^{\infty} xx'f(x)\, dx_1 \cdots dx_n - \bar{x}\bar{x}' \tag{3.15}$$

For two random vectors X and Y, the expression

$$P_{XY} = E[X - \bar{x})(Y - \bar{y})']$$

$$= \int_{-\infty}^{\infty} \cdots \int_{-\infty}^{\infty} (x - \bar{x})(y - \bar{y})'f(x,y)\, dx_1$$

$$\cdots dx_n\, dy_1 \cdots dy_m \tag{3.16}$$

defines the *cross-covariance matrix* of X and Y. In Eq. (3.16), x is an n vector, y is an m vector, $f(x,y)$ is the joint density function of X and Y, and $\bar{x} = E(X)$ and $\bar{y} = E(Y)$. It is clear that P_{xy} is $n \times m$ and that $P_{xy} = P_{yx}'$. However, we note that for $m = n$, it is not true in general that P_{xy} is symmetric or that it is positive semidefinite.

We have spoken here only of first and second moments as parameters of a probability distribution. In general, it is not possible to specify a probability distribution in terms of first and second moments alone unless the functional form of the distribution is given and all of the parameters upon which it depends can be determined from a knowledge of these first two moments. Usually, it is necessary to introduce and utilize higher-order moments, e.g., the $r = k + l$ moment of X_i and X_j which is defined by the relation

$$\mu_{kl} = \int_{-\infty}^{\infty} \cdots \int_{-\infty}^{\infty} (x_i)^k (x_j)^l f(x) \, dx_1 \cdots dx_n$$

In our work, it turns out, however, that knowledge of $\bar{x}$, P_{xx}, and P_{xy} is sufficient for all required calculations. The reason for this will become apparent as we proceed.

Example 3.3   Let us consider the situation where X is a 2 vector whose density function is the two-dimensional Rayleigh function

$$f(x) = f(x_1, x_2) = \begin{cases} 4x_1 x_2 e^{-(x_1^2 + x_2^2)} & \text{for } x_1, x_2 \geq 0 \\ 0 & \text{elsewhere} \end{cases}$$

Then,

$$E(X) = 4 \int\!\!\int_0^\infty \begin{bmatrix} x_1 \\ x_2 \end{bmatrix} x_1 x_2 e^{-(x_1^2 + x_2^2)} \, dx_1 \, dx_2$$

First, we see that

$$\begin{aligned}
E(X_1) &= 4 \int_0^\infty \int_0^\infty x_1^2 x_2 e^{-(x_1^2 + x_2^2)} \, dx_1 \, dx_2 \\
&= 4 \int_0^\infty x_2 e^{-x_2^2} \, dx_2 \int_0^\infty x_1^2 e^{-x_1^2} \, dx_1 \\
&= 4 \left(\frac{1}{2}\right) \frac{\sqrt{\pi}}{4} \\
&= \frac{\sqrt{\pi}}{2}
\end{aligned}$$

Similarly, we obtain

$$E(X_2) = \frac{\sqrt{\pi}}{2}$$

so that

$$\bar{x} = \frac{\sqrt{\pi}}{2} \begin{bmatrix} 1 \\ 1 \end{bmatrix}$$

To determine the 2×2 covariance matrix, we utilize Eq. (3.15). Then,

$$P = 4 \int\!\!\int_0^\infty \begin{bmatrix} x_1^2 & x_1 x_2 \\ x_1 x_2 & x_2^2 \end{bmatrix} x_1 x_2 e^{-(x_1^2 + x_2^2)} \, dx_1 \, dx_2 - \frac{\pi}{4} \begin{bmatrix} 1 & 1 \\ 1 & 1 \end{bmatrix}$$

and we have

$$
\begin{aligned}
p_{11} &= 4 \int\!\!\!\int_0^\infty x_1{}^3 x_2 e^{-(x_1{}^2 + x_2{}^2)} \, dx_1 \, dx_2 - \frac{\pi}{4} \\
&= 4 \int_0^\infty x_1{}^3 e^{-x_1{}^2} \, dx_1 \int_0^\infty x_2 e^{-x_2{}^2} \, dx_2 - \frac{\pi}{4} \\
&= 4 \left(\frac{1}{2}\right)\left(\frac{1}{2}\right) - \frac{\pi}{4} \\
&= 1 - \frac{\pi}{4}
\end{aligned}
$$

We note that $p_{22} = p_{11}$.

Finally,

$$
\begin{aligned}
p_{12} = p_{21} &= 4 \int\!\!\!\int_0^\infty x_1{}^2 x_2{}^2 e^{-(x_1{}^2 + x_2{}^2)} \, dx_1 \, dx_2 - \frac{\pi}{4} \\
&= 4 \frac{\sqrt{\pi}}{4} \frac{\sqrt{\pi}}{4} - \frac{\pi}{4} \\
&= 0
\end{aligned}
$$

and we have

$$
P = \begin{bmatrix} 1 - \dfrac{\pi}{4} & 0 \\[2mm] 0 & 1 - \dfrac{\pi}{4} \end{bmatrix}
$$

The significance of the zero off-diagonal terms will be discussed in Sec. 3.5.

CONDITIONAL EXPECTATION, MEAN, AND COVARIANCE

As a simple extension of the definition in Eq. (3.12), the *conditional expected value* of a scalar or vector-valued function $g(\cdot)$ of a random vector X with respect to another random vector Y is defined as

$$
E_X[g(X)|Y = y] = \int_{-\infty}^\infty \cdots \int_{-\infty}^\infty g(x) f(x|y) \, dx_1 \cdots dx_n \qquad (3.17)
$$

The subscript X on E denotes that the expected value operation, i.e., the integration, is over X. The subscript could obviously be omitted in this case without causing confusion. However, in some situations inclusion of such subscripts is essential to denote the required operations.

The *conditional mean* of X given $Y = y$ is defined as

$$
E(X|y) = \int_{-\infty}^\infty \cdots \int_{-\infty}^\infty x f(x|y) \, dx_1 \cdots dx_n \qquad (3.18)
$$

and the corresponding *conditional covariance matrix* as

$$P_{X|y} = E_X\{[X - E(X|y)][X - E(X|y)]'\}$$

$$= \int_{-\infty}^{\infty} \cdots \int_{-\infty}^{\infty} [x - E(X|y)][x - E(X|y)]'f(x|y)\, dx_1$$

$$\cdots dx_n \quad (3.19)$$

Example 3.4 Let us determine $E(X|y)$ and $P(X|y)$ for the conditional density in Example 3.2. For Eq. (3.18), we have in this case that

$$E(X|y) = \int_{-\sqrt{1-y^2}}^{\sqrt{1-y^2}} x\, \frac{1}{2\sqrt{1-y^2}}\, dx = 0$$

This result is evident from Fig. 3.5 of Example 3.2, where it is noted that each $f(x|y)$ is uniform and centered at the origin.

Equation (3.19) now assumes the form

$$P_{X|y} = \int_{-\sqrt{1-y^2}}^{\sqrt{1-y^2}} x^2\, \frac{1}{2\sqrt{1-y^2}}\, dx$$

$$= \tfrac{1}{3}(1 - y^2)$$

where $-1 \leq y \leq 1$. The conditional variance is sketched in Fig. 3.6 as a function of y^2.

Since X and Y are related, we feel that knowledge of the value of Y should permit us to say something about the corresponding value of X. Proceeding in an ad hoc fashion, we could pick $E(X|y)$ as an estimate of X under the assumption that we are told the value of Y. Moreover, since $P_{X|y}$ is a measure of the spread of X about its conditional mean, we could employ $P_{X|y}$ or its square root, the *conditional standard deviation*, as a measure of the "quality" of the estimate. For example, given $Y = y = 1$, we say that the estimate $\hat{X} = E(X|1) = 0$, with a variance of zero, or given $Y = y = \tfrac{1}{2}$, $\hat{X} = 0$ with a variance of $\tfrac{1}{4}$.

PROPERTIES OF CONDITIONAL EXPECTATION

The following properties of conditional expectation will be of use in the sequel.

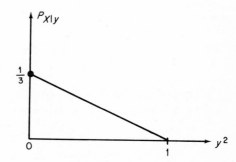

Fig. 3.6 Behavior of conditional variance as a function of the square of the conditioning variable.

Let X, Y, and Z be random vectors for which the density functions required below exist. Then,

1. $E_X[g(X)|X = x] = g(x)$
2. $E_X(AX|Y = y) = AE(X|Y = y)$, where X is n-dimensional and A is $p \times n$
3. $E_{XY}(X + Y|Z = z) = E_X(X|Z = z) + E_Y(Y|Z = z)$, where X and Y are n-dimensional and Z is m-dimensional
4. $E_Y[E_X(X|Y = y)] = E_X(X)$, where X is an n vector and Y is an m vector

The verification of these properties is relatively straightforward and is left as an exercise. We remark that it can also be shown that $E(X|y)$ is unique [5].

CHARACTERISTIC FUNCTION

The characteristic function $\phi_X(s)$ of a random n vector X is defined as

$$\phi_X(s) = E(e^{jX's}) = \int_{-\infty}^{\infty} \cdots \int_{-\infty}^{\infty} e^{jx's}f(x)\,dx_1 \cdots dx_n \qquad (3.20)$$

where $j = \sqrt{-1}$ and s is an n vector. We note that $\phi_X(\cdot)$ is a scalar-valued function of s.

The inversion formula for Eq. (3.20), which follows from Fourier transform theory, is

$$f(x) = \frac{1}{(2\pi)^n} \int_{-\infty}^{\infty} \cdots \int_{-\infty}^{\infty} e^{-js'x}\phi_X(s)\,ds_1 \cdots ds_n \qquad (3.21)$$

In a similar way, the *joint characteristic function* of the random n and m vectors X and Y, respectively, is given as

$$\phi_{XY}(s,r) = E[e^{j(X's+Y'r)}]$$
$$= \int_{-\infty}^{\infty} \cdots \int_{-\infty}^{\infty} e^{j(x's+y'r)}f(x,y)\,dx_1 \cdots dx_n\,dy_1 \cdots dy_m \qquad (3.22)$$

where s is an n vector and r is an m vector. Then

$$f(x,y) = \frac{1}{(2\pi)^{n+m}} \int_{-\infty}^{\infty} \cdots \int_{-\infty}^{\infty}$$
$$e^{-j(s'x+r'y)}\phi_{XY}(s,r)\,ds_1 \cdots ds_n\,dr_1 \cdots dr_m \qquad (3.23)$$

Obviously, Eqs. (3.22) and (3.23) can be put in the same form as Eqs. (3.20) and (3.21), respectively, by introducing the $(n + m)$ vectors

$$Z = \begin{bmatrix} X \\ Y \end{bmatrix} \qquad z = \begin{bmatrix} x \\ y \end{bmatrix} \qquad q = \begin{bmatrix} r \\ s \end{bmatrix}$$

Finally, the *conditional characteristic function* of X given $Y = y$ is written as

$$\phi_{X|y}(s) = E(e^{jX's}|Y = y)$$
$$= \int_{-\infty}^{\infty} \cdots \int_{-\infty}^{\infty} e^{jx's}f(x|y) \, dx_1 \cdots dx_n$$

and its inverse given by

$$f(x|y) = \frac{1}{(2\pi)^n} \int_{-\infty}^{\infty} \cdots \int_{-\infty}^{\infty} e^{-js'x}\phi_{X|y}(s) \, ds_1 \cdots ds_n$$

The characteristic function is useful in problems of determining the probability law of a function $g(\cdot)$ of a random vector X given the probability description of the latter. It will prove to be a very powerful tool for this purpose in Sec. 3.6.

For the present, we illustrate its use with the following simple example.

Example 3.5 Let X, x, and s be scalars and assume that X is a gaussian distributed random variable with density function

$$f(x) = \frac{1}{\sigma \sqrt{2\pi}} e^{-x^2/2\sigma^2}$$

where $\sigma > 0$. (We shall treat the gaussian distribution extensively in Sec. 3.6.) Assume that X is input into a square-law device whose output is

$$Y = \alpha X^2 \qquad \alpha > 0$$

We wish to determine the probability density function $f_Y(y)$ of the device's output.

From Eq. (3.20) and the fact that $Y = \alpha X^2$,

$$\phi_Y(s) = E(e^{jYs})$$
$$= E(e^{j\alpha X^2 s})$$
$$= \int_{-\infty}^{\infty} e^{j\alpha x^2 s} \frac{1}{\sigma \sqrt{2\pi}} e^{-x^2/2\sigma^2} \, dx$$
$$= 2 \int_{0}^{\infty} e^{j\alpha x^2 s} \frac{1}{\sigma \sqrt{2\pi}} e^{-x^2/2\sigma^2} \, dx$$

However, $y = \alpha x^2$ and $dy = 2\alpha x \, dx = 2\sqrt{\alpha y} \, dx$. Hence,

$$\phi_Y(s) = \int_{0}^{\infty} e^{jys} \frac{1}{\sigma \sqrt{2\pi}} e^{-y/2\alpha\sigma^2} \frac{dy}{\sqrt{\alpha y}}$$
$$= \int_{0}^{\infty} e^{jys} \frac{e^{-y/2\alpha\sigma^2}}{\sigma \sqrt{2\pi\alpha y}} \, dy \tag{3.24}$$

Noting that $Y = \alpha X^2$, with $\alpha > 0$, implies that $Y \geq 0$, we see by a direct comparison of Eq. (3.24) with Eq. (3.20) that

$$f_Y(y) = \begin{cases} \dfrac{e^{-y/2\alpha\sigma^2}}{\sigma \sqrt{2\pi\alpha y}} & \text{if } y \geq 0 \\ 0 & \text{elsewhere} \end{cases}$$

3.5 INDEPENDENCE AND CORRELATION

INDEPENDENT EVENTS, RANDOM VARIABLES, AND RANDOM VECTORS

Two events A, $B \in \Omega$ are said to be *independent* of each other if

$$P(A \cap B) = P(A)P(B)$$

Substituting this definition into Eq. (3.9), we see that if A and B are independent, then

$$P(A|B) = \frac{P(A)P(B)}{P(B)} = P(A)$$

In other words, the fact that B has occurred does not influence the probability law associated with A; the two events are "unrelated."

A finite or infinitely denumerable collection of sets A_1, A_2, . . . , A_n, . . . in Ω are defined as independent if

$$P[\bigcap_i A_i] = P(A_1)P(A_2) \cdots P(A_n) \cdots$$

Then, the elements X_i of a random n vector X are said to be *statistically independent*, or more briefly, *independent*, if the distribution function $F(x)$ can be expressed in the form

$$F(x) = \prod_{i=1}^{n} F_i(x_i) \tag{3.25}$$

where $F_i(x_i) = P(X_i \leq x_i)$. Thus, the collection $(X_1, \ldots, X_n)$ is a set of independent random variables.

It follows from Eq. (3.25) that the corresponding density function is

$$f(x) = \prod_{i=1}^{n} f_i(x_i) \tag{3.26}$$

where

$$f_i(x_i) = \frac{dF_i}{dx_i}$$

for $i = 1, \ldots, n$ under the assumption that the indicated derivatives exist.

Also, the corresponding characteristic function is seen to be

$$\phi_X(s) = \prod_{i=1}^{n} \phi_i(s_i) \tag{3.27}$$

where

$$\phi_i(s_i) = \int_{-\infty}^{\infty} e^{js_ix_i}f_i(x_i)\,dx_i$$

for $i = 1, \ldots, n$.

We remark that Eqs. (3.25) to (3.27) are clearly equivalent definitions of the independence of a set of n random variables.

Analogous to the definition of independent random variables, we say that two random vectors X and Y are independent if

$$F(x,y) = F_X(x)F_Y(y) \tag{3.28}$$

where $F_X(x) = P(x \leq X)$ and $F_Y(y) = P(Y \leq y)$.

It then follows that

$$f(x,y) = f_X(x)f_Y(y) \tag{3.29}$$

where

$$f_X(x) = \frac{\partial^n F_X}{\partial x_1 \cdots \partial x_n}$$

and

$$f_Y(y) = \frac{\partial^m F_Y}{\partial y_1 \cdots \partial y_m}$$

if the indicated partial derivatives exist.

The corresponding characteristic function is seen to be

$$\phi(s,r) = \phi_X(s)\phi_Y(r) \tag{3.30}$$

where

$$\phi_X(s) = \int_{-\infty}^{\infty} \cdots \int_{-\infty}^{\infty} e^{jx's}f_X(x)\,dx_1 \cdots dx_n$$

and

$$\phi_Y(r) = \int_{-\infty}^{\infty} \cdots \int_{-\infty}^{\infty} e^{jy's}f_Y(y)\,dy_1 \cdots dy_m$$

where s and r are n and m vectors, respectively.

For X and Y independent, we have

$$f(x|y) = \frac{f_X(x)f_Y(y)}{f_Y(y)} = f_X(x)$$

so that

$$E(X|y) = \int_{-\infty}^{\infty} \cdots \int_{-\infty}^{\infty} x f(x|y) \, dx_1 \cdots dx_n$$

$$= \int_{-\infty}^{\infty} \cdots \int_{-\infty}^{\infty} x f_X(x) \, dx_1 \cdots dx_n$$

$$= E(X) = \bar{x}$$

and similarly,

$$P_{X|y} = P_{XX} = E[(X - \bar{x})(X - \bar{x})']$$

Example 3.6   In Example 3.2, where

$$f(x,y) = \begin{cases} \dfrac{1}{\pi} & \text{for } x^2 + y^2 \le 1 \\ 0 & \text{elsewhere} \end{cases}$$

it is clear that the two random variables X and Y are not independent.   On the other hand, in Example 3.3, where

$$f(x_1,x_2) = \begin{cases} 4x_1 x_2 e^{-(x_1{}^2 + x_2{}^2)} & x_1, x_2 \ge 0 \\ 0 & \text{elsewhere} \end{cases}$$

we write

$$f(x_1,x_2) = f_{X_1}(x_1) f_{X_2}(x_2)$$

with

$$f_{X_1}(x_1) = \begin{cases} 2x_1 e^{-x_1{}^2} & \text{for } x_1 \ge 0 \\ 0 & \text{elsewhere} \end{cases}$$

and similarly for $f_{X_2}(x_2)$.   It is easily verified that

$$f_{X_1}(x_1) \ge 0$$

and

$$\int_0^{\infty} f_{X_1}(x_1) \, dx_1 = 1$$

and similarly for $f_{X_2}(x_2)$.

CORRELATION

Two random variables X_i and Y_j are said to be *uncorrelated* if

$$E_{X_i Y_j}(X_i Y_j) = E_{X_i}(X_i) E_{Y_j}(Y_j) \tag{3.31}$$

and correlated otherwise.

If X is a random n vector whose components are uncorrelated, then

$$E[(X_i - \bar{x}_i)(X_j - \bar{x}_j)] = E(X_i X_j) - \bar{x}_i \bar{x}_j = \bar{x}_i \bar{x}_j - \bar{x}_i \bar{x}_j = 0$$

for all $i \neq j; i, j = 1, \ldots, n$. From this it follows that the corresponding covariance matrix

$$P = E[(X - \bar{x})(X - \bar{x})']$$

is diagonal.

If X and Y are random vectors with components X_i, $i = 1, \ldots, n$, and Y_j, $j = 1, \ldots, m$, and if Eq. (3.31) is satisfied for all i and j, X and Y are termed *uncorrelated random vectors*. In this case,

$$E_{XY}(XY') = E(X)E(Y') = \bar{x}\bar{y}'$$

From Eq. (3.16),

$$P_{XY} = E[(X - \bar{x})(Y - \bar{y})'] = E(XY') - \bar{x}\bar{y}' = 0$$

Thus the cross-covariance matrix of two uncorrelated random vectors is zero.

The two matrices

$$\Psi_{XX} = E(XX')$$

and

$$\Psi_{XY} = E(XY')$$

are called the *correlation matrices* of X, and of X and Y, respectively.

For two random variables X_i and Y_j, the quantity

$$\rho_{ij} = \frac{E[(X_i - \bar{x}_i)(Y_j - \bar{y}_j)]}{\sqrt{E[(X_i - \bar{x}_i)^2] \cdot E[(Y_j - \bar{y}_j)^2]}}$$

is called the *correlation coefficient* of X_i and Y_j. It is obviously zero if X_i and Y_j are uncorrelated.

We assert that $|\rho_{ij}| \leq 1$. To see this, it is sufficient to consider the case where $\bar{x}_i = \bar{y}_j = 0$. Then for c = scalar constant, we note that

$$E[(cX_i - Y_j)^2] = E(X_i{}^2)c^2 - 2E(X_iY_j)c + E(Y_j{}^2) \geq 0$$

However, since $E(X_i{}^2)c^2 - 2E(X_iY_j)c + E(Y_j{}^2)$ is quadratic in c, necessary and sufficient conditions for it to be nonnegative are

$$E(X_i{}^2) \geq 0 \quad \text{and} \quad E(Y_j{}^2) - \frac{[E(X_iY_j)]^2}{E(X_i{}^2)} \geq 0$$

The first condition is obviously satisfied. The second gives

$$1 - \frac{[E(X_iY_j)]^2}{E(X_i{}^2)E(Y_j{}^2)} \geq 0$$

or, equivalently, $1 - \rho^2 \geq 0$, which verifies the assertion.

Finally, we take note of the fact that two independent random vectors are uncorrelated, but that the converse is not true in general.

If X and Y are independent, then obviously

$$E_{XY}(XY') = \int_{-\infty}^{\infty} \cdots \int_{-\infty}^{\infty} xy'f_X(x)f_Y(y)\,dx_1 \cdots dx_n\,dy_1 \cdots dy_m$$

$$= \int_{-\infty}^{\infty} \cdots \int_{-\infty}^{\infty} xf_X(x)\,dx_1 \cdots dx_n \int_{-\infty}^{\infty} \cdots$$

$$\int_{-\infty}^{\infty} y'f_Y(y)\,dy_1 \cdots dy_m$$

$$= E(X)E(Y')$$

To verify that the converse is not generally true, it is sufficient to take X and Y as scalars. Assume that $f_X(x)$ is symmetric with $\bar{x} = 0$ and take $Y = X^2$. Then,

$$E(XY) = E(X^3) = \int_{-\infty}^{\infty} x^3 f(x)\,dx = 0$$

since the integrand is an odd function of x. Since $\bar{X} = 0$, we have

$$E(XY) = \bar{X} \cdot \bar{Y} = 0$$

The two random variables are obviously uncorrelated, but by the relation $Y = X^2$, not independent.

3.6 GAUSSIAN DISTRIBUTION

One, but by no means the only, probability distribution which has played a significant role in the study of random phenomena in nature is the so-called *gaussian* or *normal* distribution. This has been especially true in the areas of communication theory, estimation theory, and control theory since the early 1940s.

There are basically two reasons why the gaussian distribution has enjoyed this prominence in applications. First, it has been found through experience that the gaussian distribution provides a model which is a reasonable approximation to observed random behavior in certain physical systems. In addition, a partial justification for the use of the gaussian distribution in describing random phenomena stems from the central limit theorem which is examined below.

Secondly, by virtue of its mathematical form, the gaussian distribution is analytically and computationally tractable. This is due to the fact that it is completely specified by its first and second moments, namely, its mean and covariance matrix.

PROBABILITY DESCRIPTION

At this point, it is convenient for us to introduce a simple notational change. We do this to facilitate our work in this section and in Chaps. 4

through 10. In particular, we shall hereafter use the lowercase letters x, y, z, ξ, and ζ to denote random vectors.

A random n vector is said to be *gaussian distributed* if its characteristic function is

$$\phi_x(s) = \exp\left(j\bar{x}'s - \tfrac{1}{2}s'Ps\right) \tag{3.32}$$

where s is an n vector, $\bar{x} = E(x)$, and $P = E[(x - \bar{x})(x - \bar{x})']$.

We shall show later in this section that the corresponding density function is

$$f(x) = \frac{1}{\sqrt{(2\pi)^n|P|}} \exp\left[-\tfrac{1}{2}(x - \bar{x})'P^{-1}(x - \bar{x})\right] \tag{3.33}$$

where $|P|$ is the determinant of P. We note that $f(x)$ does not exist if P is singular. For this reason, the gaussian distribution is usually defined by its characteristic function. In the sequel, we shall assume that P is positive definite, and therefore nonsingular, when working with the gaussian density function.

Two random vectors, x, n-dimensional, and y, m-dimensional, are *jointly gaussian distributed*, or, more simply, *jointly gaussian*, if their joint characteristic function is

$$\phi_{xy}(s,r) = \exp\left\{j\begin{bmatrix}\bar{x}\\\bar{y}\end{bmatrix}'\begin{bmatrix}s\\r\end{bmatrix} - \tfrac{1}{2}\begin{bmatrix}s\\r\end{bmatrix}'P\begin{bmatrix}s\\r\end{bmatrix}\right\}$$

$$= \exp\left\{j(\bar{x}'s + \bar{y}'r) - \tfrac{1}{2}\begin{bmatrix}s\\r\end{bmatrix}'P\begin{bmatrix}s\\r\end{bmatrix}\right\} \tag{3.34}$$

where s is an n vector, r an m vector, $\bar{x} = E(x)$, $\bar{y} = E(y)$, and P is the $(n + m) \times (n + m)$ matrix

$$P = \begin{bmatrix}P_{xx} & P_{xy}\\P_{yx} & P_{yy}\end{bmatrix}$$

with

$$P_{xx} = E[(x - \bar{x})(x - \bar{x})'] \qquad P_{xy} = E[(x - \bar{x})(y - \bar{y})']$$

$$P_{yx} = P'_{xy} \qquad P_{yy} = E[(y - \bar{y})(y - \bar{y})']$$

It is clear that by defining the two $(n + m)$ vectors

$$z = \begin{bmatrix}x\\y\end{bmatrix} \quad\text{and}\quad q = \begin{bmatrix}r\\s\end{bmatrix}$$

Eq. (3.34) can be written in the same form as Eq. (3.32). The joint density function of x and y would then be

$$f(x,y) = f(z) = \frac{1}{\sqrt{(2\pi)^{n+m}|P|}} \exp\left[-\tfrac{1}{2}(z - \bar{z})'P^{-1}(z - \bar{z})\right] \tag{3.35}$$

for P positive definite.

When dealing with the density function of two jointly gaussian random vectors, it is often convenient to have an explicit expression for P^{-1} in terms of P_{xx}, P_{xy}, and P_{yy}. This can be obtained by defining

$$P^{-1} = \begin{bmatrix} A & B \\ B' & C \end{bmatrix}$$

where A, B, and C are $n \times n$, $n \times m$, and $m \times m$ matrices, respectively, and are determined so that $P^{-1}P = I$, the $(n + m) \times (n + m)$ identity matrix. The details are left as an exercise; the results are:

$$A = (P_{xx} - P_{xy}P_{yy}^{-1}P_{yx})^{-1} = P_{xx}^{-1} + P_{xx}^{-1}P_{xy}CP_{yx}P_{xx}^{-1} \quad (3.36)$$

$$B = -AP_{xy}P_{yy}^{-1} = -P_{xx}^{-1}P_{xy}C \quad (3.37)$$

$$C = (P_{yy} - P_{yx}P_{xx}^{-1}P_{xy})^{-1} = P_{yy}^{-1} + P_{yy}^{-1}P_{yx}AP_{xy}P_{yy}^{-1} \quad (3.38)$$

The expression $P^{-1}P = I$ of course requires that P be positive definite. Since P is also symmetric, this means that P^{-1} will be symmetric and positive definite.

Under the assumption that the indicated matrix inverses in Eqs. (3.36) to (3.38) exist, we now utilize these relations to develop an expression for the conditional gaussian density $f(x|y)$.

From Eqs. (3.35) and (3.33),

$$f(x,y) = \frac{1}{\sqrt{(2\pi)^{n+m}|P|}} \exp\left\{-\tfrac{1}{2} \begin{bmatrix} x - \bar{x} \\ y - \bar{y} \end{bmatrix}' \begin{bmatrix} A & B \\ B' & C \end{bmatrix} \begin{bmatrix} x - \bar{x} \\ y - \bar{y} \end{bmatrix}\right\}$$

and

$$f(y) = \frac{1}{\sqrt{(2\pi)^m|P_{yy}|}} \exp\left[-\tfrac{1}{2}(y - \bar{y})'P_{yy}^{-1}(y - \bar{y})\right]$$

respectively. Then from Bayes' rule,

$$f(x|y) = \frac{1}{\sqrt{(2\pi)^n \dfrac{|P|}{|P_{yy}|}}} \exp\left\{-\tfrac{1}{2} \begin{bmatrix} x - \bar{x} \\ y - \bar{y} \end{bmatrix}' \begin{bmatrix} A & B \\ B' & C - P_{yy}^{-1} \end{bmatrix} \begin{bmatrix} x - \bar{x} \\ y - \bar{y} \end{bmatrix}\right\} \quad (3.39)$$

Expanding the quadratic form in the exponential, substituting from Eqs. (3.37) and (3.38), and rearranging terms, we have

$$\begin{bmatrix} x - \bar{x} \\ y - \bar{y} \end{bmatrix}' \begin{bmatrix} A & B \\ B' & C - P_{yy}^{-1} \end{bmatrix} \begin{bmatrix} x - \bar{x} \\ y - \bar{y} \end{bmatrix}$$

$$= (x - \bar{x})'A(x - \bar{x}) + 2(x - \bar{x})'B(y - \bar{y})$$
$$\qquad + (y - \bar{y})'(C - P_{yy}^{-1})(y - \bar{y})$$

$$= (x - \bar{x})'A(x - \bar{x}) - 2(x - \bar{x})'AP_{xy}P_{yy}^{-1}(y - \bar{y})$$
$$\qquad + (y - \bar{y})'P_{yy}^{-1}P_{yx}AP_{xy}P_{yy}^{-1}(y - \bar{y})$$

$$= [x - \bar{x} - P_{xy}P_{yy}^{-1}(y - \bar{y})]'A[x - \bar{x} - P_{xy}P_{yy}^{-1}(y - \bar{y})]$$

Defining

$$m = \bar{x} + P_{xy}P_{yy}^{-1}(y - \bar{y})$$

and

$$Q = A^{-1} = P_{xx} - P_{xy}P_{yy}^{-1}P_{yx}$$

the quadratic form becomes

$$(x - m)'Q^{-1}(x - m)$$

Noting that

$$P = \begin{bmatrix} P_{xx} & P_{xy} \\ P_{yx} & P_{yy} \end{bmatrix} = \begin{bmatrix} P_{xx} - P_{xy}P_{yy}^{-1}P_{yx} & P_{xy} \\ 0 & P_{yy} \end{bmatrix} \begin{bmatrix} I_n & 0 \\ P_{yy}^{-1}P_{xy} & I_m \end{bmatrix}$$

where I_n is the $n \times n$ identity matrix and I_m is the $m \times m$ identity matrix, we have

$$|P| = |P_{xx} - P_{xy}P_{yy}^{-1}P_{yx}| \cdot |P_{yy}|$$

Hence,

$$\frac{|P|}{|P_{yy}|} = |P_{xx} - P_{xy}P_{yy}^{-1}P_{yx}| = |Q|$$

Equation (3.39) can now be written

$$f(x|y) = \frac{1}{\sqrt{(2\pi)^n|Q|}} \exp\left[-\tfrac{1}{2}(x - m)'Q^{-1}(x - m)\right] \qquad (3.40)$$

Comparing this result with Eq. (3.33), we see that $f(x|y)$ is a gaussian density function with mean m and covariance Q. The conditional mean and covariance matrix are then evidently

$$m = E(x|y) = \bar{x} + P_{xy}P_{yy}^{-1}(y - \bar{y}) \qquad (3.41)$$

and

$$Q = P_{x|y} = P_{xx} - P_{xy}P_{yy}^{-1}P_{yx} \qquad (3.42)$$

respectively. Also, it is clear that the appropriate conditional characteristic function is

$$\phi_{x|y}(s) = \exp\left(jm's - \tfrac{1}{2}s'Qs\right) \qquad (3.43)$$

where s is an n vector. We note again that the probability description in terms of the characteristic function requires only that the covariance matrix be positive semidefinite; its inverse need not exist.

One important property of the gaussian distribution is that it is completely characterized by its mean and covariance matrix. In particular, it can be shown that all moments of a gaussian distributed random variable can be expressed in terms of its first two moments [4].

CENTRAL LIMIT THEOREM [1–5]

As remarked earlier, partial justification for using the gaussian distribution in practice is the central limit theorem. The theorem, which we state here without proof, is as follows.

Theorem 3.1 *Let x^i, $i = 1, \ldots, r$, be a set of independent, identically distributed random n vectors with finite means $\bar{x}^i$ and covariance matrices P^i. Let y^r be the random n vector*

$$y^r = \sum_{i=1}^{r} x^i$$

and z^r the random n vector

$$z^r = [P^r]^{-1}(y - \bar{y}^r)$$

where

$$\bar{y}^r = \sum_{i=1}^{r} \bar{x}^i$$

and

$$P^r = \prod_{i=1}^{r} P^i$$

Then,

$$\lim_{r \to \infty} f(z^r) = \frac{1}{\sqrt{(2\pi)^n}} \exp\left(-\tfrac{1}{2}z'z\right)$$

That is, as $r \to \infty$, z^r becomes a zero mean gaussian random n vector whose covariance matrix is the identity matrix. This means that if the random phenomenon which we observe at the macroscopic level is the superposition of an arbitrarily large number of independent random phenomena which occur at the microscopic level, we are justified in describing the former phenomenon in terms of the gaussian distribution.

LINEAR TRANSFORMATIONS AND LINEAR COMBINATIONS

Much of our work will involve linear transformations on and linear combinations of gaussian random vectors. We now show that such operations lead to gaussian random vectors.

Suppose that x is a gaussian random n vector with mean $\bar{x}$ and covariance matrix P_{xx}, and let A be an $m \times n$ matrix. We assert that the random m vector $y = Ax$ is gaussian with mean $\bar{y} = A\bar{x}$ and covariance matrix $P_{yy} = AP_{xx}A'$. The proof follows easily with the aid of characteristic functions. From Eq. (3.32), the characteristic function for x is

$$\phi_x(s) = E(e^{jx's}) = \exp\left(j\bar{x}'s - \tfrac{1}{2}s'Ps\right) \tag{3.44}$$

The characteristic function for y is, by definition,

$$\phi_y(r) = E(e^{jy'r}) \tag{3.45}$$

where r is an m vector.

Since $y = Ax$,

$$\begin{aligned}
\phi_y(r) &= E[\exp j(Ax)'r] \\
&= E[\exp jx'(A'r)] \\
&= \phi_x(A'r) \\
&= \exp\left[j\bar{x}'A'r - \tfrac{1}{2}r'APA'r\right] = \exp\left[j(A\bar{x})'r - \tfrac{1}{2}r'(APA')r\right]
\end{aligned}$$

and the assertion is proved.

Now suppose that x and y are gaussian random n and m vectors, respectively, whose joint characteristic function is given by Eq. (3.34). Let A and B be real $p \times n$ and $p \times m$ matrices, respectively. We shall show that the random p vector $z = Ax + By$ is gaussian with mean $\bar{z} = A\bar{x} + B\bar{y}$ and covariance matrix

$$P_{zz} = AP_{xx}A' + AP_{xy}B' + BP_{yx}A' + BP_{yy}B'$$

The derivation essentially parallels the one given above. We let ξ be the random $(n + m)$ vector defined as

$$\xi = \begin{bmatrix} x \\ y \end{bmatrix}$$

Also, we let q be an $(n + m)$ vector, and let C denote the $p \times (n + m)$ matrix $[A \quad B]$. Then, $z = C\xi$.

From Eq. (3.32),

$$\begin{aligned}
\phi_\xi(q) &= E(e^{j\xi'q}) \\
&= \exp\left(j\bar{\xi}'q - \tfrac{1}{2}q'Pq\right) \tag{3.46}
\end{aligned}$$

where

$$P = \begin{bmatrix} P_{xx} & P_{xy} \\ P_{yx} & P_{yy} \end{bmatrix}$$

Also,

$$\phi_z(s) = E(e^{jz's}) \tag{3.47}$$

where s is a p vector.

Since $z = C\xi$, it follows from Eqs. (3.47) and (3.46) that

$$
\begin{aligned}
\phi_z(s) &= E[\exp j(C\xi)'s] = E[\exp j\xi'(C's)] \\
&= \phi_\xi(C's) \\
&= \exp\left[j\bar{\xi}'(C's) - \tfrac{1}{2}(C's)'P(Cs)\right] \\
&= \exp\left[j(C\bar{\xi})'s - \tfrac{1}{2}s'(CPC')s\right]
\end{aligned}
$$

This means that z is a gaussian distributed random p vector with mean

$$
\bar{z} = C\bar{\xi} = [A \quad B]\begin{bmatrix} \bar{x} \\ \bar{y} \end{bmatrix} = A\bar{x} + B\bar{y}
$$

and covariance matrix

$$
\begin{aligned}
P_{zz} &= CPC' \\
&= [A \quad B]\begin{bmatrix} P_{xx} & P_{xy} \\ P_{yx} & P_{yy} \end{bmatrix}\begin{bmatrix} A' \\ B' \end{bmatrix} \\
&= AP_{xx}A' + AP_{xy}B' + BP_{yx}A' + BP_{yy}B'
\end{aligned}
$$

as claimed.

DERIVATION OF THE DENSITY FUNCTION

We now utilize the inversion formula, Eq. (3.21), to show that Eq. (3.33) is the density function which corresponds to the gaussian characteristic function, Eq. (3.32).

Let x be a gaussian random n vector whose characteristic function is given by Eq. (3.32), and assume that P is positive definite. The matrix P then has a unique positive definite square root $A = P^{1/2}$; that is, $A^2 = P$. (See Bellman [6] and page 22.)

Now consider the random n vector

$$
y = P^{-1/2}(x - \bar{x}) \tag{3.48}
$$

where $P^{-1/2}$ is the inverse of $P^{1/2}$. This random vector is obviously gaussian, since it is obtained from x via a linear transformation. It is also clear that

$$
E(y) = 0
$$

and

$$
P_{yy} = E(yy') = P^{-1/2}PP^{-1/2} = I
$$

Hence,

$$
\phi_y(s) = \exp\left(-\tfrac{1}{2}s's\right) = \exp\left(-\tfrac{1}{2}\sum_{i=1}^{n} s_i^2\right)
$$

Substituting this result into Eq. (3.21), we have

$$f(y) = \frac{1}{(2\pi)^n} \int_{-\infty}^{\infty} \cdots \int_{-\infty}^{\infty} \left[\exp \sum_{i=1}^{n} (-jy_is_i - \tfrac{1}{2}s_i{}^2) \right] ds_1 \cdots ds_n$$

Since

$$\int_{-\infty}^{\infty} \exp(-jx_is_i - \tfrac{1}{2}s_i{}^2)\, dx_i = \int_{-\infty}^{\infty} \exp(-jy_is_i) \cdot \exp(-\tfrac{1}{2}s_i{}^2)\, ds_i$$

$$= \int_{-\infty}^{\infty} \cos(y_is_i) \exp(-\tfrac{1}{2}s_i{}^2)\, ds_i$$

$$= \sqrt{2\pi}\, e^{-y_i{}^2/2}$$

it follows that

$$f(y) = \frac{1}{\sqrt{(2\pi)^n}} \exp(-\tfrac{1}{2}y'y)$$

From Eq. (3.48),

$$y'y = (x - \bar{x})'P^{-\frac{1}{2}}P^{-\frac{1}{2}}(x - \bar{x})$$

$$= (x - \bar{x})'P^{-1}(x - \bar{x})$$

and we have

$$f(x) = \frac{J}{\sqrt{(2\pi)^n}} \exp[-\tfrac{1}{2}(x - \bar{x})'P^{-1}(x - \bar{x})]$$

$$= \frac{1}{\sqrt{(2\pi)^n|P|}} \exp[-\tfrac{1}{2}(x - \bar{x})'P^{-1}(x - \bar{x})]$$

where $J = |P|^{-\frac{1}{2}}$ is the Jacobian of the transformation which is defined by Eq. (3.48).

INDEPENDENCE AND CORRELATION

In Sec. 3.5, we showed that two independent random vectors are uncorrelated, but that the converse is not generally true. However, if the two random vectors are gaussian distributed, the converse is true. This useful and important property is easily established.

Let x and y be n and m-dimensional gaussian random vectors, respectively, which are uncorrelated. Hence, $P_{xy} = 0$, and Eq. (3.34) becomes

$$\phi_{xy}(s,r) = \exp\left\{ j(\bar{x}'s + \bar{y}'r) - \tfrac{1}{2} \begin{bmatrix} s \\ r \end{bmatrix}' \begin{bmatrix} P_{xx} & 0 \\ 0 & P_{yy} \end{bmatrix} \begin{bmatrix} s \\ r \end{bmatrix} \right\}$$

$$= \exp[j(\bar{x}'s + \bar{y}'r) - \tfrac{1}{2}(s'P_{xx}s + r'P_{yy}r)]$$

$$= \exp(j\bar{x}'s - \tfrac{1}{2}s'P_{xx}s) \exp(j\bar{y}'r - \tfrac{1}{2}r'P_{yy}r)$$

$$= \phi_x(s)\phi_y(r)$$

thereby establishing the result. This property can also be derived by working with the joint density function $f(x,y)$ under the assumption that P_{xx} and P_{yy} are positive definite.

At this point, we choose to summarize two of the principal results which we have obtained above.

Theorem 3.2

    a. *Independent gaussian random vectors are uncorrelated, and conversely.*

    b. *Linear transformation on, and linear combinations of, gaussian random vectors are gaussian random vectors.*

PROPERTIES OF GAUSSIAN CONDITIONAL EXPECTATION

We conclude this section with a derivation of certain properties of conditional expectation involving gaussian random vectors. These properties are of fundamental importance for our work in estimation.

1. $E(x|y)$ is a gaussian random vector which is a linear combination of the elements of y.

This property follows from Eq. (3.41) and part 2 of Theorem 3.2.

2. $x - E(x|y)$ is independent of the random vector obtained by any linear transformation on y.

To show this, let x be n-dimensional, y r-dimensional, and A any $m \times r$ matrix. The two random vectors $\xi = x - E(x|y)$ and $\zeta = Ay$ are obviously gaussian. First, we see that

$$\bar{\xi} = E(\xi) = \bar{x} - E[E_x(x|y)] = 0$$

and that $\bar{\zeta} = E(\zeta) = A\bar{y}$.

It is then clear that

$$
\begin{aligned}
P_{\xi\zeta} &= E[(\xi - \bar{\xi})(\zeta - \bar{\zeta})'] \\
&= E\{[x - E(x|y)](y - \bar{y})'A'\} \\
&= E\{[(x - \bar{x}) - P_{xy}P_{yy}^{-1}(y - \bar{y})](y - \bar{y})'\}A' \\
&= (P_{xy} - P_{xy})A' \\
&= 0
\end{aligned}
$$

and the result follows from part 1 of Theorem 3.2.

3. If y and z are independent, where z is a random m vector, then

$$E(x|y,z) = E(x|y) + E(x|z) - \bar{x} \tag{3.49}$$

To show this, let ξ denote the random $(r + m)$ vector

$$\xi = \begin{bmatrix} y \\ z \end{bmatrix}$$

Then, from Eq. (3.41),

$$E(x|y,z) = E(x|\xi) = \bar{x} + P_{x\xi}P_{\xi\xi}^{-1}(\xi - \bar{\xi})$$

Now

$$P_{x\xi} = E\left((x - \bar{x})\begin{bmatrix} y - \bar{y} \\ z - \bar{z} \end{bmatrix}'\right) = [P_{xy} \quad P_{xz}]$$

and

$$P_{\xi\xi}^{-1} = \left\{E\left(\begin{bmatrix} y - \bar{y} \\ z - \bar{z} \end{bmatrix}[(y - \bar{y})'(z - \bar{z})']\right)\right\}^{-1}$$

$$= \begin{bmatrix} P_{yy} & 0 \\ 0 & P_{zz} \end{bmatrix}^{-1}$$

$$= \begin{bmatrix} P_{yy}^{-1} & 0 \\ 0 & P_{zz}^{-1} \end{bmatrix}$$

where we have utilized the fact that $P_{yz} = 0$, since y and z are independent. Consequently,

$$E(x|y,z) = \bar{x} + [P_{xy} \quad P_{xz}]\begin{bmatrix} P_{yy}^{-1} & 0 \\ 0 & P_{zz}^{-1} \end{bmatrix}\begin{bmatrix} y - \bar{y} \\ z - \bar{z} \end{bmatrix}$$

$$= \bar{x} + P_{xy}P_{yy}^{-1}(y - \bar{y}) + P_{xz}P_{zz}^{-1}(z - \bar{z})$$

$$= E(x|y) + E(x|z) - \bar{x}$$

4. For y and z not necessarily independent, we have

$$E(x|y,z) = E(x|y,\tilde{z}) \tag{3.50}$$

where $\tilde{z} = z - E(z|y)$ and

$$E(x|y,\tilde{z}) = E(x|y) + E(x|\tilde{z}) - \bar{x} \tag{3.51}$$

Equation (3.51) follows immediately from properties 2 and 3 above. To verify Eq. (3.50), we begin with the result in Eq. (3.41) and see that

$$E(x|y,z) = \bar{x} + [P_{xy} \quad P_{xz}]\begin{bmatrix} P_{yy} & P_{yz} \\ P_{zy} & P_{zz} \end{bmatrix}^{-1}\begin{bmatrix} y - \bar{y} \\ z - \bar{z} \end{bmatrix}$$

From Eqs. (3.36) to (3.38), we have

$$\begin{bmatrix} P_{yy} & P_{yz} \\ P_{zy} & P_{zz} \end{bmatrix}^{-1} = \begin{bmatrix} P_{yy}^{-1} + P_{yy}^{-1}P_{yz}CP_{zy}P_{yy}^{-1} & -P_{yy}^{-1}P_{yz}C \\ -CP_{zy}P_{yy}^{-1} & C \end{bmatrix}$$

where

$$C = (P_{zz} - P_{zy}P_{yy}^{-1}P_{yz})^{-1}$$

Hence,

$$\begin{aligned}
E(x|y,z) &= \bar{x} + (P_{xy}P_{yy}^{-1} + P_{xy}P_{yy}^{-1}P_{yz}CP_{zy}P_{yy}^{-1} \\
&\quad - P_{xz}CP_{zy}P_{yy}^{-1})(y - \bar{y}) \\
&\quad\quad + (P_{xz}C - P_{xy}P_{yy}^{-1}P_{yz}C)(z - \bar{z}) \\
&= \bar{x} + [P_{xy}P_{yy}^{-1} + (P_{xy}P_{yy}^{-1}P_{yz} \\
&\quad - P_{xz})CP_{zy}P_{yy}^{-1}](y - \bar{y}) \\
&\quad\quad + (P_{xz} - P_{xy}P_{yy}^{-1}P_{yz})C(z - \bar{z}) \quad (3.52)
\end{aligned}$$

On the other hand,

$$\begin{aligned}
E(x|y,\tilde{z}) &= E(x|y) + E(y|\tilde{z}) - \bar{x} \\
&= \bar{x} + P_{xy}P_{yy}^{-1}(y - \bar{y}) + P_{x\tilde{z}}P_{\tilde{z}\tilde{z}}^{-1}[\tilde{z} - E(\tilde{z})]
\end{aligned}$$

But, $E(\tilde{z}) = E[z - E(z|y)] = 0$, and

$$\begin{aligned}
P_{x\tilde{z}} &= E\{(x - \bar{x})[\tilde{z} - E(\tilde{z})]'\} \\
&= E\{(x - \bar{x})[z - \bar{z} - P_{zy}P_{yy}^{-1}(y - \bar{y})]'\} \\
&= P_{xz} - P_{xy}P_{yy}^{-1}P_{yz}
\end{aligned}$$

Also,

$$\begin{aligned}
P_{\tilde{z}\tilde{z}} &= E\{[(z - \bar{z}) - P_{zy}P_{yy}^{-1}(y - \bar{y})][(z - \bar{z}) - P_{zy}P_{yy}^{-1}(y - \bar{y})]'\} \\
&= P_{zz} - P_{zy}P_{yy}^{-1}P_{yz}
\end{aligned}$$

Hence,

$$\begin{aligned}
E(x|y,\tilde{z}) &= \bar{x} + P_{xy}P_{yy}^{-1}(y - \bar{y}) \\
&\quad + (P_{xz} - P_{xy}P_{yy}^{-1}P_{yz})(P_{zz} - P_{zy}P_{yy}^{-1}P_{yz})^{-1}[z - E(z|y)] \\
&= \bar{x} + P_{xy}P_{yy}^{-1}(y - \bar{y}) \\
&\quad + (P_{xz} - P_{xy}P_{yy}^{-1}P_{yz})C[(z - \bar{z}) - P_{zy}P_{yy}^{-1}(y - \bar{y})] \\
&= \bar{x} + [P_{xy}P_{yy}^{-1} + (P_{xy}P_{yy}^{-1}P_{yz} - P_{xz})CP_{zy}P_{yy}^{-1}](y - \bar{y}) \\
&\quad + (P_{xz} - P_{xy}P_{yy}^{-1}P_{yz})C(z - \bar{z}) \quad (3.53)
\end{aligned}$$

Comparing Eqs. (3.52) and (3.53), we see that Eq. (3.50) follows.

We note that many of the manipulations which we carried out above would have been easier if all the random vectors had zero means. Since the mean value of a random vector is the average value about which the amplitudes of the random vector are distributed, it is often convenient to imagine a translation of coordinates, viz., $\xi = x - \bar{x}$, and to perform all manipulations with respect to the zero mean random vector ξ. The nonzero mean can then be reintroduced as desired. In general, we shall proceed along these lines in the sequel.

PROBLEMS

3.1. Verify Eq. (3.3).

3.2. What is $f(x|y)$ for $y = +1$ and for $y = -1$ in Example 3.2?

3.3. Sketch $f(x_1,x_2)$ for $x_1 = x_2$ in Example 3.3 and locate $\bar{x}$ on this sketch.

3.4. Verify the four general properties of conditional expectation which are given on page 86.

3.5. Verify the results in Eqs. (3.36) to (3.38).

3.6. Let x be a gaussian random n vector with mean $\bar{x}$ and covariance matrix P_{xx}. Show that $y = Ax + b$, where A is an $m \times n$ matrix and b is a constant m vector, is a gaussian random m vector, and determine its mean and covariance matrix.

3.7. Given the gaussian probability density function in Eq. (3.33), utilize the definitions of $E(x)$ and $E[(x - \bar{x})(x - \bar{x})']$ to show that $E(x) = \bar{x}$ and $E[(x - \bar{x})(x - \bar{x})'] = P$, respectively.

3.8. Let x be a zero mean gaussian random variable with variance $P = \sigma^2 > 0$. Show that the kth moment of x, defined by the relation $\mu_k = E(x^k)$, where $k = 1, 2, \ldots$, is

$$\mu_k = \begin{cases} 1 \cdot 3 \cdots (k-1)\sigma^k & \text{for } k \text{ even} \\ 0 & \text{for } k \text{ odd} \end{cases}$$

Hint: Work with a general characteristic function for a random variable first and establish the result

$$\left. \frac{d^k \phi}{ds^k} \right|_{s=0} = j^k \mu_k$$

where $j = \sqrt{-1}$. This latter result, which can be extended to the n-dimensional case, is called the *moment theorem* [4].

3.9. If x is a gaussian random n vector, does it follow that a k vector, which comprises any k elements of x, $k = 1, 2, \ldots, n - 1$, is also a gaussian random vector? Give a reason for your answer.

3.10. Show by working with the density function in Eq. (3.35) that two uncorrelated gaussian random vectors are independent. Assume that P_{xx} and P_{yy} are positive definite.

The remaining four problems are intended to illustrate the application of certain results in this chapter to obtain "optimal" estimates of a random vector.

3.11. Let x and y be zero mean random n and m vectors, respectively. Assume that y is given, and let $\hat{x} = h(y)$ denote an estimate of x where h is a fixed n-dimensional vector-valued function of y; that is, the value of $h(\cdot)$ is known whenever the actual value of y is given. Since the estimation error, defined here as $\tilde{x} = x - \hat{x}$, will not be zero in general, it is desired to determine $\hat{x}$ in a manner such that some measure of $\tilde{x}$ is minimized. One possible measure is

$$L = E(\tilde{x}'\tilde{x}) = E\left[\sum_{i=1}^{n} \tilde{x}_i^2 \right]$$

which is called the mean square error. Other such measures, commonly termed loss functions, penalty functions, indices or criteria of performance, etc., are, of course, also possible. Show that the value of $\hat{x}$ that minimizes the mean square error L defined above is given by the expression

$$\hat{x} = E(x|y)$$

Hint: Note that $E_x[g(x)] = E_y\{E_x[g(x)|y]\}$ and set $g(x) = \tilde{x}'\tilde{x} = (x - \hat{x})'(x - \hat{x})$.

The result here shows that for a quadratic performance measure, the minimizing or *optimal* estimate is the conditional expectation.

3.12. What is the expression for the optimal estimate in Prob. 3.11 if x and y are independent?

3.13. Let x and y be zero mean n- and m-dimensional random vectors, respectively, with known joint probability density function $f(x,y)$. Assume that y is given and that it is desired to determine an estimate of x in the form $\hat{x} = A^0 y$ where A^0 is an $n \times m$ matrix. Assume further that A^0 is to be determined from the set of all possible $n \times m$ matrices A such that the performance measure

$$L = E[(x - Ay)'(x - Ay)]$$

is minimized.

Show that the optimal estimate $\hat{x} = A^0 y$ of x is

$$\hat{x} = E(xy')[E(yy')]^{-1}y$$

i.e., that

$$A^0 = E(xy')[E(yy')]^{-1} = P_{xy}P_{yy}^{-1}$$

under the assumption that P_{yy} is nonsingular.

Show also that the covariance matrix of the estimation error $\tilde{x} = x - \hat{x}$ is

$$P_{\tilde{x}\tilde{x}} = P_{xx} - P_{xy}P_{yy}^{-1}P_{yx}$$

3.14. What is the significance of the result in Probs. 3.11 and 3.13 if $f(x,y)$ is a gaussian density function?

REFERENCES

1. Feller, W., "An Introduction to Probability and Its Applications," vol. I, John Wiley & Sons, Inc., New York, 1950.
2. ———, "An Introduction to Probability and Its Applications," vol. II, John Wiley & Sons, Inc., New York, 1966.
3. Gnedenko, B. V., "The Theory of Probability," Chelsea Publishing Company, New York, 1962.
4. Papoulis, A., "Probability, Random Variables, and Stochastic Processes," McGraw-Hill Book Company, New York, 1965.
5. Parzen, E., "Modern Probability Theory and Its Applications," John Wiley & Sons, Inc., New York, 1960.
6. Bellman, R., "Introduction to Matrix Analysis," McGraw-Hill Book Company, New York, 1960.

4
Elements of the Theory of Stochastic Processes and Development of System Models

4.1 INTRODUCTION

Our goal in this chapter is to develop the system models for which we shall formulate and solve estimation and control problems in the sequel. Basically, our approach consists in combining certain results which were given in Chaps. 2 and 3. In particular, we attach a probabilistic description to the initial condition, system disturbance, and measurement error vectors of the discrete and continuous linear system models of Chap. 2. Although we could attempt to do this directly, we choose to proceed more methodically by first examining certain results from the theory of stochastic processes. Via this route, we shall form a good foundation for the development of the system models. Hopefully, this will lead to a better understanding and appreciation of our final results than is possible by the more direct approach.

Very simply, the theory of stochastic or random processes deals with the study of phenomena which are governed by probabilistic laws in which time or some other suitable variable is a parameter. Although a complete, rigorous treatment of the subject is beyond the scope of this book, we are adequately prepared at this point to deal formally with those aspects of the theory which are essential to our pursuits. The reader who is interested in a considerably more detailed treatment of the theory of stochastic processes at approximately the same level as in this chapter will find, for example, the textbooks of Parzen [1] and Papoulis [2] most helpful. More traditional developments from an engineering point of view may be found in Laning and Battin [3], Davenport and Root [4], Lee [5], and James et al. [6], for example.

Following our study of stochastic processes in Sec. 4.2, we develop the two system models in Secs. 4.3 and 4.4. These models, termed the *Gauss-Markov sequence* for discrete linear systems and the *Gauss-Markov process* for continuous linear systems, have experienced considerable success in engineering practice in the modeling of systems which are subject to random initial conditions, system disturbances, and measurement errors. The reason for this is that the models have proven to be sufficiently complete to give meaningful results, yet simple enough to be analytically and computationally tractable.

4.2 ELEMENTS OF THE THEORY OF STOCHASTIC PROCESSES

STOCHASTIC PROCESS

Let us consider a phenomenon which is evolving in time in a manner governed by probabilistic laws, e.g., the motion of a charged particle in a randomly fluctuating magnetic field or the deviation of an interplanetary space probe from its desired trajectory as a result of random errors in its guidance and propulsion systems. (In both of these cases, the state vector is six-dimensional, viz., three position and three velocity coordinates in a suitably specified coordinate frame.) The collection of all possible evolutions in time of any such phenomenon is termed a stochastic process. Viewed rather simply, a stochastic process is a collection of random vectors in which the notion of time plays a role. To be more precise, we introduce the following definition.

DEFINITION

A stochastic process is a family of random vectors $\{x(t),\ t \in I\}$ indexed by a parameter t all of whose values lie in some appropriate index set I.

The definition does not preclude the case where x is a scalar. However, we shall commonly be concerned with those situations where x is a vector.

The index set I can be any abstract set. However, we have already introduced the notion of time as the index and shall proceed accordingly. In particular, we shall need only the following two time index sets. The first is the set of discrete time instants $I = \{t_k : k = 0, 1, \ldots\}$, where $t_k < t_{k+1}$ and the time instants are not necessarily uniformly spaced. To simplify the notation, we replace t_k by k so that our index set becomes $I = \{k : k = 0, 1, \ldots\}$. In this case, we have a *discrete-time stochastic process*.

The second time index set that we need is one which involves intervals on the time axis such as $I = \{t : 0 \leq t \leq T\}$ or $I = \{t : t \geq t_0\}$. Here, the process is called a *continuous-time stochastic process*.

The phrase "family of random vectors" in the definition of a stochastic process is especially significant. It means that a stochastic process is composed of a collection or *ensemble* of random vectors over the index set. The ensemble may contain either a countable or nondenumerable number of elements. In our work, only the latter situation is of consequence.

The combination of the notions of time and ensemble implies that a stochastic process is actually a function of two variables. This is easily indicated by the notation $\{x(t,\omega),\ t \in I$ and $\omega \in \Omega\}$, where Ω is an appropriately defined sample space for the experiment under consideration. For a fixed value of t, $x(t,\cdot)$ is a vector-valued function on the sample space Ω; that is, it is a random vector. If, on the other hand, ω is fixed, then $x(\cdot,\omega)$ is a vector-valued function of time which is one possible *realization* or *sample function* of the process.

If the sample space Ω is discrete, the term *chain* is commonly used instead of the word *process*. Thus, one may speak of a discrete-time stochastic chain or a continuous-time stochastic chain. We note that if the ensemble of a stochastic process contains a countable number of elements, then we are speaking of a chain. Since chains play no role in the sequel, we shall not consider them further.

Let us now give some examples of stochastic processes.

Example 4.1 Consider first the scalar stochastic process $\{x(t),\ t \geq 0\}$ where $x(t) = V_0 \sin t$ and V_0 is a continuous random variable which is uniformly distributed between ± 1; that is,

$$f(V_0) = \begin{cases} \tfrac{1}{2} & -1 \leq V_0 \leq 1 \\ 0 & \text{elsewhere} \end{cases}$$

The ensemble contains a nondenumerable number of elements, and $\{x(t), t \geq t_0\}$ is obviously a continuous-time stochastic process. The process's ensemble may be represented as shown in Fig. 4.1.

For some fixed value of $t = t_1 \geq 0$, $x(t_1)$ is clearly a random variable. Also, any one of the sinusoids is a sample function of the process.

If we introduce sampling into the process, and, say, consider only the set of time points $\{t_k : t_k = k\pi/4, k = 0, 1, \ldots\}$, we would have a discrete-time stochastic process.

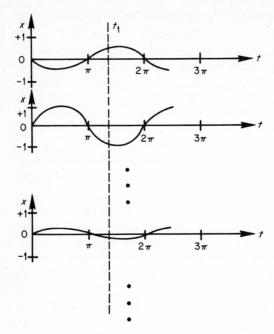

Fig. 4.1 Ensemble representation of a continuous-time stochastic process.

Example 4.2 As our second example we consider the discrete-time scalar stochastic process $\{y(k),\ k = 0,\ 1,\ \ldots\}$, where, for each k, $y(k)$ is assumed to be a zero mean gaussian random variable. We allow the possibility that the variance of $y(k)$ may depend upon k by writing the probability density function as

$$f(y,k) = \frac{1}{\sigma(k)\sqrt{2\pi}}\ e^{-y^2/2\sigma^2(k)}$$

where $\sigma^2(k) = E[y^2(k)]$. We assume that the value of $y(k)$ for a given k is statistically independent of its value at all other time points in the index set.

Two typical sample functions of the process are indicated in Fig. 4.2.

This process is called an independent gaussian sequence. The adjective gaussian is used to denote the fact that for a given k, the distribution of amplitudes "across the ensemble" is gaussian, whereas the use of the word independent denotes the statistical independence, in time, of the elements in all the

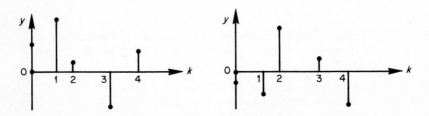

Fig. 4.2 Sample functions for a discrete-time stochastic process.

sample functions. The term sequence follows from the fact that any sample function is simply a sequence of random variables which occur at discrete instants in time.

We shall have occasion in the sequel to examine this type of process in some detail in connection with the characterization of system disturbances and measurement errors in discrete linear systems.

DESCRIPTION OF STOCHASTIC PROCESSES

It is clear from our discussion above that the description of a stochastic process involves a real (or imagined) picture of the ensemble and a probability law which governs the ensemble. We turn now to the question of how one might go about expressing the probability law to which the ensemble is subject.

To begin, suppose that we have a discrete-time stochastic process $\{x(k),\ k \in I\}$, where $I = \{k: k = 1, \ldots, N\}$ and N is some fixed positive integer. A rather natural description in this case can be given by specifying the joint probability distribution function of the N random n vectors $x(1), \ldots, x(N)$ for all n vectors $x^1, \ldots, x^N$, namely,

$$F(x^1, \ldots, x^N) = P[x(1) \leq x^1, \ldots, x(N) \leq x^N] \tag{4.1}$$

Equivalent descriptions could also be given in terms of the joint probability density function

$$f(x^1, \ldots, x^N) = \frac{\partial^{nN} F}{\partial x_1{}^1 \cdots \partial x_n{}^1 \cdots \partial x_1{}^N \cdots \partial x_n{}^N} \tag{4.2}$$

where $x_i{}^j,\ i = 1, \ldots, n$ and $j = 1, \ldots, N$, is the ith component of the n vector x^j under the assumption that the indicated nN partial derivatives exist; or, in terms of the joint characteristic function

$$\phi(s^1, \ldots, s^N) = E[\exp j \sum_{i=1}^{N} x'(i)s^i] \tag{4.3}$$

for all n vectors $s^i,\ i = 1, \ldots, N$, where $j = \sqrt{-1}$.

If the index set is discrete and infinite, i.e., $I = \{k: k = 1\ 2, \ldots\}$ or continuous and either finite or infinite, i.e., $I = \{t: 0 \leq t \leq T\}$ or $I = \{t;\ t \geq 0\}$, respectively, a natural description in the sense of the one given above is not obvious. However, from a practical point of view, it seems reasonable that a stochastic process $\{x(t),\ t \in I\}$, where I is any one of the above three types, can be adequately represented by considering its behavior at a finite number of time points $t_1, \ldots, t_m$ in I. Then the process can be described by specifying the joint probability distribution function of the m random n vectors $x(t_1), \ldots, x(t_m)$ for all integers m and m times points $t_1, \ldots, t_m$ in I. This can be written as

$$F(x^1, \ldots, x^m) = P[x(t_1) \leq x^1, \ldots, x(t_m) \leq x^m] \tag{4.4}$$

for all n vectors $x^1, \ldots, x^m$. The expression is, of course, analogous to Eq. (4.1), but far more involved, since we must consider all integers m and, correspondingly, all sets of m time points $t_1, \ldots, t_m$ in I.

Equivalent descriptions in terms of the joint probability density and joint characteristic functions then follow by analogy to Eqs. (4.2) and (4.3), respectively. In the first case, we write

$$f(x^1, \ldots, x^m) = \frac{\partial^{mn} F}{\partial x_1^{~1} \cdots \partial x_n^{~1} \cdots \partial x_1^{~m} \cdots \partial x_n^{~m}} \qquad (4.5)$$

assuming that the indicated mn partial derivatives exist. In the second, we have

$$\phi(s^1, \ldots, s^m) = E[\exp j \sum_{i=1}^{m} x'(t_i) s^i] \qquad (4.6)$$

for all n vectors s^i, $i = 1, \ldots, m$.

In Eqs. (4.1) through (4.6), it is sometimes convenient to include time explicitly on the left-hand side. This can be done, for example, by writing $f(x^1, \ldots, x^m)$ as either $f(x^1, t_1; \ldots ; x^m, t_m)$ or $f[x(t_1), \ldots, x(t_m)]$, and similarly for F and ϕ. We shall make use of all three of these forms in the sequel.

The notion of conditional probability can also be introduced in describing a stochastic process, but we defer this until later.

The approach given above and defined by either Eqs. (4.4) to (4.6) is the most natural way of describing the probability law that governs a stochastic process. In practice, however, it has been found more expedient to describe certain stochastic processes in terms of their first and second moments, which are called the mean value function and the covariance kernel, respectively. We consider these quantities next.

MEAN VALUE FUNCTION AND COVARIANCE KERNEL

The *mean value function* of a stochastic process $\{x(t), t \in I\}$ is defined for all $t \in I$ by the expression

$$\bar{x}(t) = E[x(t)] \qquad (4.7)$$

This function is also called the *mean value* or the *mean* of $\{x(t) \ t \in I\}$. If it is zero for all $t \in I$ then we say that $\{x(t), t \in I\}$ is a zero mean stochastic process.

The *covariance kernel* of $\{x(t), t \in I\}$ is defined for all t and $\tau \in I$ by the relation

$$P(t,\tau) = E\{[x(t) - \bar{x}(t)] [x(\tau) - \bar{x}(\tau)]'\} \qquad (4.8)$$

For two stochastic processes $\{x(t), t \in I\}$ and $\{y(t), t \in I\}$ with $E[x(t)] = \bar{x}(t)$ and $E[y(t)] = \bar{y}(t)$, the function

$$P_{xy}(t,\tau) = E\{[x(t) - \bar{x}(t)][y(\tau) - \bar{y}(\tau)]'\} \tag{4.9}$$

is called the *cross-covariance kernel* of the two processes for all $t, \tau \in I$. It is clear that x and y need not both have the same number of elements.

When $t = \tau$, we denote $P(t,t)$ by $P(t)$ and call it the *covariance matrix* of $\{x(t), t \in I\}$. Similarly, we define $P_{xy}(t) = P_{xy}(t,t)$ and term it the *cross-covariance matrix* of the two processes.

We note immediately that

$$P(t) = E[x(t)x'(t)] - \bar{x}(t)\bar{x}'(t)$$

and

$$P_{xy}(t) = E[x(t)y'(t)] - \bar{x}(t)\bar{y}'(t)$$

Also, it is clear that $P'(t) = P(t)$ but that $P'_{xy}(t) \neq P_{xy}(t)$, in general. However, $P'_{xy}(t) = P_{yx}(t)$ does hold in all instances.

The use of the mean value function and the covariance kernel in characterizing a stochastic process is, of course, analogous to the use of the mean and covariance matrix in characterizing a random vector, the difference being that time is now a parameter in the description.

Example 4.3 For the continuous-time stochastic process in Example 4.1, it is clear that

$$\bar{x}(t) = E(V_0 \sin t) = 0$$

for all $t \geq 0$, since V_0 is uniformly distributed between -1 and $+1$ and is therefore a zero mean random variable.

Also,

$$
\begin{aligned}
P(t,\tau) &= E(V_0^2 \sin t \sin \tau) \\
&= E(V_0^2) \sin t \sin \tau \\
&= \tfrac{1}{3} \sin t \sin \tau
\end{aligned}
$$

from which it follows that

$$P(t) = \tfrac{1}{3} \sin^2 t$$

Example 4.4 For the independent gaussian sequence in Example 4.2, we have $\bar{y}(k) = 0$ and $P(j,k) = 0$ for all $j, k = 0, 1, \ldots$, when $j \neq k$. For $j = k$, obviously, $P(k) = \sigma^2(k)$.

Assuming that the time points $k = 0, 1, \ldots$, for the stochastic process in Example 4.2 coincide with the time points $t_k = k\pi/4, k = 0, 1, \ldots$, in the discrete-time stochastic process of Example 4.1, and that V_0 is statistically independent of $y(k)$ for all $k = 0, 1, \ldots$, it follows that

$$P_{xy}(j,k) = E[x(j)y(k)] = 0$$

for all $j, k = 0, 1, \ldots$ for these two processes.

INDEPENDENT AND CORRELATED STOCHASTIC PROCESSES

The notions of independence and correlation carry over from probability theory to the theory of stochastic processes in a simple manner, just as the notions of mean and covariance did above.

First, we say that a stochastic process $\{x(t),\ t \in I\}$ is *independent* if, for any m time points $t_1,\ \dots,\ t_m$ in I, where m is any integer,

$$P[x(t_1) \leq x^1,\ \dots,\ x(t_m) \leq x^m] = \prod_{i=1}^{m} P_i[x(t_i) \leq x^i]$$

or alternately,

$$F(x^1,\ \dots,\ x^m) = \prod_{i=1}^{m} F_i(x^i) \tag{4.10}$$

for all n vectors $x^1,\ \dots,\ x^m$.

In terms of the relevant probability density functions, assuming that they exist, the definition can be expressed by

$$f(x^1,\ \dots,\ x^m) = \prod_{i=1}^{m} f_i(x^i) \tag{4.11}$$

where

$$f(x^1,\ \dots,\ x^m) = \frac{\partial^{mn} F}{\partial x_1{}^1 \cdots \partial x_n{}^1 \cdots \partial x_1{}^m \cdots \partial x_n{}^m}$$

and

$$f_j(x^j) = \frac{\partial^n F_j}{\partial x_1{}^j \cdots \partial x_n{}^j} \qquad j = 1,\ \dots,\ m$$

The definition in terms of characteristic function follows in the same way.

Now let $\{x(t),\ t \in I\}$ and $\{y(t),\ t \in I\}$ be two stochastic processes, where x is an n vector and y is a p vector. The two processes are said to be *independent* of each other if, for any m time points $t_1,\ \dots,\ t_m$ in I, where m is any integer,

$$P[x(t_1) \leq x^1,\ \dots,\ x(t_m) \leq x^m, y(t_1) \leq y^1,\ \dots,\ y(t_m) \leq y^m]$$
$$= P_1[x(t_1) \leq x^1,\ \dots,\ x(t_m) \leq x^m] P_2[y(t_1) \leq y^1,\ \dots,\ y(t_m) \leq y^m]$$

for all n vectors $x^1,\ \dots,\ x^m$ and all p vectors $y^1,\ \dots,\ y^m$.

Relations analogous to Eqs. (4.10) and (4.11) then follow trivially, and we write

$$F(x^1,\ \dots,\ x^m, y^1,\ \dots,\ y^m) = F_1(x^1,\ \dots,\ x^m) F_2(y^1,\ \dots,\ y^m)$$

and

$$f(x^1,\ \dots,\ x^m, y^1,\ \dots,\ y^m) = f_1(x^1,\ \dots,\ x^m) f_2(y^1,\ \dots,\ y^m),$$

respectively.

Clearly, one must exercise care in distinguishing between an "independent stochastic process" and "independent stochastic processes" in discussions involving two or more stochastic processes.

In regard to Examples 4.1 and 4.2, it is obvious that the stochastic process in the latter example is an independent one (by hypothesis, actually), whereas neither of the two processes in the former one is. Going one step further, if the time points in the discrete-time stochastic process in Example 4.1 are assumed to be coincident with those of $\{y(k),\ k = 0,\ 1,\ \ldots\}$ in Example 4.2 and if V_0 is independent of $x(k)$ for all k, then these two processes are independent of each other.

We turn next to the notion of *correlation* in the description of stochastic processes. In particular, the *correlation matrix* of a stochastic process is defined for all $t,\ \tau \in I$ by the expression

$$\Psi(t,\tau) = E[x(t)x'(\tau)] \tag{4.12}$$

Expanding the right-hand side of Eq. (4.8), we have

$$P(t,\tau) = E[x(t)x'(\tau)] - 2\bar{x}(t)\bar{x}'(\tau) + \bar{x}(t)\bar{x}'(\tau)$$
$$= \Psi(t,\tau) - \bar{x}(t)\bar{x}'(\tau) \tag{4.13}$$

If $\{x(t),\ t \in I\}$ is a zero mean process, then obviously

$$P(t,\tau) = \Psi(t,\tau)$$

for all $t,\ \tau \in I$.

We observe that for $t \neq \tau$, the correlation matrix, Eq. (4.12), and the covariance kernel, Eq. (4.8), give a measure of "time correlation" for the stochastic process $\{x(t),\ t \in I\}$ as distinguished from the "space correlation" that results when $t = \tau$ in these expressions.

We now say that the stochastic process $\{x(t),\ t \in I\}$ is *uncorrelated* if

$$\Psi(t,\tau) = E[x(t)]E[x'(\tau)] \tag{4.14}$$

for all t and τ in I, $t \neq \tau$.

If a stochastic process is independent, then for $t,\ \tau \in I$, $t \neq \tau$, $f(x^1,t;x^2,\tau) = f_1(x^1,t)f_2(x^2,\tau)$, and it follows that

$$\Psi(t,\tau) = \int_{-\infty}^{\infty} \cdots \int_{-\infty}^{\infty} x^1 x^{2\prime} f_1(x^1,t) f_2(x^2,\tau)\ dx_1{}^1 \cdots dx_n{}^1\ dx_1{}^2 \cdots$$
$$dx_n{}^2$$
$$= E[x(t)]E[x'(\tau)]$$

Hence, an independent stochastic process is also an uncorrelated one. If, in addition, the process has mean zero, then

$$P(t,\tau) = \Psi(t,\tau) = 0$$

for all $t,\ \tau \in I$, $t \neq \tau$.

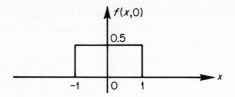

Fig. 4.3 Density function for initial value of discrete-time stochastic process.

We now show by means of a simple example that an uncorrelated stochastic process need not also be an independent one.

Example 4.5 Consider the scalar stochastic process $\{x(k),\ k = 0, 1\}$ for which $x(1) = \alpha x^2(0)$, where α is a nonzero scalar. Obviously this is not an independent process. Suppose that $x(0)$ is uniformly distributed between -1 and 1, as shown in Fig. 4.3.

We have first that

$$\Psi(1,0) = E[x(1)x(0)] = \alpha E[x^3(0)]$$

$$= \frac{\alpha}{2} \int_{-1}^{1} \xi^3\, d\xi$$

$$= 0$$

On the other hand, it is clear that $E[x(0)] = 0$ and that

$$E[x(1)] = \alpha E[x^2(0)]$$

$$= \frac{\alpha}{2} \int_{-1}^{1} \xi^2\, d\xi$$

$$= \frac{\alpha}{3}$$

from which it follows that

$$E[x(1)]E[x(0)] = 0 = \Psi(1,0)$$

Similarly, $\Psi(0,1) = E[x(0)]E[x(1)]$, and we have here a case of an uncorrelated process which is not independent.

The *cross-correlation matrix* of two stochastic processes $\{x(t),\ t \in I\}$ and $\{y(t),\ t \in I\}$ is denoted by $\Psi_{xy}(t,\tau)$ and defined for all t and τ in I by the relation

$$\Psi_{xy}(t,\tau) = E[x(t)y'(\tau)] \tag{4.15}$$

If Eq. (4.15) has the property that

$$\Psi_{xy}(t,\tau) = E[x(t)]E[y'(\tau)] \tag{4.16}$$

for all t and τ in I, the two stochastic processes are termed *uncorrelated*. We note here that the restriction $t \neq \tau$ used in the definition of a single uncorrelated stochastic process is absent. We remark also that care must be exercised in distinguishing between the above two types of "uncorrelatedness."

We leave as an exercise the task of showing that two stochastic processes which are independent of each other are also uncorrelated with each other, but that the converse is not necessarily true.

STATIONARY AND NONSTATIONARY PROCESSES

For some studies involving stochastic processes, it is often convenient to separate stochastic processes into two classes based on the nature of the mechanism producing the process. Stated rather simply, a *stationary stochastic process* is one in which the probability laws governing the mechanism producing the process remain time-invariant as the process evolves in time. On the other hand, if these probability laws are time-variant during the process's evolution, the process is termed *nonstationary*.

A stochastic process $\{x(t), t \in I\}$ is said to be *strict-sense stationary* if, for any two sets of m time points each $(t_1 + \tau, \ldots, t_m + \tau)$ and $(t_1, \ldots, t_m)$ in I where m is any integer and τ = constant,

$$F(x^1, t_1 + \tau; \ldots; x^m, t_m + \tau) = F(x^1, t_1; \ldots; x^m, t_m) \qquad (4.17)$$

for all n vectors $x^1, \ldots, x^m$. Definitions in terms of the probability density and characteristic functions are obvious.

If Eq. (4.17) is satisfied, then $f(x,t) = f(x,0) = f(x)$, and it follows that

$$\bar{x}(t) = E[x(t)] = \int_{-\infty}^{\infty} \cdots \int_{-\infty}^{\infty} xf(x)\, dx = \bar{x} = \text{constant}$$

and

$$P(t) = E\{[x(t) - \bar{x}(t)][x(t) - \bar{x}(t)]'\}$$
$$= \int_{-\infty}^{\infty} \cdots \int_{-\infty}^{\infty} (x - \bar{x})(x - \bar{x})'f(x)\, dx = P = \text{constant}$$

Also, $f(x^1, t_1; x^2, t_2) = f(x^1, 0; x^2, t_2 - t_1)$, so that

$$P(t_1, t_2) = \int_{-\infty}^{\infty} \cdots \int_{-\infty}^{\infty} (x^1 - \bar{x})(x^2 - \bar{x})f(x^1, 0; x^2, t_2 - t_1)\, dx^1\, dx^2$$
$$= P(0, t_2 - t_1)$$

That is, the covariance kernel depends only upon the time difference $t_2 - t_1$. This is commonly denoted by writing

$$P(t_1, t_2) = P(t_2 - t_1)$$

In this case, $\{x(t), t \in I\}$ is said to be *wide-sense* or *covariance stationary*.

It is clear that a strict-sense stationary process is covariance stationary, but not necessarily vice versa.

We observe from Eq. (4.13) that

$$\Psi(t,\tau) = P(t,\tau) + \bar{x}(t)\bar{x}'(\tau)$$

For a covariance stationary process, $\bar{x}(t) = \bar{x}(\tau) = \bar{x}$ and

$$P(t,\tau) = P(t - \tau)$$

and it follows that

$$\Psi(t,\tau) = P(t - \tau) + \bar{x}\bar{x}'$$

Let us illustrate the above notions with an example.

Example 4.6   Let $\{x(t), t \geq 0\}$ be the scalar stochastic process which is defined by the differential equation $\dot{x} = -ax + w$, where a is a positive constant and w is a zero mean random variable which is independent of the initial condition $x(0)$ and has a constant variance $\sigma_w^2 > 0$.   For simplicity, we take $x(0) = 0$.

Any sample function of the process is of the form

$$x(t) = \frac{w}{a}(1 - e^{-at})$$

for all $t \geq 0$.   It is clear that $\bar{x}(t) = 0$ and

$$P(t) = E[x^2(t)]$$

$$= \frac{\sigma_w^2}{a^2}(1 - e^{-at})^2$$

Also,

$$P(t_1,t_2) = E[x(t_1)x(t_2)]$$

$$= \frac{\sigma_w^2}{a^2}(1 - e^{-at_1})(1 - e^{-at_2})$$

and the process is nonstationary.

However, for sufficiently large t_1 and t_2,

$$P(t_1,t_2) \rightarrow \frac{\sigma_w^2}{a^2}$$

and the process becomes covariance stationary in the "steady state."

GAUSS-MARKOV STOCHASTIC PROCESS

We focus our attention now on the description of an important and useful type of stochastic process which is termed the *Gauss-Markov process*. This process enjoys essentially the same prominence in the theory of stochastic processes that the gaussian distribution does in probability theory.   The comments which were made in Sec. 3.6 in reference to the latter's utility in practice are equally applicable to the former here.

We present first the definition of a gaussian or normal stochastic process.

DEFINITION

A stochastic process $\{x(t), t \in I\}$ is defined to be gaussian or normal if, for any m time points $t_1, \ldots, t_m$ in I, where m is any integer, the set of m random n vectors $x(t_1), \ldots, x(t_m)$ is jointly gaussian distributed.

In terms of the joint characteristic function, the definition requires that

$$\phi(s^1, \ldots, s^m) = \exp\left(j \sum_{i=1}^{m} \bar{x}^{i\prime} s^i - \tfrac{1}{2} \sum_{i=1}^{m} \sum_{k=1}^{m} s^{i\prime} P^{ik} s^k\right) \tag{4.18}$$

for all n vectors $s^1, \ldots, s^m$, where

$$\bar{x}^i = E[x(t_i)] \qquad i = 1, \ldots, m \tag{4.19}$$

and

$$P^{ik} = E\{[x(t_i) - \bar{x}^i][x(t_k) - \bar{x}^k]'\} \qquad i, k = 1, \ldots, m \tag{4.20}$$

In terms of the probability density function, we write

$$f(x^*) = \frac{1}{\sqrt{(2\pi)^{nm}|P^*|}} \exp\left[-\tfrac{1}{2}(x^* - \bar{x}^*)' P^{*-1}(x^* - \bar{x}^*)\right] \tag{4.21}$$

where x^* is the nm vector

$$x^* = \begin{bmatrix} x^1 \\ \cdot \\ \cdot \\ \cdot \\ x^m \end{bmatrix}$$

and P^* is the $nm \times nm$ matrix

$$P^* = [P^{ik}] \qquad i, k = 1, \ldots, m$$

with the P^{ik} given by Eq. (4.20).

The description in terms of the probability density function in Eq. (4.21) is, of course, valid only if P^* is positive definite. If P^* is positive semidefinite, then Eq. (4.18) must be employed in the description.

In any event, we see that the probability law of a gaussian stochastic process is completely specified by the mean $\bar{x}(t)$ for all $t \in I$ and the covariance kernel $P(t, \tau)$ for all t and τ in I.

A particular type of gaussian stochastic process that will concern us in the sequel is the gaussian white process whose definition follows.

DEFINITION

A stochastic process $\{x(t), t \in I\}$ is said to be a gaussian white process if, for any m time points $t_1, \ldots, t_m$ in I, where m is any integer, the m random n vectors $x(t_1), \ldots, x(t_m)$ are independent gaussian random vectors.

The adjective "white" is used here instead of independent in keeping with traditional terminology [3–5]. The term arose as a consequence of the fact that certain stationary, independent, stochastic processes have a power spectral density which is constant over a wide range of frequencies just as white light does.

As a result of independence, the joint characteristic function of a gaussian white process is

$$\phi(s^1, \ldots, s^m) = \prod_{i=1}^{m} \phi_i(s^i)$$

where

$$\phi_i(s^i) = \exp\left(j\bar{x}^{i'}s^i - \tfrac{1}{2}s^{i'}P^{ii}s^i\right) \qquad i = 1, \ldots, m$$

for all n vectors $s^1, \ldots, s^m$; and where $\bar{x}^i$ and P^{ii}, $i = 1, \ldots, m$, are given by Eqs. (4.19) and (4.20), respectively.

We observe that, for a gaussian white process, the covariance kernel $P(t,\tau) = 0$ for all t and τ in I, where $t \neq \tau$.

It is clear that the probability law of a gaussian white process is completely specified by the mean value function $\bar{x}(t)$ and the covariance matrix $P(t)$ for all $t \in I$.

Finally, as a matter of convenience in the sequel, we shall call a discrete-time gaussian white process a *gaussian white sequence* and refer to a continuous-time gaussian white process more simply as a *gaussian white process*.

Let us consider next the formal definition of a Markov process. We say that a stochastic process $\{x(t),\ t \in I\}$ is a *Markov process* if, for any m time points $t_1 < t_2 < \cdots < t_m$ in I, where m is any integer, the conditional probability distribution function of $x(t_m)$ for given values of $x(t_1), \ldots, x(t_{m-1})$ has the property that

$$P[x(t_m) \leq x^m | x(t_{m-1}) = x^{m-1}, \ldots, x(t_1) = x^1]$$
$$= P[x(t_m) \leq x^m | x(t_{m-1}) = x^{m-1}]$$

for all n vectors $x^1, \ldots, x^m$.

In terms of the conditional probability density function, the definition means that

$$f(x^m | x^{m-1}, \ldots, x^1) = f(x^m | x^{m-1})$$

for all n vectors $x^1, \ldots, x^m$. Equivalently, we write also

$$f[x(t_m) | x(t_{m-1}), \ldots, x(t_1)] = f[x(t_m) | x(t_{m-1})]$$

From the definition, we see that if we consider t_{m-1} to be the present time, and $t_{m-2}, \ldots, t_1$ to be the past, then the probability law describing the process in the future, i.e., at time t_m, depends only on the present value that the process assumes and is completely independent of the process's behavior in the past. This feature is termed the "Markov property."

A Markov process is described by its *transition probability distribution function* $F[x(t) \leq \xi | x(\tau) = \zeta]$, which denotes the conditional probability that $x(t) \leq \xi$, where ξ is any n vector given that $x(\tau) = \zeta$, where ζ is a specified n vector, and t and τ are elements of the process's index set with $t > \tau$. The description can also be given by the transition probability density function, denoted $f[x(t)|x(\tau)]$, under the assumption that it exists.

We now exhibit a particular way of describing a Markov process that will be of use to us. Let $\{x(t), t \in I\}$ be a Markov process for which $f[x(t)|x(\tau)]$ is given for all t and τ in I, $t > \tau$. One way of describing the process is to specify the joint probability density function of the m random n vectors $x(t_1), \ldots, x(t_m)$ for all integers m and m time points $t_1 < t_2 < \cdots < t_m$ in I and all n vectors $x^1, \ldots, x^m$.

Let $t_1 < t_2 < \cdots < t_m$ be any m time points in I, where m is any integer. Then, from Bayes' rule,

$$f(x^m, \ldots, x^1) = f(x^m | x^{m-1}, \ldots, x^1)f(x^{m-1}, \ldots, x^1) \qquad (4.22)$$

However, since the process is Markov,

$$f(x^m | x^{m-1}, \ldots, x^1) = f(x^m | x^{m-1})$$

which means that Eq. (4.22) can be simplified to

$$f(x^m, \ldots, x^1) = f(x^m | x^{m-1})f(x^{m-1}, \ldots, x^1) \qquad (4.23)$$

Again, from Bayes' rule and the fact that the process is Markov,

$$f(x^{m-1}, \ldots, x^1) = f(x^{m-1} | x^{m-2}, \ldots, x^1)f(x^{m-2}, \ldots, x^1)$$
$$= f(x^{m-1} | x^{m-2})f(x^{m-2}, \ldots, x^1)$$

Substitution of this result into Eq. (4.23) gives

$$f(x^m, \ldots, x^1) = f(x^m | x^{m-1})f(x^{m-1} | x^{m-2})f(x^{m-2}, \ldots, x^1)$$

Continuing in this fashion by reapplying Bayes' rule and the Markov property of the process, we are led finally to the result that

$$f(x^m, \ldots, x^1) = f(x^m | x^{m-1})f(x^{m-1} | x^{m-2}) \cdots f(x^2 | x^1)f(x^1) \qquad (4.24)$$

Each term on the right-hand side of Eq. (4.24) except $f(x^1) = f[x(t_1)]$ is known, since the transition probability density function $f[x(t)|x(\tau)]$ is given for all t and τ in I, $t > \tau$. Hence, the joint probability density function of the process is completely determined if the probability density function $f[x(t_1)]$ is specified. In particular, if t_1 is the initial time in the index set I, we see that a Markov process is completely specified by the probability density function of the process at the initial time and the transition probability density function.

In carrying out the above argument, we have intentionally omitted subscripts on the probability density functions for the sake of simplicity. This does not mean, for example, that the conditional probability density function for $x(t_m)$ given $x(t_{m-1})$ is the same as that for $x(t_{m-1})$ given $x(t_{m-2})$, etc.

Also, we have assumed that all the indicated probability density functions exist. It is, of course, clear that the development could also be conducted utilizing either probability distribution functions or characteristic functions.

We present next two properties of Markov processes which are of some importance.

1. Let $\{x(t), t \in I\}$ be a Markov process and suppose that I_1 is a subset of I. Then $\{x(t), t \in I_1\}$ is also a Markov process. The verification of this property is left as an exercise.

2. A Markov process is also reverse-time Markov. In other words, for $t_m < t_{m+1} < \cdots < t_{m+k}$ in I, where m and k are any two integers, we have the result

$$f(x^m | x^{m+1}, \ldots, x^{m+k}) = f(x^m | x^{m+1})$$

To verify this, we first utilize Bayes' rule and see that

$$f(x^m | x^{m+1}, \ldots, x^{m+k}) = \frac{f(x^m, x^{m+1}, \ldots, x^{m+k})}{f(x^{m+1}, \ldots, x^{m+k})}$$

However, by rearranging elements and utilizing the fact that the process is Markov, we have

$$f(x^m, x^{m+1}, \ldots, x^{m+k}) = f(x^{m+k}, x^{m+k-1}, \ldots, x^m)$$
$$= f(x^{m+k} | x^{m+k-1}) \cdots f(x^{m+1} | x^m) f(x^m)$$

where the second line follows from Eq. (4.24).

Similarly, it is clear that

$$f(x^{m+1}, \ldots, x^{m+k}) = f(x^{m+k}, x^{m+k-1}, \ldots, x^{m+1})$$
$$= f(x^{m+k} | x^{m+k-1}) \cdots f(x^{m+2} | x^{m+1}) f(x^{m+1})$$

Dividing these last two relations and canceling the common terms, it follows that

$$f(x^m | x^{m+1}, \ldots, x^{m+k}) = \frac{f(x^{m+1} | x^m) f(x^m)}{f(x^{m+1})}$$

Utilizing Bayes' rule once more, we have

$$\frac{f(x^{m+1}|x^m)f(x^m)}{f(x^{m+1})} = \frac{f(x^{m+1},x^m)}{f(x^{m+1})}$$

$$= \frac{f(x^m,x^{m+1})}{f(x^{m+1})}$$

$$= f(x^m|x^{m+1})$$

and the result is established.

The concept of a Markov process can be easily extended to include what are termed higher-order Markov processes. For example, we consider a stochastic process $\{x(t),\ t \in I\}$, which, for any m time points $t_1 < t_2 < \cdots < t_m$, where m is any integer, has the property that

$$P[x(t_m) \le x^m | x(t_{m-1}) = x^{m-1},\ \ldots,\ x(t_1) = x^1]$$

$$= P[x(t_m) \le x^m | x(t_{m-1}) = x^{m-1},\ x(t_{m-2}) = x^{m-2}]$$

for all n vectors $x^1,\ \ldots,\ x^m$. Such a process is said to be *Markov-2*. The extension to Markov-k processes for other integers is obvious.

Our concern is primarily with Markov-1 processes, which we introduced first and which we shall continue to refer to simply as Markov processes. However, we shall encounter Markov-2 processes in our study of optimal data smoothing and refer to them specifically as such.

By analogy to the description of a Markov process, a Markov-2 process is described by its transition probability distribution function $F[x(t) \le \xi | x(\tau_1) = \zeta^1,\ x(\tau_2) = \zeta^2]$, where ξ it any n vector, ζ^1 and ζ^2 are two specified n vectors, and $t,\ \tau_1,\ \tau_2 \in I$, with $t > \tau_1 > \tau_2$. The corresponding transition probability density function is, of course, $f[x(t)|x(\tau_1),\ x(\tau_2)]$.

The two properties of Markov processes which we discussed above also pertain to Markov-2 processes.

We conclude this section by introducing and illustrating the notions of Gauss-Markov and Gauss-Markov-2 processes. The definitions follow trivially from the names.

DEFINITION

A stochastic process $\{x(t),\ t \in I\}$ is Gauss-Markov if and only if it is both Gaussian and Markov.

DEFINITION

A stochastic process $\{x(t)\ t \in I\}$ is Gauss-Markov-2 if and only if it is both Gaussian and Markov-2.

Again, as a matter of convenience in the sequel, we shall call a discrete-time Gauss-Markov process a *Gauss-Markov sequence* and use the term *Gauss-Markov process* when speaking of a continuous-time Gauss-Markov process, and similarly for the Markov-2 case.

It is clear that the gaussian nature of these processes dictates the amplitude distribution, while the Markov nature governs the process's evolution in time.

We remark that the two properties of Markov and Markov-2 processes which we noted above carry over when the processes are also gaussian.

For a Gauss-Markov sequence whose index set is $I = \{k: k = 0, 1, \ldots\}$ we note that the process is completely specified if the gaussian probability density functions $f[x(0)]$ and $f[x(k + 1)|x(k)]$ are given, the latter for all $k \in I$. By analogy to Eq. (4.24), we can then write

$$f[x(k + 1), \ldots, x(0)] = f[x(k + 1)|x(k)]f[x(k)|x(k - 1)] \cdots$$
$$f[x(1)|x(0)]f[x(0)] \quad (4.25)$$

for any $k \in I$, which gives us the joint probability density function for the process.

Finally, we note that a gaussian white process, either discrete or continuous-time, can be viewed as a Gauss-Markov process, discrete or continuous time, respectively, for which $f[x(t)|x(\tau)] = f[x(t)]$ for all t and τ in I with $t > \tau$.

Example 4.7 Consider the scalar process $\{x(t), t \geq 0\}$ which is defined by the differential equation

$$\dot{x} = -\frac{1}{t + 1} x$$

where $x(0)$ is a gaussian random variable with mean zero and variance $\sigma_0^2 > 0$.
By direct integration,

$$x(t) = \frac{x(0)}{t + 1}$$

for all $t \geq 0$. Since $x(0)$ is gaussian distributed, it follows immediately that, for any m time points $t_1, \ldots, t_m$ in $I = [0, \infty)$, $f(x^1, \ldots, x^m)$ is a gaussian density function.

In addition, for some $t_2 > t_1 \geq 0$, it is easily shown that

$$x(t_2) = \frac{t_1 + 1}{t_2 + 1} x(t_1)$$

As a result, for any ordered set of time points $t_m, \ldots, t_1 \in [0, \infty)$,

$$f(x^m|x^{m-1}, \ldots, x^1) = f(x^m|x^{m-1})$$

Consequently, the process is Gauss-Markov.

We know that

$$f[x(0)] = f[x,0] = \frac{1}{\sqrt{2\pi}\,\sigma_0}\, e^{-x^2/2\sigma_0^2}$$

Let us determine $f[x(t)|x(\tau)] = f(x,t|x,\tau)$. Since $\{x(t),\ t \geq 0\}$ is Gauss-Markov, we know that $f(x,t|x,\tau)$ is a gaussian density function, and is therefore completely specified in terms of its conditional mean and variance.

Since

$$x(t) = \frac{\tau+1}{t+1}\, x(\tau)$$

for $t > \tau$, then obviously

$$E[x(t)|x(\tau)] = \frac{\tau+1}{t+1}\, x(\tau)$$

and

$$E[\{x(t) - E[x(t)|x(\tau)]\}^2] = E[x(t) - \frac{\tau+1}{t+1}\, x(\tau)]^2 = 0$$

This simply means that $f(x,t|x,\tau)$ is a unit Dirac delta function. The result is not surprising, since a knowledge of $x(\tau)$ permits us to determine $x(t)$ exactly for all $t \geq \tau \geq 0$.

Since $x(t)$ is a gaussian random variable, we can characterize it in terms of its mean and variance. In particular,

$$\bar{x}(t) = E\left[\frac{x(0)}{t+1}\right] = 0$$

and

$$P(t) = E[x(t) - \bar{x}(t)]^2$$
$$= E\left[\frac{x^2(0)}{(t+1)^2}\right] = \frac{\sigma_0^2}{(t+1)^2}$$

from which it follows that

$$f(x,t) = \frac{(t+1)}{\sqrt{2\pi}\,\sigma_0}\, e^{-(t+1)^2 x^2/2\sigma_0^2}$$

Example 4.8 Suppose that $\{x_1(t),\ t \geq 0\}$ is a scalar process with $\ddot{x}_1 = 0$, where $x_1(0)$ and $\dot{x}_1(0)$ are jointly gaussian distributed. We note that

$$x_1(t) = x_1(0) + \dot{x}_1(0)t$$

and $\{x_1(t),\ t \geq 0\}$ is obviously a gaussian process.

Let us now consider three time points $t_3 > t_2 > t_1 \geq 0$ and show that the process is Markov-2. We have

$$x_1(t_3) = x_1(0) + \dot{x}_1(0)t_3$$
$$x_1(t_2) = x_1(0) + \dot{x}_1(0)t_2$$
$$x_1(t_1) = x_1(0) + \dot{x}_1(0)t_1$$

Subtracting the second relation from the first, we obtain

$$x_1(t_3) - x_1(t_2) = (t_3 - t_2)\dot{x}_1(0)$$

or

$$x_1(t_3) = x_1(t_2) + (t_3 - t_2)\dot{x}_1(0)$$

Similarly, we obtain

$$x_1(t_2) - x_1(t_1) = (t_2 - t_1)\dot{x}_1(0)$$

or

$$\dot{x}_1(0) = \frac{x_1(t_2) - x_1(t_1)}{t_2 - t_1}$$

from the last two relations.

From these results, it follows that

$$x_1(t_3) = x_1(t_2) + \frac{t_3 - t_2}{t_2 - t_1}[x_1(t_2) - x_1(t_1)]$$

We see that $x_1(t_3)$ is specified if both $x_1(t_2)$ and $x_1(t_1)$ are given. In general then,

$$f(x^m|x^{m-1}, \ldots, x^1) = f(x^m|x^{m-1}, x^{m-2})$$

and the process is Markov-2.

The fact that $\{x_1(t), t \geq 0\}$ is not simply Markov follows because its defining relation $\ddot{x}_1 = 0$ is a second-order differential equation, and therefore requires two constants of integration. In other words, in order to determine the future values of $x(t)$, we must have knowledge of its past values for two previous time points.

Let us now consider the stochastic process $\{x(t), t \geq 0\}$, where

$$x(t) = \begin{bmatrix} x_1(t) \\ \dot{x}_1(t) \end{bmatrix} \triangleq \begin{bmatrix} x_1(t) \\ x_2(t) \end{bmatrix}$$

That is, our process is now two-dimensional and consists of the "position" and "velocity" coordinates of the above process. Clearly,

$$x(t) = \begin{bmatrix} x_1(t) \\ x_2(t) \end{bmatrix} = \begin{bmatrix} x_1(0) + \dot{x}_1(0)t \\ \dot{x}_1(0) \end{bmatrix}$$

Since $x(t)$ is obviously a gaussian 2 vector, the process is gaussian. In addition, for $t_2 > t_1 \geq 0$

$$x(t_2) = \begin{bmatrix} x_1(0) + \dot{x}_1(0)t_2 \\ \dot{x}_1(0) \end{bmatrix} = \begin{bmatrix} x_1(t_2) \\ x_2(t_2) \end{bmatrix}$$

and

$$x(t_1) = \begin{bmatrix} x_1(0) + \dot{x}_1(0)t_1 \\ \dot{x}_1(0) \end{bmatrix} = \begin{bmatrix} x_1(t_1) \\ x_2(t_1) \end{bmatrix}$$

Hence, given $x(t_1)$, it follows that

$$x(t_2) = \begin{bmatrix} x_1(t_1) + (t_2 - t_1)x_2(t_1) \\ x_2(t_1) \end{bmatrix}$$

That is, $x(t_2)$ can be determined exactly from a knowledge of $x(t_1)$, which means that $\{x(t), t \geq 0\}$ is Markov in addition to being gaussian. We have given here an example of how a Gauss-Markov-2 process can be "reduced" to a Gauss-Markov process by "extending" the dimension of the state vector. The procedure will prove useful to us on a number of occasions.

4.3 GAUSS-MARKOV SEQUENCE MODEL

We are now ready to develop the first of our system models. We consider the description of the system dynamics and measurement scheme separately, and in that order. Following the development of each of the two parts of the model, we shall discuss how that part pertains to certain physical situations. We then conclude this section with a brief treatment of the Gauss-Markov-2 sequence model.

SYSTEM DYNAMICS

Let $\{w(k), k \in I\}$ where $I = \{k: k = 0, 1, \ldots\}$ be a p-dimensional gaussian white sequence whose mean

$$E[w(k)] = \bar{w}(k)$$

and covariance

$$E\{[w(j) - \bar{w}(j)][w(k) - \bar{w}(k)]'\} = Q(k)\delta_{jk} \tag{4.26}$$

are given for all j, $k = 0, 1, \ldots$, where δ_{jk} is the Kronecker delta and $Q(k)$ is a positive semidefinite $p \times p$ matrix. Further, let $x(0)$ be a gaussian random n vector with known mean

$$E[x(0)] = \bar{x}(0)$$

and covariance matrix

$$E\{[x(0) - \bar{x}(0)][x(0) - \bar{x}(0)]'\} = P(0)$$

where $P(0)$ is a positive semidefinite $n \times n$ matrix. Also, assume that $\{w(k), k \in I\}$ is independent of $x(0)$, so that

$$E\{[x(0) - \bar{x}(0)][w(k) - \bar{w}(k)]'\} = 0 \tag{4.27}$$

for all $k \in I$.

As our model, we take the system

$$x(k + 1) = \Phi(k + 1, k)x(k) + \Gamma(k + 1, k)w(k) \tag{4.28}$$

for $k \in I$, where x is an n vector, the state; $\Phi(k + 1, k)$ is an $n \times n$ matrix, the state transition matrix; w is a p vector, the system disturb-

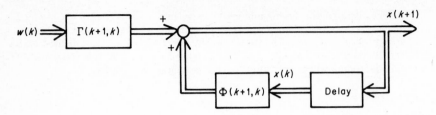

Fig. 4.4　Discrete linear system dynamics model.

ance; $\Gamma(k + 1, k)$ is an $n \times p$ matrix, the disturbance transition matrix; and $\{w(k), k = 0, 1, \ldots\}$ and $x(0)$ have the properties listed above.

For simplicity, we ignore for the present the fact that the model may include a control input term, viz., $\Psi(k + 1, k)u(k)$, where u is an r vector, the control, and $\Psi(k + 1, k)$ is an $n \times r$ matrix, the control transition matrix. We shall, however, return to this point later and refine the model accordingly.

The block diagram for the model is shown in Fig. 4.4, where it is understood that on any one "run" of the system, the input $w(k)$, $k = 0, 1, \ldots$, is any sample function of the gaussian white sequence $\{w(k), k = 0, 1, \ldots\}$ and the initial condition vector $x(0)$ is any sample drawn from the given gaussian distributed population of initial condition vectors. The model is seen to be the same as that in Sec. 2.4 except that $x(0)$ and $w(k)$ now have a particular stochastic description.

Evidently then, $\{x(k), k \in I\}$ is a stochastic process. In particular, we assert that it is a Gauss-Markov sequence.

We now establish this assertion and give two ways of describing the process.

We show first that the process is Markov. Let $t_1 < t_2 < \cdots < t_m$ be any m time points in I, where m is any integer. Also, let k and j be the integers in I that correspond to the time points t_m and t_{m-1}, respectively. The situation may be considered as shown in Fig. 4.5, where it is clear that there will, in general, be points in I that lie between the time points $t_1 < t_2 < \cdots < t_m$ as well as outside the interval of time spanned by them.

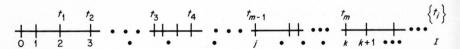

Fig. 4.5　Index set I and arbitrary set of ordered time points $\{t_i : i = 1, \ldots, m; t_1 < t_2 < \cdots < t_m\}$.

From Eq. (4.28), it is clear that

$$x(j + 1) = \Phi(j + 1, j)x(j) + \Gamma(j + 1, j)w(j)$$

and

$$x(j + 2) = \Phi(j + 2, j + 1)x(j + 1) + \Gamma(j + 2, j + 1)w(j + 1)$$

Substituting the former into the latter and rearranging terms, we have

$$\begin{aligned}
x(j + 2) &= \Phi(j + 2, j + 1)[\Phi(j + 1, j)x(j) + \Gamma(j + 1, j)w(j)] \\
&\qquad + \Gamma(j + 2, j + 1)w(j + 1) \\
&= \Phi(j + 2, j)x(j) + \Phi(j + 2, j + 1)\Gamma(j + 1, j)w(j) \\
&\qquad + \Gamma(j + 2, j + 1)w(j + 1) \\
&= \Phi(j + 2, j)x(j) + \sum_{i=j+1}^{j+2} \Phi(j + 2, i)\Gamma(i, i - 1)w(i - 1)
\end{aligned}$$

where $\Phi(j + 2, j) \triangleq \Phi(j + 2, j + 1)\Phi(j + 1, j)$. Continuing in this manner, we obtain the general relationship

$$x(j + n) = \Phi(j + n, j)x(j) + \sum_{i=j+1}^{j+n} \Phi(j + n, i)\Gamma(i, i - 1)w(i - 1)$$

$$(4.29)$$

where $n = 1, 2, \ldots$, with

$$\Phi(j + n, i) = \Phi(j + n, j + n - 1) \cdots \Phi(i + 1, i)$$

for $i = j, j + 1, \ldots, j + n$.

For $j + n = k$, it is obvious that

$$x(k) = \Phi(k,j)x(j) + \sum_{i=j+1}^{k} \Phi(k,i)\Gamma(i, i - 1)w(i - 1) \qquad (4.30)$$

Recalling that k corresponds to t_m and j to t_{m-1}, it is clear from Eq. (4.30) that the conditional probability distribution function of $x(t_m)$ for any given set of values of $x(t_{m-1})$, $x(t_{m-2})$, $\ldots$, $x(t_2)$, $x(t_1)$ depends only upon $x(t_{m-1})$; that is, the conditioning is dependent only on $x(t_{m-1})$. Since this is true for any m time points $t_1 < t_2 < \cdots < t_m$, where m is any integer, $\{x(k), k \in I\}$ is obviously Markov.

It is trivial to show that the process is gaussian. Setting $j = 0$ in Eq. (4.30),

$$x(k) = \Phi(k,0)x(0) + \sum_{i=1}^{k} \Phi(k,i)\Gamma(i, i - 1)w(i - 1) \qquad (4.31)$$

Since $x(0)$ and each $w(i - 1)$, $i = 1, \ldots, k$ are gaussian by hypothesis, it follows that $x(k)$ is also gaussian for each $k = 0, 1, \ldots$, since it is merely the sum of gaussian random vectors. Consequently, for any integer m and any set of time points $t_1, t_2, \ldots, t_m \in I$, the set of random n vectors $x(t_1), \ldots, x(t_m)$ is jointly gaussian distributed, and the assertion is proved.

We now give two ways of describing the process. From Eq. (4.25), we know that the joint probability density function of a Gauss-Markov sequence $\{x(k), k \in I\}$, where $I = \{k: k = 0, 1, \ldots\}$ is completely determined for all $k \in I$ if its gaussian probability density functions $f[x(0)]$ and $f[x(k+1)|x(k)]$ are given, the latter for all $k \in I$. Since $x(0)$ is a given gaussian random n vector with mean $\bar{x}(0)$ and covariance matrix $P(0)$, $f[x(0)]$ is determined.† For a given $x(k)$, it follows from Eq. (4.28) that

$$E[x(k+1)|x(k)] = \Phi(k+1, k)x(k) + \Gamma(k+1, k)\bar{w}(k) \qquad (4.32)$$

Then,

$$\begin{aligned}
x(k+1) - E[x(k+1)|x(k)] &= \Phi(k+1, k)x(k) + \Gamma(k+1, k)w(k) \\
&\quad - \Phi(k+1, k)x(k) - \Gamma(k+1, k)\bar{w}(k) \\
&= \Gamma(k+1, k)[w(k) - \bar{w}(k)]
\end{aligned}$$

Hence, the conditional covariance matrix is

$$\begin{aligned}
E[\{x(k+1) &- E[x(k+1)|x(k)]\}\{x(k+1) - E[x(k+1)|x(k)]\}'] \\
&= E\{\Gamma(k+1, k)[w(k) - \bar{w}(k)][w(k) - \bar{w}(k)]'\Gamma'(k+1, k)\} \\
&= \Gamma(k+1, k)Q(k)\Gamma'(k+1, k) \qquad (4.33)
\end{aligned}$$

and the conditional probability density function is determined with the conditional mean given by Eq. (4.32) and the conditional covariance matrix by Eq. (4.33). If $\Gamma(k+1, k)Q(k)\Gamma'(k+1, k)$ is singular, then the description is given in terms of the conditional characteristic function.

In the second description, we also take advantage of both the gaussian and Markov properties of the process to obtain a very simple and useful description. In particular, we derive expressions for the mean and covariance matrix of the process.

Letting $E[x(k)] = \bar{x}(k)$, we have from Eq. (4.28) that

$$\bar{x}(k+1) = \Phi(k+1, k)\bar{x}(k) + \Gamma(k+1, k)\bar{w}(k) \qquad (4.34)$$

for all $k \in I$. Since $\bar{x}(0)$ and $\bar{w}(k)$, $k \in I$, are given, Eq. (4.34) is a recursive relation for the mean of the sequence.

---

† This requires that $P(0)$ be positive definite. If it is not, then the description must be given in terms of the characteristic function.

Next, letting $P(k) = E\{[x(k) - \bar{x}(k)][x(k) - \bar{x}(k)]'\}$, we have from Eqs. (4.28), (4.34), and (4.26) that

$$
\begin{aligned}
P(k+1) &= E\{[x(k+1) - \bar{x}(k+1)][x(k+1) - \bar{x}(k+1)]'\} \\
&= E(\{\Phi(k+1, k)[x(k) - \bar{x}(k)] \\
&\quad + \Gamma(k+1, k)[w(k) - \bar{w}(k)]\} \\
&\quad \{\Phi(k+1, k)[x(k) - \bar{x}(k)] \\
&\quad + \Gamma(k+1, k)[w(k) - \bar{w}(k)]\}') \\
&= \Phi(k+1, k)P(k)\Phi'(k+1, k) \\
&\quad + \Phi(k+1, k)E\{[x(k) - \bar{x}(k)][w(k) \\
&\quad - \bar{w}(k)]'\}\Gamma'(k+1, k) \\
&\quad + \Gamma(k+1, k)E\{[w(k) - \bar{w}(k)][x(k) \\
&\quad - \bar{x}(k)]'\}\Phi'(k+1, k) \\
&\quad + \Gamma(k+1, k)Q(k)\Gamma'(k+1, k) \quad (4.35)
\end{aligned}
$$

where the middle two terms remain to be evaluated. Since one of these terms is the transpose of the other, it is sufficient to consider only the first one.

Subtracting Eq. (4.34) from Eq. (4.28), we obtain

$$
\begin{aligned}
x(k+1) - \bar{x}(k+1) &= \Phi(k+1, k)[x(k) - \bar{x}(k)] \\
&\quad + \Gamma(k+1, k)[w(k) - \bar{w}(k)]
\end{aligned}
$$

which, after defining $\tilde{x}(k) = x(k) - \bar{x}(k)$ and $\tilde{w}(k) = w(k) - \bar{w}(k)$, can be written as

$$
\tilde{x}(k+1) = \Phi(k+1, k)\tilde{x}(k) + \Gamma(k+1, k)\tilde{w}(k)
$$

for $k = 0, 1, \ldots$.

By direct analogy to Eq. (4.31), we have

$$
\tilde{x}(k) = \Phi(k,0)\tilde{x}(0) + \sum_{i=1}^{k} \Phi(k,i)\Gamma(i, i-1)\tilde{w}(i-1) \quad (4.36)
$$

Postmultiplying in Eq. (4.36) by $\tilde{w}'(k)$ and taking the expected value of the result, we obtain

$$
\begin{aligned}
E[\tilde{x}(k)\tilde{w}'(k)] &= \Phi(k,0)E[\tilde{x}(0)\tilde{w}'(k)] \\
&\quad + \sum_{i=1}^{k} \Phi(k,i)\Gamma(i, i-1)E[\tilde{w}(i-1)\tilde{w}'(k)]
\end{aligned}
$$

Clearly, $\{\tilde{w}(k),\ k = 0, 1, \ldots\}$ is a zero mean gaussian white sequence which is independent of the zero mean gaussian random n vector

$\bar{x}(0)$. This means that

$$E[\bar{w}(i-1)\bar{w}'(k)] = 0$$

for all $i - 1 \neq k$ and that

$$E[\bar{x}(0)\bar{w}'(k)] = 0$$

for all $k = 0, 1, \ldots$. Hence,

$$E[\bar{x}(k)\bar{w}'(k)] = E\{[x(k) - \bar{x}(k)][w(k) - \bar{w}(k)]'\} = 0$$

for all $k = 0, 1, \ldots$, which means that the second and third terms in Eq. (4.35) are zero, and we have

$$P(k+1) = \Phi(k+1, k)P(k)\Phi'(k+1, k)$$
$$+ \Gamma(k+1, k)Q(k)\Gamma'(k+1, k) \quad (4.37)$$

for $k = 0, 1, \ldots$. Since $P(0)$ and $Q(k)$ are given, the latter for all $k = 0, 1, \ldots$, Eq. (4.37) is a recursive matrix equation for the $n \times n$ covariance matrix of the sequence.

For each value of k, we can express the characteristic function of the sequence in the form

$$\phi_x(s,k) = \exp\left[j\bar{x}'(k)s - \tfrac{1}{2}s'P(k)s\right]$$

where s is an n vector.

The use of $P(k)$ in our description appears to be an oversimplification. Since $\{x(k), k \in I\}$ is gaussian, we actually need the covariance kernel $P(k,j)$ instead of $P(k)$. To justify our present description, we now show that knowledge of $P(k)$ is sufficient to determine $P(k,j)$.

By analogy to Eq. (4.30), we have

$$\bar{x}(k) = \Phi(k,j)\bar{x}(j) + \sum_{i=j+1}^{k} \phi(k,i)\Gamma(i, i-1)\bar{w}(i-1)$$

for $k > j$. From this we see that

$$P(k,j) = E[\bar{x}(k)\bar{x}'(j)]$$
$$= \Phi(k,j)E[\bar{x}(j)\bar{x}'(j)]$$
$$+ \sum_{i=j+1}^{k} \Phi(k,i)\Gamma(i, i-1)E[\bar{w}(i-1)\bar{x}'(j)]$$

Replacing k by j in Eq. (4.36), taking the transpose of the result, premultiplying by $\bar{w}(i-1)$, and taking the expected value, we have

$$E[\bar{w}(i-1)\bar{x}'(j)] = E[\bar{w}(i-1)\bar{x}'(0)]\Phi'(j,0)$$
$$+ \sum_{l=1}^{j} E[\bar{w}(i-1)\bar{w}'(l-1)]\Gamma'(l, l-1)\Phi'(j,l)$$

As above, $E[\tilde{w}(i-1)\tilde{x}'(0)] = 0$ for all $i = 1, 2, \ldots$. Also, the indexing on i is $i = j + 1, j + 2, \ldots, k$. Hence,

$$E[\tilde{w}(i-1)\tilde{w}'(l-1)] = 0$$

for the relevant values of i and l, and we have

$$E[\tilde{w}(i-1)\tilde{x}'(j)] = 0$$

This means that the expression for $P(k,j)$ reduces to

$$P(k,j) = \Phi(k,j)P(j)$$

clearly indicating that it is sufficient to know the covariance matrix in order to describe the sequence.

In retrospect, the gaussian nature of $\{x(k),\ k \in I\}$ requires knowledge of the pair $\{\bar{x}(k),\ P(k,j),\ k,\ j \in I\}$ to describe the sequence. The Markov property permits the requirement to be reduced to specification of $\{\bar{x}(k),\ P(k),\ k \in I\}$.

A very simple interpretation of Eq. (4.37) is that it describes how uncertainty in the system propagates in time. This is due to the fact that the covariance matrix gives a measure of the spread of the distribution of $x(k)$ about its mean.

Throughout our work, the second of the above two descriptions will play the dominant role primarily because of the above interpretation.

Our model is easily extended to include a control input, viz.,

$$x(k+1) = \Phi(k+1, k)x(k) + \Gamma(k+1, k)w(k) + \Psi(k+1, k)u(k)$$

where the term $\Psi(k+1, k)u(k)$ was defined earlier. Under the assumption that the control sequence $\{u(k),\ k = 0, 1, \ldots\}$ is known or can be specified as desired, the stochastic process $\{x(k),\ k = 0, 1, \ldots\}$ is still a Gauss-Markov sequence. This follows from the fact that the presence of a known control input merely adds a deterministic component to the system output, and, therefore, affects only the mean of $\{x(k),\ k = 0, 1, \ldots\}$. In particular, we have

$$\bar{x}(k+1) = \Phi(k+1, k)\bar{x}(k) + \Gamma(k+1, k)\bar{w}(k) + \Psi(k+1, k)u(k)$$

for $k = 0, 1, \ldots$. Also,

$$x(k+1) - \bar{x}(k+1) = \Phi(k+1, k)[x(k) - \bar{x}(k)] \\ + \Gamma(k+1, k)[w(k) - \bar{w}(k)]$$

from which it is clear that the covariance matrix of the sequence is the same regardless of whether or not a given control input is present.

In determining the conditional mean for the sequence, we make use of the fact that the control input is known to us to obtain the result

$$E[x(k + 1)|x(k),u(k)] = \Phi(k + 1, k)x(k) + \Gamma(k + 1, k)\bar{w}(k)$$
$$+ \Psi(k + 1, k)u(k)$$

where $k = 0, 1, \ldots$. The conditional covariance matrix is then the same as before.

We observe that our model, Eq. (4.28), is that of a discrete linear system with a gaussian white sequence forcing function which is independent of the system's initial state. There are four assumptions involved here that merit some comment.

First, we have assumed that the model is linear. The justifications for this are that (1) the use of linear models in engineering studies has proved fruitful and (2) the techniques of linear systems analysis are well developed, whereas those for nonlinear systems are not, in general.

Second, we have assumed that the forcing function, viz., the disturbance vector stochastic process, is independent of the initial state. This is reasonable for many systems from physical considerations if the mechanism producing the disturbance vector is unrelated to, or does not interact with, that which produces the initial state. For example, we can argue that the random internal disturbances present in a supersonic transport's inertial navigation system (due to imperfect gyros, accelerometers, and associated electronics) are independent of the initial coordinates (latitude, longitude, and altitude) that are input to the navigation system just prior to takeoff. Unfortunately, situations in which this assumption of independence is not valid also arise in practice. As a case in point here, consider the relationship between the random wind gusts experienced by a spacecraft during liftoff and boost in relation to the initial conditions, viz., the launch site coordinates. Obviously, the wind patterns are related to the location of the launch site. Nevertheless, we shall restrict our studies to cases of the first type described.

Third, the assumption that the amplitudes of the elements of the disturbance vector are, at each time point of interest, jointly gaussian distributed is based on Theorem 3.1, the central limit theorem. Implicit here is the further assumption that the system disturbance, a macroscopic stochastic process, is the sum of a "large" number of independently acting microscopic stochastic processes. The same is true of the initial state $x(0)$. The justification here is that this assumption has proven to be a useful idealization for modeling many physical phenomena.

Finally, the fourth assumption which we have made is that the disturbance vector stochastic process is "white," i.e., independent. This assumption is commonly employed in practice in the absence of any analyses or experimental data to the contrary. The relaxation of this assumption to the case where the process is correlated will be considered in Chap. 5 after the filtering problem has been solved under the assumption of independence. The task of formulating the system model for this purpose is treated below in our discussion of the Gauss-Markov-2 sequence.

MEASUREMENTS

We turn next to the formulation of the measurement model which is assumed to be of the form

$$z(k + 1) = H(k + 1)x(k + 1) + v(k + 1) \tag{4.38}$$

for $k \in I$ where z is an m vector, the measurement vector; H is an $m \times n$ matrix that relates the state vector to the measurement vector; and v is an m vector, the measurement error vector. Our real concern here is the specification of the model for the measurement error vector.

We assume that the measurement error process can be modeled as an m-dimensional gaussian white sequence $\{v(k + 1), \ k \in I\}$, whose mean

$$E[v(k + 1)] = \bar{v}(k + 1)$$

and covariance

$$E\{[v(j + 1) - \bar{v}(j + 1)][v(k + 1) - \bar{v}(k + 1)]'\} = R(k + 1)\delta_{jk} \tag{4.39}$$

are given for all j and k in I, where δ_{jk} is the Kronecker delta and $R(k + 1)$ is a positive semidefinite $m \times m$ matrix for all $k \in I$.

We also assume that $\{v(k + 1), \ k \in I\}$ is independent of $x(0)$ for all $k \in I$. Hence,

$$E\{[x(0) - \bar{x}(0)][v(k + 1) - \bar{v}(k + 1)]'\} = 0 \tag{4.40}$$

for all $k = 0, 1, \ldots$.

We allow the possibility that $\{w(k), k \in I\}$ and $\{v(k + 1), k \in I\}$ may be correlated with each other by writing

$$E\{[w(j + 1) - \bar{w}(j + 1)][v(k + 1) - \bar{v}(k + 1)]'\} = S(k + 1)\delta_{jk} \tag{4.41}$$

for $j, \ k = 0, 1, \ldots$, where $S(k + 1)$ is a $p \times m$ matrix, the cross-covariance matrix. Most of our work will be concerned with the case where $S(k + 1) = 0$ for all k. However, in certain parts of our development, as well as in the problems, we shall consider the more general situation. In either case, the measurement model has the block diagram shown in Fig. 4.6.

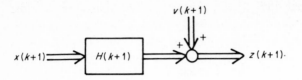

Fig. 4.6 Discrete linear system measurement model.

Our reasons for selecting the linear structure of Eq. (4.38) for our measurement model are the same as those for our choice of Eq. (4.28) to model the system dynamics.

Similarly, we assume that the measurement error process can be modeled as a gaussian white sequence for the same reasons that we assumed the disturbance process could be considered gaussian and white. Our model, then, presupposes that measurement errors made at one time are independent of those at any other time.   After we have solved the filtering problem in Chap. 5 under this assumption, we shall show how the theory can be extended to handle problems where the measurement error vector process is correlated.

The justification for the assumed independence of $x(0)$ and $\{v(k + 1), k \in I\}$ is that, physically, we expect the mechanism from which the measurement errors arise to be independent of the one leading to the initial state.   Generally, the measurement system for a given system is peripheral to the system dynamics, so that the hardware and electronic component imperfections, which cause measurement errors, function independently of the initial state.

Also, from basic implementation considerations, it is plausible to expect that mechanisms producing system disturbances are unrelated to those that generate measurement errors in many cases.

As an example wherein the assumption that $\{v(k + 1), k \in I\}$ is independent of both $x(0)$ and $\{w(k), k \in I\}$ is reasonable, let us consider once more the supersonic transport from above.   Let us assume that it is enroute from Los Angeles to New York and that a doppler navigation fix is being made over St. Louis.   Since the operation of the doppler system electronics is in no way affected by the aircraft's coordinates at takeoff, we are clearly justified in assuming that the initial state and the doppler measurement errors made in taking the navigation fix are independent.   In addition, since the doppler system electronics are peripheral to the inertial navigation system, it is reasonable to assume that the measurement errors are independent of the random disturbances within the navigation system.

On the other hand, it is possible to envision situations in which the system disturbance and the measurement error are correlated.   For example, in a ground-based tracking radar, a position sensor may be coupled to the radar dish drive shaft through a suitable gear train. Then random wind gusts impinging on the antenna are not only a system disturbance but also appear in the output of the position sensor as measurement noise.

As is clear from Eq. (4.38), we have assumed that no measurement is made at the initial time; that is, there is no $z(0)$. This is actually no restriction, and the removal of the assumption is trivial.

For analytical ease, it is convenient to assume that $x(0)$, $\{w(k)$ $k \in I\}$, and $\{v(k+1),\ k \in I\}$ all have zero mean in specifying the system model of Eqs. (4.28) and (4.38). There is no loss of generality in so doing, since the means $\bar{x}(k+1)$ and $\bar{z}(k+1)$ are easy to compute, using the relations

$$\bar{x}(k+1) = \Phi(k+1, k)\bar{x}(k) + \Gamma(k+1, k)\bar{w}(k)$$

and

$$\bar{z}(k+1) = H(k+1)\bar{x}(k+1) + \bar{v}(k+1)$$

respectively, for $k = 0, 1, \ldots$, when $x(0)$, $\{w(k), k \in I\}$, and $\{v(k+1), k \in I\}$ have nonzero means. We shall make the above assumption for the system model when we consider the estimation problem in Chaps. 5 and 6 and the control problem in Chap. 9.

The basic idea here is that nonzero mean values enter into the system description as deterministic elements which are readily computed whenever desired. Hence, one is justified in making them subordinate (but not necessarily ignoring them) in analyses.

GAUSS-MARKOV-2 SEQUENCE

Let us examine the stochastic process $\{x(k),\ k = 0, 1, \ldots\}$ which is defined by the second-order linear vector difference equation

$$x(k+2) = \Phi(k+2, k+1)x(k+1) + \theta(k+2, k)x(k)$$
$$+ \Gamma(k+2, k)w(k) \quad (4.42)$$

for $k = 0, 1, \ldots$. We assume that $x(0)$ and $x(1)$ are jointly gaussian distributed with

$$E[x(0)] = \bar{x}(0) \qquad E[x(1)] = \bar{x}(1)$$
$$E\{[x(0) - \bar{x}(0)][x(0) - \bar{x}(0)]'\} = P_{00}$$
$$E\{[x(1) - \bar{x}(1)][x(1) - \bar{x}(1)]'\} = P_{11}$$

and

$$E\{[x(0) - \bar{x}(0)][x(1) - \bar{x}(1)]'\} = P_{01}$$

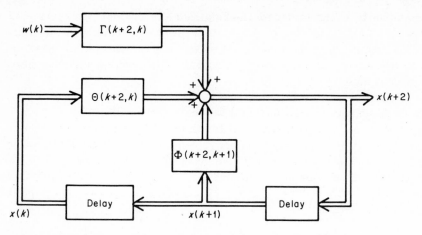

Fig. 4.7 Block diagram of Gauss-Markov-2 system dynamics model.

given. We assume further that $\{w(k),\ k = 0,\ 1,\ \ldots\}$ is a gaussian white sequence with known mean and covariance and that this sequence is independent of $x(0)$ and $x(1)$.

In Eq. (4.42), x is an n vector, w is a p vector, $\Phi(k + 2,\ k + 1)$ and $\theta(k + 2,\ k)$ are both $n \times n$ matrices, and $\Gamma(k + 2,\ k)$ is an $n \times p$ matrix.

Because of the gaussian nature of $x(0)$, $x(1)$, and $\{w(k),\ k = 0,\ 1,\ \ldots\}$ and the linearity of Eq. (4.42), $\{x(k),\ k = 0,\ 1,\ \ldots\}$ is obviously a gaussian process. Moreover, because of the second-order nature of Eq. (4.42), the conditional probability distribution function of the state x at any time point given the values of the state at an arbitrary number of previous time points depends upon only two previous values; i.e., the process is Markov-2.

The block diagram for the system is shown in Fig. 4.7, where we note the presence of two delay operations, which are the result, of course, of the second-order nature of Eq. (4.42).

Rather than working directly with this Gauss-Markov-2 sequence, it is convenient to convert it into a Gauss-Markov sequence and take advantage of the previously developed results. To do this, we first introduce a new n vector which is defined by the relation

$$x(k + 1) = y(k) \tag{4.43}$$

Utilizing this definition, we then write Eq. (4.42) as

$$y(k + 1) = \Phi(k + 2,\ k + 1)y(k) + \theta(k + 2,\ k)x(k)$$
$$+ \Gamma(k + 2,\ k)w(k) \tag{4.44}$$

Next, letting x^* denote the $2n$ vector

$$x^* = \begin{bmatrix} x \\ y \end{bmatrix}$$

we combine Eqs. (4.43) and (4.44) into one relation

$$x^*(k + 1) = \begin{bmatrix} 0 & I \\ \theta(k + 2, k + 1) & \Phi(k + 2, k + 1) \end{bmatrix} x^*(k)$$
$$+ \begin{bmatrix} 0 \\ \Gamma(k + 2, k) \end{bmatrix} w(k)$$

where I is the $n \times n$ identity matrix.

Defining

$$\Phi^*(k + 1, k) = \begin{bmatrix} 0 & I \\ \theta(k + 2, k + 1) & \Phi(k + 2, k + 1) \end{bmatrix}$$

and

$$\Gamma^*(k + 1, k) = \begin{bmatrix} 0 \\ \Gamma(k + 2, k) \end{bmatrix}$$

we can write

$$x^*(k + 1) = \Phi^*(k + 1, k)x^*(k) + \Gamma^*(k + 1, k)w(k) \tag{4.45}$$

for $k = 0, 1, \ldots$.

From its definition,

$$x^*(0) = \begin{bmatrix} x(0) \\ y(0) \end{bmatrix} = \begin{bmatrix} x(0) \\ x(1) \end{bmatrix}$$

is obviously a gaussian random $2n$ vector with mean

$$\bar{x}^*(0) = \begin{bmatrix} \bar{x}(0) \\ \bar{x}(1) \end{bmatrix}$$

and covariance matrix

$$P(0) = \begin{bmatrix} P_{00} & P_{01} \\ P'_{01} & P_{11} \end{bmatrix}$$

Since $\{w(k), k = 0, 1, \ldots\}$ is a gaussian white sequence which is independent of $x^*(0)$ and Eq. (4.45) is of the same form as Eq. (4.28), it follows that $\{x^*(k), k = 0, 1, \ldots\}$ is a Gauss-Markov sequence.

The above procedure, we note, is identical to that which we utilized in the second part of Example 4.8.

The price which we have paid to convert $\{x(k), k = 0, 1, \ldots\}$ into a Gauss-Markov sequence is that of introducing a new state vector which has twice as many elements as the original state vector. The alternative would have been to work with a second-order probability description. As noted earlier, we choose instead to take advantage of results already at our disposal.

We treat next a slightly different situation which involves a Gauss-Markov-2 sequence. Suppose now that the system is described by the two first-order difference equations

$$x(k + 1) = \Phi(k + 1, k)x(k) + \Gamma(k + 1, k)w(k) \tag{4.46}$$

and

$$w(k + 1) = \theta(k + 1, k)w(k) + \Delta(k + 1, k)\xi(k) \tag{4.47}$$

for $k = 0, 1, \ldots$, where x is an n vector, w is a p vector, and ξ is a q vector. The matrices $\Phi(k + 1, k)$, $\Gamma(k + 1, k)$, $\theta(k + 1, k)$, and $\Delta(k + 1, k)$ are $n \times n$, $n \times p$, $p \times p$, and $p \times q$, respectively. We take x to be the primary state vector of interest and assume that w is the system disturbance.

We assume that $x(0)$ and $w(0)$ are jointly gaussian distributed with

$$E[x(0)] = \bar{x}(0) \qquad E[w(0)] = \bar{w}(0)$$
$$E\{[x(0) - \bar{x}(0)][x(0) - \bar{x}(0)]'\} = P_{xx}(0)$$
$$E\{[w(0) - \bar{w}(0)][w(0) - \bar{x}(0)]'\} = P_{ww}(0)$$

and

$$E\{[x(0) - \bar{x}(0)][w(0) - \bar{w}(0)]'\} = P_{xw}(0)$$

In addition, we assume that $\{\xi(k), k = 0, 1, \ldots\}$ is a gaussian white sequence which is independent of both $x(0)$ and $w(0)$ and has known mean and covariance.

Obviously, $\{w(k), k = 0, 1, \ldots\}$ is a Gauss-Markov sequence. Hence, Eq. (4.46), which is of the same form as Eq. (4.28), models a system in which the system disturbance is a Gauss-Markov sequence instead of a gaussian white sequence. In other words, our interest here is in extending the model of Eq. (4.28) to include cases where the system disturbance is a particular type of correlated stochastic process.

A simple way of observing that $\{x(k), k = 0, 1, \ldots\}$ is Gauss-Markov-2 is to consider the block diagram for the system, which is defined by Eqs. (4.46) and (4.47) and is shown in Fig. 4.8. Between its input $\xi(k)$ and its output $x(k + 1)$, the system is obviously second-order. Then, since $x(0)$ and $w(0)$ are jointly gaussian distributed, and $\{\xi(k), k = 0, 1, \ldots\}$ is a Gaussian white sequence which is independent of the initial conditions, the stochastic process $\{x(k), k = 0, 1, \ldots\}$ has the same properties as the one defined by Eq. (4.42); viz., it is a Gauss-Markov-2 sequence.

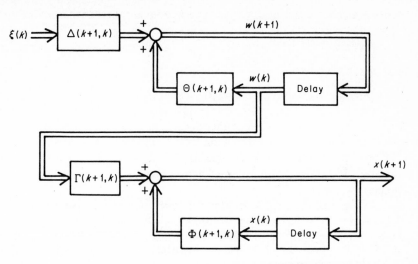

Fig. 4.8   Block diagram for system of Eqs. (4.46) and (4.47).

Finally, it is a trivial matter to show that $\{x^*(k),\ k = 0, 1, \ldots\}$ where x^* is the $(n + p)$ vector

$$x^* = \begin{bmatrix} x \\ w \end{bmatrix}$$

is a Gauss-Markov sequence.

The technique of extending the state vector to reduce higher-order Markov processes to simple Markov processes is called the augmented state vector approach [7].

4.4 GAUSS–MARKOV PROCESS MODEL

In this section, we develop the second of our system models, the one for continuous linear systems.   The development is analogous to the one for discrete linear systems.

We begin with a formulation of the notion of the gaussian white process which we then use to develop the system model.   As in Sec. 4.3, we conclude with a consideration of the Markov-2 version of the system dynamics.

GAUSSIAN WHITE PROCESS

Proceeding directly by analogy to the gaussian white sequence, we postulate a p-dimensional stochastic process $\{w(t),\ t \geq t_0\}$ to be a gaussian white process if it is an independent gaussian process with given mean

$$E[w(t)] = \bar{w}(t)$$

and covariance kernel

$$E\{[w(t) - \bar{w}(t)][w(\tau) - \bar{w}(\tau)'\} = Q(t)\delta(t - \tau)$$

where t_0 is a given initial time, $t, \tau \geq t_0$, $Q(t)$ is a continuous positive semidefinite $p \times p$ matrix, and $\delta(t - \tau)$ is the Dirac delta function. The latter is introduced by analogy to the Kronecker delta, which appears in the covariance kernel expression for the gaussian white sequence.

We give now a plausibility argument for the above formulation. The argument, which is due to Kalman [7], is based on consideration of the limiting behavior of a piecewise constant gaussian white sequence in which the frequency of occurrence of event points is made arbitrarily large within a given time interval.

Let $\{w(k), k = 0, 1, \ldots\}$ be a zero mean gaussian white sequence with covariance $E[w(j)w'(k)] = Q(k)\delta_{jk}, j, k = 0, 1, \ldots$, for which the spacing between successive time points in the time index set is some number $\Delta t > 0$. Let t denote continuous time, t_0 correspond to $k = 0$, and t_1 correspond to $k = n$. Then, $t_1 = t_0 + n\,\Delta t$, or equivalently, $n\,\Delta t = t_1 - t_0$. For some given value of n, let $\{w^{(n)}(t), t_0 \leq t \leq t_1\}$ denote the piecewise constant gaussian white sequence for which one component is indicated in Fig. 4.9. Note that the value over each subinterval is defined from the left.

We remark that our choice of zero mean for $\{w(k), k = 0, 1, \ldots\}$ is simply for convenience.

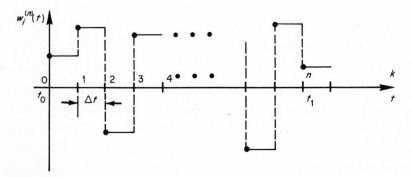

Fig. 4.9 Component of piecewise constant gaussian white sequence sample function.

Keeping t_1 fixed while increasing n such that

$$n \, \Delta t = t_1 - t_0 = \text{constant}$$

we examine the nature of $\{w^{(n)}(t), \; t_0 \leq t \leq t_1\}$. In the limit, we are concerned with the situation $n \to \infty$ and $\Delta t \to 0$, that is, where the switching points occur at an arbitrarily high rate.

Now suppose that $\{w^{(n)}(t), \; t_0 \leq t \leq t_1\}$ is the input to some dynamic system. For example, suppose that it is the accelerating force which acts on a system of p particles, each of unit mass, and that each component of the process acts on one and only one particle. Assuming that each particle has zero initial velocity and is constrained to rectilinear motion, we have

$$\dot{\nu} = w^{(n)}(t)$$

for $t \geq t_0$; ν is the p vector of the particles' velocities, with $\nu(t_0) = 0$. It follows that

$$\nu(t_1) = \int_{t_0}^{t_1} w^{(n)}(t) \, dt$$

is a random p vector. Since the accelerating force is piecewise constant, we see that

$$\nu(t_1) = \sum_{i=0}^{n-1} w^{(n)}(t_0 + i \, \Delta t) \, \Delta t$$

In addition, since each $w^{(n)}(t_0 + i \, \Delta t)$ is a gaussian random vector, it follows from this last expression that $\nu(t_1)$ is also gaussian distributed.

Clearly,

$$E[\nu(t_1)] = 0$$

since $E[w^{(n)}(t_0 + i \, \Delta t)] = 0$ for each $i = 0, 1, \ldots, n - 1$ by hypothesis.

The covariance matrix of the final velocity is seen to be

$$
\begin{aligned}
E[\nu(t_1)\nu'(t_1)] &= E\left[\int_{t_0}^{t_1} w^{(n)}(\sigma) \, d\sigma \int_{t_0}^{t_1} w^{(n)}(\tau) \, d\tau \right] \\
&= E\left[\sum_{i=0}^{n-1} w^{(n)}(t_0 + i \, \Delta t) \, \Delta t \sum_{j=0}^{n-1} w^{(n)}(t_0 + j \, \Delta t) \, \Delta t \right] \\
&= \sum_{i=0}^{n-1} Q(t_0 + i \, \Delta t)(\Delta t)^2 \qquad (4.48)
\end{aligned}
$$

where the last line is a consequence of the fact that $\{w^{(n)}(t), \; t_0 \leq t < t_1\}$ is an independent process.

If Q is constant over each subinterval, we see from this equation that in the limit as $n \to \infty$ and $\Delta t \to 0$, such that $n\, \Delta t = t_1 - t_0 = $ constant,

$$E[\nu(t_1)\nu'(t_1)] = nQ(\Delta t)^2 = (t_1 - t_0)Q\, \Delta t \to 0$$

Along with the fact that $E[\nu(t_1)] = 0$, this means that $\nu(t_1)$, the final velocity, is a deterministic quantity, viz., zero, in the limit, a result which does not make physical sense.

On the other hand, if we replace $Q(t_0 + i\, \Delta t)$ in Eq. (4.48) by $Q(t_0 + i\, \Delta t)/\Delta t$, the covariance matrix of $v(t_1)$ is nonzero, which agrees with our intuition.

We now define a gaussian white process $\{w(t),\ t_0 \le t \le t_1\}$ to be the limit of the gaussian white sequence

$$\{w(t),\ t_0 \le t \le t_1\} = \lim_{n \to \infty} \{w^{(n)}(t),\ t_0 \le t \le t_1\}$$

as described above, where the covariance matrix $Q(t_0 + i\, \Delta t) = Q(i)$ is to be replaced by $Q(t)/\Delta t$ in taking limits, where t corresponds to the time point i and $Q(t) = Q(i)$.

We see that a sample function of a gaussian white process can be thought of as a time function which is composed of the superposition of an arbitrarily large number of independent pulses of very brief duration which have gaussian distributed amplitudes. As such, it is a useful idealization of many physical phenomena such as random wind gusts, electronic and atmospheric noise, and measurement errors in data sensors.

Because the process is used to model extraneous and undesirable effects in systems, it is more commonly called a *gaussian white noise*. In the sequel, we shall use this latter terminology.

Finally, we give an argument for the presence of the Dirac delta function in the covariance expression for the gaussian white noise.

We note that

$$\lim_{\Delta t \to 0} \frac{Q(t)}{\Delta t}$$

does not make sense. However, the quantity with which we are dealing is defined over an interval of width Δt, and we can envision the situation as depicted in Fig. 4.10. Then in the limit as $\Delta t \to 0$, the function which

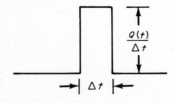

Fig. 4.10 Representation of covariance matrix for piecewise constant gaussian white sequence.

is $1/\Delta t$ over the interval Δt and zero elsewhere becomes the Dirac delta function. For this reason, we write the covariance expression for the gaussian white noise as

$$E\{[w(t) - \bar{w}(t)][w(\tau) - \bar{w}(\tau)]'\} = Q(t)\delta(t - \tau)$$

for all $t, \tau \geq t_0$.

SYSTEM DYNAMICS

Consider the continuous linear system

$$\dot{x} = F(t)x + G(t)w(t) \tag{4.49}$$

for $t \geq t_0$, where x is an n vector, the state; w is a p vector, the disturbance; $F(t)$ and $G(t)$ are continuous $n \times n$ and $n \times r$ matrices, respectively; and the dot denotes the time derivative. The initial condition $x(t_0)$ is arbitrary for the present, as is $w(t)$.

For two time instants t and $t + \Delta t$, $t \geq t_0$ and $\Delta t > 0$, we know from Eq. (2.32) that

$$x(t + \Delta t) = \Phi(t + \Delta t, t)x(t) + \int_t^{t+\Delta t} \Phi(t + \Delta t, \tau)G(\tau)w(\tau)\,d\tau$$

$$= \Phi(t + \Delta t, t)x(t) + \Gamma(t + \Delta t, t)w(t) \tag{4.50}$$

under the assumption that $w(\tau) = w(t) = \text{constant}$ for $t \leq \tau < t + \Delta t$. In Eq. (4.50),

$$\Gamma(t + \Delta t, t) = \int_t^{t+\Delta t} \Phi(t + \Delta t, \tau)G(\tau)\,d\tau$$

and Φ is the $n \times n$ state transition matrix which satisfies the relations

$$\dot{\Phi}(t,\tau) = F(t)\Phi(t,\tau)$$

and

$$\Phi(\tau,\tau) = I$$

for all $t, \tau \geq t_0$.

Now let $w(t) = w^{(n)}(t)$ and recall from Eqs. (2.34) and (2.35) that

$$\Gamma(t + \Delta t, t) = G(t)\,\Delta t + 0(\Delta t^2)$$

and

$$\Phi(t + \Delta t, t) = I + F(t)\,\Delta t + 0(\Delta t^2)$$

Equation (4.50) can now be written

$$x(t + \Delta t) = [I + F(t)\,\Delta t + 0(\Delta t^2)]x(t) + G(t)w^{(n)}(t)\,\Delta t + 0(\Delta t^2)$$

or, equivalently, as

$$x(t + \Delta t) - x(t) = F(t)x(t)\,\Delta t + G(t)w^{(n)}(t)\,\Delta t + 0(\Delta t^2) \tag{4.51}$$

Dividing through by Δt and taking the limit as $\Delta t \rightarrow 0$, we have

$$\dot{x} = F(t)x + G(t)w(t) \tag{4.52}$$

for $t \geq t_0$, where $\{w(t),\ t \geq t_0\}$ is now a gaussian white noise with given mean $\bar{w}(t)$ and covariance matrix

$$E\{[w(t) - \bar{w}(t)][w(\tau) - \bar{w}(\tau)]'\} = Q(t)\delta(t - \tau)$$

Letting $x(t_0)$ be a gaussian random n vector which is independent of $\{w(t),\ t \geq t_0\}$ and has mean $\bar{x}(t_0)$ and positive semidefinite $n \times n$ covariance matrix,

$$E\{[x(t_0) - \bar{x}(t_0)][x(t_0) - \bar{x}(t_0)]'\} = P(t_0)$$

completes the specification of the model for the system dynamics. As a consequence of the assumption of independence, it is clear that

$$E\{[x(t_0) - \bar{x}(t_0)][w(t) - \bar{w}(t)]'\} = 0$$

for all $t \geq t_0$.

Let us examine the nature of $\{x(t),\ t \geq t_0\}$. The process is obviously Markov, since the solution of Eq. (4.52) can be written

$$x(t_m) = \Phi(t_m,t_{m-1})x(t_{m-1}) + \int_{t_{m-1}}^{t_m} \Phi(t_m,\tau)G(\tau)w(\tau)\,d\tau$$

where $t_m > t_{m-1} \geq t_0$.

Also, for any $t \geq t_0$, let $x^{(n)}(t)$ be given by

$$
\begin{aligned}
x^{(n)}(t) &= \Phi(t,t_0)x(t_0) + \int_{t_0}^{t} \Phi(t,\tau)G(\tau)w^{(n)}(\tau)\,d\tau \\
&= \Phi(t,t_0)x(t_0) \\
&\quad + \sum_{i=0}^{n-1} \Phi(t,\ t_0 + i\,\Delta t)G(t_0 + i\,\Delta t)w^{(n)}(t_0 + i\,\Delta t)\,\Delta t
\end{aligned} \tag{4.53}
$$

where $\{w^{(n)}(\tau),\ t_0 \leq \tau \leq t\}$ is the same as above and the interval $[t_0,t]$ has been subdivided into n subintervals, each of length $\Delta t = (t - t_0)/n$.

For $\{x(t),\ t \geq t_0\}$ we know that

$$x(t) = \Phi(t,t_0)x(t_0) + \int_{t_0}^{t} \Phi(t,\tau)G(\tau)w(\tau)\,d\tau$$

Since

$$\lim_{n \rightarrow \infty} \{w^{(n)}(\tau),\ t_0 \leq \tau \leq t\} = \{w(\tau),\ t_0 \leq \tau \leq t\}$$

it follows that

$$\lim_{n \to \infty} \{x^{(n)}(t), t \geq 0\} = \{x(t), t \geq t_0\}$$

Referring now to Eq. (4.53), it is clear that the term $\Phi(t,t_0)x(t_0)$ is a gaussian random n vector for all $t \geq t_0$, since $x(t_0)$ is a gaussian random n vector. Moreover, for each $n = 1, 2, \ldots$, it is clear that

$$\sum_{i=0}^{n-1} \Phi(t, t_0 + i \, \Delta t)G(t_0 + i \, \Delta t)w^{(n)}(t_0 + i \, \Delta t) \, \Delta t$$

is also a gaussian random n vector, being simply the sum of a set of n linear transformations on the set of n independent gaussian random n vectors $w^{(n)}(t_0 + i \, \Delta t)$, $i = 0, 1, \ldots, n - 1$.

It follows immediately that for any $n = 1, 2, \ldots$, $x^{(n)}(t)$ is a gaussian random n vector for all $t \geq t_0$. Hence, $\{x(t), t \geq t_0\}$ is a Gauss-Markov process. Its mean follows by formally taking the expected value in Eq. (4.52) to obtain

$$\dot{\bar{x}} = F(t)\bar{x} + G(t)\bar{w}(t) \tag{4.54}$$

for $t \geq t_0$, subject to the initial condition $\bar{x}(t_0)$. Alternately, we take the expected value in Eq. (4.51), divide the result by Δt, and take the limit as $\Delta t \to 0$ to arrive at the same expression.

To complete the characterization of $\{x(t), t \geq t_0\}$, we determine a relation for its covariance matrix

$$P(t) = E\{[x(t) - \bar{x}(t)][x(t) - \bar{x}(t)]'\}$$

We approach this task by considering first the limiting behavior of Eq. (4.37), which we repeat here for convenience:

$$P(k + 1) = \Phi(k + 1, k)P(k)\Phi'(k + 1, k)$$
$$+ \Gamma(k + 1, k)Q(k)\Gamma'(k + 1, k)$$

This equation gives the evolution of the covariance matrix for the discrete version of the stochastic process, which is defined by Eq. (4.52) wherein the system disturbance is $\{w^{(n)}(t), t \geq t_0\}$.

Let t correspond to the time point k and $t + \Delta t$ to $k + 1$ where $\Delta t > 0$. Then, making use of the fact that $Q(k)$ is replaced by $Q(t)/\Delta t$, we have

$$P(t + \Delta t) = \Phi(t + \Delta t, t)P(t)\Phi'(t + \Delta t, t)$$
$$+ \Gamma(t + \Delta t, t) \frac{Q(t)}{\Delta t} \Gamma'(t + \Delta t, t)$$

Substituting for $\Phi(t + \Delta t, t)$ and $\Gamma(t + \Delta t, t)$ and expanding, we obtain

$$P(t + \Delta t) = [I + F(t) \Delta t + 0(\Delta t^2)]P(t)[I + F(t) \Delta t + 0(\Delta t^2)]'$$

$$+ [G(t) \Delta t + 0(\Delta t^2)] \frac{Q(t)}{\Delta t} [G(t) \Delta t + 0(\Delta t^2)]'$$

$$= P(t) + F(t)P(t) \Delta t + P(t)F'(t) \Delta t$$

$$+ G(t)Q(t)G'(t) \Delta t + 0(\Delta t^2)$$

Transposing $P(t)$ to the left-hand side of the equation, dividing through by Δt, and taking the limit as $\Delta t \to 0$, we have the matrix differential equation

$$\dot{P} = F(t)P + PF'(t) + G(t)Q(t)G'(t) \tag{4.55}$$

for $t \geq t_0$. The initial condition is obviously $P(t_0)$.

Equation (4.55), in analogy to Eq. (4.37), describes how uncertainty propagates in the system dynamics.

Since P is $n \times n$, it contains n^2 elements. However, P is a covariance matrix and is, therefore, symmetric. This means that only $n(n + 1)/2$ of the equations in the system of Eq. (4.55) are independent relations.

We see that Eq. (4.55) is linear. Hence, its solution consists of the homogeneous solution, which depends upon the initial condition $P(t_0)$, and the particular solution, which depends upon the forcing function $Q(t)$. This permits us to determine the contribution to the total uncertainty $P(t)$ of the uncertainty associated with $x(t_0)$ and that associated with $w(t)$, respectively.

Once Eqs. (4.54) and (4.55) are solved to obtain $\bar{x}(t)$ and $P(t)$ for $t \geq t_0$, the Gauss-Markov process $\{x(t),\ t \geq t_0\}$ can be characterized in terms of its characteristic function

$$\phi_x(s,t) = \exp{[j\bar{x}'(t)s - \tfrac{1}{2}s'P(t)s]}$$

where s is a real n vector.

An alternate derivation of Eq. (4.55) can be carried out by working directly with the solution of Eq. (4.52) and the definition of $P(t)$. First we see that

$$x(t) = \Phi(t,t_0)x(t_0) + \int_{t_0}^{t} \Phi(t,\tau)G(\tau)w(\tau) \, d\tau$$

and

$$\bar{x}(t) = \Phi(t,t_0)\bar{x}(t_0) + \int_{t_0}^{t} \Phi(t,\tau)G(\tau)\bar{w}(\tau) \, d\tau$$

from which it follows trivially that

$$x(t) - \bar{x}(t) = \Phi(t,t_0)[x(t_0) - \bar{x}(t_0)] + \int_{t_0}^{t} \Phi(t,\tau)G(\tau)[w(\tau) - \bar{w}(\tau)] \, d\tau$$

Then from the definition of $P(t)$,

$$
\begin{aligned}
P(t) = \ & E\left(\left\{\Phi(t,t_0)[x(t_0) - \bar{x}(t_0)] + \int_{t_0}^{t} \Phi(t,\tau)G(\tau)[w(\tau) - \bar{w}(\tau)]\, d\tau\right\}\right.\\
& \left.\left\{\Phi(t,t_0)[x(t_0) - \bar{x}(t_0)] + \int_{t_0}^{t} \Phi(t,\sigma)G(\sigma)[w(\sigma) - \bar{w}(\sigma)]\, d\sigma\right\}'\right)\\
= \ & \Phi(t,t_0)E\{[x(t_0) - \bar{x}(t_0)][x(t_0) - \bar{x}(t_0)]'\}\Phi'(t,t_0)\\
& + \Phi(t,t_0)\int_{t_0}^{t} E\{[x(t_0) - \bar{x}(t_0)][w(\sigma) - \bar{w}(\sigma)]'\}G'(\sigma)\Phi'(t,\sigma)\, d\sigma\\
& + \int_{t_0}^{t} \Phi(t,\tau)G(\tau)E[w(\tau) - \bar{w}(\tau)][x(t_0) - \bar{x}(t_0)]'\}\, d\tau\ \Phi'(t,t_0)\\
& + \iint_{t_0}^{t} \Phi(t,\tau)G(\tau)E\{[w(\tau) - \bar{w}(\tau)][w(\sigma)\\
& \hspace{4cm} - \bar{w}(\sigma)\,']\}G'(\sigma)\Phi'(t,\sigma)\, d\tau\, d\sigma
\end{aligned}
$$

Since $x(t_0)$ is independent of $\{u(t), t \geq t_0\}$, the middle two terms on the right-hand side vanish. Also,

$$
E\{[w(\tau) - \bar{w}(\tau)][w(\sigma) - \bar{w}(\sigma)]'\} = Q(\tau)\delta(\tau - \sigma)
$$

and we can carry out the integration in the last term with respect to σ. Consequently,

$$
P(t) = \Phi(t,t_0)P(t_0)\Phi'(t,t_0) + \int_{t_0}^{t} \Phi(t,\tau)G(\tau)Q(\tau)G'(\tau)\Phi'(t,\tau)\, d\tau \quad (4.56)
$$

Equation (4.56) gives the desired covariance matrix.

To reconcile this result with Eq. (4.55), we differentiate it with respect to t to obtain

$$
\begin{aligned}
\dot{P}(t) = \ & \Phi(t,t_0)P(t_0)\dot{\Phi}'(t,t_0) + \dot{\Phi}(t,t_0)P(t_0)\Phi'(t,t_0)\\
& + \Phi(t,t)G(t)Q(t)G'(t)\Phi'(t,t) + \int_{t_0}^{t} \Phi(t,\tau)G(\tau)Q(\tau)G'(\tau)\dot{\Phi}'(t,\tau)\, d\tau\\
& + \int_{t_0}^{t} \dot{\Phi}(t,\tau)G(\tau)Q(\tau)G'(\tau)\Phi'(t,\tau)\, d\tau\\
= \ & \Phi(t,t_0)P(t_0)\Phi'(t,t_0)F'(t) + F(t)\Phi(t,t_0)P(t_0)\Phi'(t,t_0)\\
& + G(t)Q(t)G'(t) + \int_{t_0}^{t} \Phi(t,\tau)G(\tau)Q(\tau)G'(\tau)\Phi'(t,\tau)\, d\tau\, F'(t)\\
& + F(t)\int_{t_0}^{t} \Phi(t,\tau)G(\tau)Q(\tau)G'(\tau)\, d\tau\\
= \ & F(t)\left[\Phi(t,t_0)P(t_0)\Phi'(t,t_0) + \int_{t_0}^{t} \Phi(t,\tau)G(\tau)Q(\tau)G'(\tau)\, d\tau\right]\\
& + \left[\Phi(t,t_0)P(t_0)\Phi'(t,t_0) + \int_{t_0}^{t} \Phi(t,\tau)G(\tau)Q(\tau)G'(\tau)\, d\tau\right]F'(t)\\
& + G(t)Q(t)G'(t) \quad (4.57)
\end{aligned}
$$

where we have made use of the fact that $\dot{\Phi}(t,\tau) = F(t)\Phi(t,\tau)$ and $\Phi(\tau,\tau) = I$ for all $t,\ \tau \geq t_0$. Finally, substituting into the first two terms on the right-hand side of Eq. (4.57) from Eq. (4.56), we have

$$\dot{P} = F(t)P + PF'(t) + G(t)Q(t)G'(t)$$

which is identically Eq. (4.55).

Equation (4.56) is obviously the solution of Eq. (4.55). However, it is of little use for numerical computation of $P(t)$, in general, because it requires that $\Phi(t,\tau)$ be determined first. For computational purposes, it is more expedient to determine $P(t)$ by numerical integration of Eq. (4.55).

As in the discrete-time case, the model is easily extended to include a control input by writing

$$\dot{x} = F(t)x + G(t)w(t) + C(t)u(t) \tag{4.58}$$

where u is an r vector and $C(t)$ is a continuous $n \times r$ matrix. If $u(t)$, $t \geq t_0$, is a known control input, then clearly $\{x(t),\ t \geq t_0\}$ is a Gauss-Markov process. The only modification which is necessary in the process's description is that

$$\dot{\bar{x}} = F(t)\bar{x} + G(t)\bar{w}(t) + C(t)u(t)$$

is the differential equation for the mean.

Our justification for the model in this section is the same as in the discrete-time case.

MEASUREMENTS

Analogous to the discrete-time model, we choose

$$z(t) = H(t)x(t) + v(t) \tag{4.59}$$

for $t \geq t_0$ as the measurement model for continuous linear systems. In Eq. (4.59), z is an m vector, the measurement; $H(t)$ is a continuous $m \times n$ matrix; and v, the measurement error, is also an m vector. We assume that the measurement error process, $\{v(t),\ t \geq t_0\}$ is an m-dimensional gaussian white noise with known mean

$$E[v(t)] = \bar{v}(t)$$

and covariance

$$E\{[v(t) - \bar{v}(t)][v(\tau) - \bar{v}(\tau)]'\} = R(t)\delta(t - \tau)$$

for all $t,\ \tau \geq t_0$. We assume that $R(t)$, which is $m \times m$, is continuous and positive definite for all $t \geq t_0$. This is in contrast to $R(k + 1)$, which was required to be positive semidefinite for $k = 0, 1, \ldots$. The reason for this difference will become clear in our study of the estimation problem.

We assume further that $\{v(t),\, t \geq t_0\}$ is independent of $x(t_0)$ but that it may be correlated with $\{w(t), t \geq t_0\}$. Hence, we have

$$E\{[x(t_0) - \bar{x}(t_0)][v(t) - \bar{v}(t)]'\} = 0$$

for all $t \geq t_0$ and

$$E\{[w(t) - \bar{w}(t)][v(\tau) - \bar{v}(\tau)]\} = S(t)\delta(t - \tau)$$

for all $t, \tau \geq t_0$, where $S(t)$ is $p \times m$ and continuous.

Again, our justification for the above model is the same as for the discrete-time one, as discussed earlier.

The block diagram for the complete model is the one which we gave in Fig. 2.3, the distinction being that $x(t_0)$, $w(t)$, and $v(t)$, $t \geq t_0$, now have the descriptions which we have given above.

GAUSS-MARKOV-2 PROCESS

The development here is entirely analogous to that for the Gauss-Markov-2 sequence model. Hence, for simplicity, we assume zero mean values throughout.

Consider the stochastic process $\{x(t), t \geq t_0\}$, which is defined by

$$\ddot{x} = A(t)\dot{x} + B(t)x + G(t)w(t) \tag{4.60}$$

for $t \geq t_0$. We assume that $x(t_0)$ and $\dot{x}(t_0)$ are zero mean jointly gaussian distributed n-vectors, with $E[x(t)x'(t)] = P_{00}$, $E[x(t)\dot{x}'(t)] = P_{01}$, and $E[\dot{x}(t)\dot{x}'(t)] = P_{11}$. We assume that $A(t)$ and $B(t)$ are continuous $n \times n$ matrices and that $G(t)$ and $\{w(t), t \geq t_0\}$ are the same as above with the exception that $\bar{w}(t) = 0$ for all $t \geq t_0$. Also, $\{w(t), t \geq t_0\}$ is independent of both $x(0)$ and $\dot{x}(0)$.

Since Eq. (4.60) is second-order, it is clear that for $t_m > t_{m-1} > \cdots > t_1 \geq t_0$, the probability distribution function of $x(t_m)$ given $x(t_{m-1}), \ldots, x(t_1)$ depends upon both $x(t_{m-1})$ and $x(t_{m-2})$. Hence $\{x(t), t \geq t_0\}$ is Markov-2.

Defining the n vector y as

$$\dot{x} = y$$

we write Eq. (4.60) as

$$\dot{y} = B(t)x + A(t)y + G(t)w(t)$$

Letting x^* be the $2n$ vector

$$x^* = \begin{bmatrix} x \\ y \end{bmatrix}$$

we combine the two equations to obtain

$$\dot{x}^* = \begin{bmatrix} 0 & I \\ B(t) & A(t) \end{bmatrix} x^* + \begin{bmatrix} 0 \\ G(t) \end{bmatrix} w(t) \tag{4.61}$$

for $t \geq t_0$.

Now let

$$\Phi(t,\tau) = \begin{bmatrix} \Phi_{11}(t,\tau) & \Phi_{12}(t,\tau) \\ \Phi_{21}(t,\tau) & \Phi_{22}(t,\tau) \end{bmatrix}$$

denote the $2n \times 2n$ state transition matrix of the system in Eq. (4.61), where each of the $\Phi_{ij}(t,\tau)$, $i, j = 1, 2$, is $n \times n$. Clearly,

$$x(t) = \Phi_{11}(t,t_0)x(t_0) + \Phi_{12}(t,t_0)y(t_0) + \int_{t_0}^{t} \Phi_{12}(t,\tau)G(\tau)w(\tau)\,d\tau$$

Since $x(t_0)$, $y(t_0) = \dot{x}(t_0)$, and $\{w(\tau),\ \tau \geq t_0\}$ are gaussian, so is $x(t)$ for all $t \geq t_0$. Hence, $\{x(t),\ t \geq t_0\}$ is Gauss-Markov-2.

On the other hand, $\{x^*(t),\ t \geq t_0\}$, which is defined by Eq. (4.61), is obviously Gauss-Markov. This follows by noting that $x^*(t_0)$ is a zero mean gaussian random $2n$ vector with

$$E[x^*(t_0)x^{*\prime}(t_0)] = \begin{bmatrix} P_{00} & P_{01} \\ P_{01}' & P_{11} \end{bmatrix}$$

and is independent of the zero mean gaussian white noise $\{w(t),\ t \geq t_0\}$. The model is then the same as that in Eq. (4.52), except that the state vector now has $2n$ instead of n elements.

We note from above that a Gauss-Markov-2 process results when a gaussian white noise is input to a second-order system wherein the initial conditions, which are independent of this noise input, are jointly gaussian distributed.

With this in mind, we consider the process $\{x(t),\ t \geq t_0\}$, which is defined by the relation

$$\dot{x} = F(t)x + G(t)w(t) \tag{4.62}$$

where the system disturbance is itself a Gauss-Markov process which is governed by the expression

$$\dot{w} = A(t)w + B(t)\xi(t) \tag{4.63}$$

for $t \geq t_0$ [cf. Eqs. (4.46) and (4.47), respectively].

In these equations x is an n vector, w a p vector, and ξ a q vector; and $F(t)$, $G(t)$, $A(t)$, and $B(t)$ are continuous $n \times n$, $n \times p$, $p \times p$, and $p \times q$ matrices, respectively. We assume that $x(t_0)$ and $w(t_0)$ are gaussian random vectors with known means and covariance matrices, not

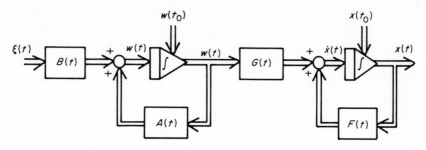

Fig. 4.11 Block diagram for system of Eqs. (4.62) and (4.63).

necessarily independent of each other. We assume further that $\{\xi(t),$ $t \geq t_0\}$ is a gaussian white noise which is independent of both $x(t_0)$ and $w(t_0)$ and whose mean and covariance matrix are given.

The block diagram for the process $\{x(t),\ t \geq t_0\}$, with $\{\xi(t),\ t \geq t_0\}$ as its input, is shown in Fig. 4.11.   Between input $\xi(t)$ and output $x(t)$, the system is obviously second-order, the initial conditions are independent of the input and gaussian distributed, and its input is a gaussian white noise.   Hence, $\{x(t),\ t \geq t_0\}$ is a Gauss-Markov-2 process.

As in the discrete-time case, it is also a simple matter to show that $\{x^*(t),\ t \geq t_0\}$, where x^* is the $(n + p)$ vector

$$x^* = \begin{bmatrix} x \\ w \end{bmatrix}$$

is a Gauss-Markov process.

Discussion   The two system models which we have developed in this chapter are generalizations of the system models which were introduced in classical communication and control theory [3–6].   The classical theory is based on the independent work of Wiener [8] in this country and Kolomogorov [9] in the Soviet Union.

In the classical work, the model was that of a time-invariant linear system whose input was a *stationary* white noise and in which the effect of initial conditions was assumed negligible.   The time index set was chosen to be $-\infty < t < \infty$, and the resulting stochastic process was stationary.   These simplifications permitted analyses to be conducted in the frequency domain in a straightforward manner.   Numerous examples are given in the references [3–6].

The extensions here are apparent: the system is, in general, time-varying, the noise input need not be stationary, and the effects of uncertainty in the initial conditions are included.

No claim is made that the current models are a panacea. Rather, they are refinements of earlier models, and are themselves subject to refinement in the future. They obviously do not account for numerous physical phenomena which we would like to model for purposes of system analysis and design. They are, however, reasonable models for many phenomena, some of which we mentioned earlier.

PROBLEMS

4.1. Consider the sequence of independent random variables $x(1)$, $x(2)$, . . . , which are identically gaussian distributed with mean zero and variance σ^2. Clearly $\{x(k), k = 1, 2, \ldots\}$ and $\{y(k), k = 1, 2, \ldots\}$, where $y(k) = x(1) + x(2) + \cdots + x(k)$ are discrete-time stochastic processes. For any integer k, specify the characteristic function for each process.

4.2. Consider the discrete-time stochastic process $\{x(k), k = 0,1, \ldots\}$ defined by

$$x(k + 1) = \frac{-1}{k + 1}\, x(k)$$

where $x(0)$ is a random n vector with mean $\bar{x}(0)$ and covariance matrix $P(0)$. Determine the mean value function, covariance kernel, and correlation matrix for the process.

4.3. Assuming that three stochastic proceses $\{x(t), t \in I\}$, $\{y(t), t \in I\}$, and $\{z(t), t \in I\}$ are pairwise independent, show that they are not necessarily triplewise independent.

4.4. Show that two stochastic processes which are independent of each other are uncorrelated with each other, but that the converse is not true in general.

4.5. Show that a covariance stationary gaussian stochastic process is also strict-sense stationary.

4.6. If $\{x(t), t \in I\}$ is a Markov process, show that $\{x(t), t \in I_1\}$, where I_1 is a subset of I, is also a Markov process.

4.7. If the stochastic process $\{x(t), t \in I\}$ has the property that, for every $t \le t_1 < t_2$ in I, $x(t_2) - x(t_1)$ is independent of $x(t)$, show that the process is Markov.

4.8. For the Gauss-Markov model of Eq. (4.28), consider the case where x, Φ, and w are scalars and

$$x(k + 1) = e^{-k}x(k) + w(k) \qquad k = 0, 1, \ldots$$

Assume that $\bar{x}(0) = \bar{w}(k) = 0$, $E[x^2(0)] = \sigma_0^2$, $E[w^2(k)] = \sigma_w^2$, and $E[x(0)w(k)] = 0$ for all k. Determine the expression for $P(k + 1) \triangleq \sigma^2(k + 1) = E[x^2(k + 1)]$ and consider its behavior as $k \to \infty$. Compare this result with the behavior of $P(k + 1)$ if e^{-k} is replaced by $(1 - e^{-k})$ in the equation that defines the process. For arbitrarily large k, is the process covariance stationary in either of these cases? Strict-sense stationary?

4.9. If in the system model of Eq. (4.28), $x(0)$ is not independent of $\{w(k), k = 0, 1, \ldots\}$, is $\{x(k), k = 0, 1, \ldots\}$ still Gauss-Markov? Explain why or why not.

4.10. If an additive control input $\Psi(k + 1, k)u(k)$ where, $u(k) = \mu[x(k),k]$ with μ a known r-dimensional vector-valued function of $x(k)$ and k is present in Eq. (4.28), show that $\{x(k), k = 0, 1, \ldots\}$ is Markov, but not necessarily gaussian. Give one form for μ for which $\{x(k), k = 0, 1, \ldots\}$ is Gauss-Markov.

4.11. Let $\{x(k), k = 0, 1, \ldots\}$ be a Gauss-Markov sequence of the type described by Eq. (4.28). The covariance matrix of the sequence is determined recursively using the relation $P(k + 1) = \Phi(k + 1, k)P(k)\Phi'(k + 1, k) + \Gamma(k + 1, k)Q(k)\Gamma'(k + 1, k)$, $k = 0, 1, \ldots$.

(a) In some cases, $P(0)$ and $Q(k)$ are not known exactly and are subject to errors $\Delta P(0)$ and $\Delta Q(k)$, respectively. Develop a recursive error covariance equation which shows how these errors propagate in time.

(b) In other cases, $P(0)$ and $Q(k)$ are known reasonably well, but the state transition matrix $\Phi(k + 1, k)$ is subject to an error $\Delta\Phi(k + 1, k)$. Develop a recursive error covariance equation that describes how these errors affect the covariance matrix computation.

4.12. Show that the stochastic process $\{x^*(k), k = 0, 1, \ldots\}$, which is discussed at the end of Sec. 4.3, is Gauss-Markov.

4.13. Consider the system model $x(k + 1) = \Phi(k + 1, k)x(k) + \Gamma(k + 1, k)w(k)$ and $z(k + 1) = H(k + 1)x(k + 1) + v(k + 1)$ for $k = 0, 1, \ldots$, where all the terms are as defined in Sec. 4.3 with the following exception. Namely, $\{v(k + 1), k = 0, 1, \ldots\}$ is now a correlated process, in particular one that can be modeled by the relation

$$v(k + 1) = \Theta(k + 1, k)v(k) + \xi(k)$$

for $k = 0, 1, \ldots$, where $\Theta(k + 1, k)$ is $m \times m$. Assume that $v(0)$ is a gaussian random m vector which is independent of $x(0)$, $\{w(k), k = 0, 1, \ldots\}$ and $\{\xi(k), k = 0, 1, \ldots\}$, and has mean $\bar{v}(0)$ and covariance $V(0)$. Also assume that $\{\xi(k), k = 0, 1, \ldots\}$ is independent of $x(0)$, $v(0)$, and $\{w(k), k = 0, 1, \ldots\}$ and is a gaussian white sequence with mean $\bar{\xi}(k)$ and covariance $E\{[\xi(k) - \bar{\xi}(k)][\xi(k) - \bar{\xi}(k)]'\} = Z(k)\delta_{jk}$ for all $j, k = 0, 1, \ldots$.

(a) Utilize the augmented state vector approach to develop a Gauss-Markov model of the form in Eq. (4.28) for the system.

(b) As an alternative to the procedure in (a), suppose that a "new" measurement is formed as a linear combination of the two successive measurements $z(k + 1)$ and $z(k)$; namely, let $\zeta(k) = z(k + 1) + A(k)z(k)$ for $k = 1, 2, \ldots$, where $A(k)$ is $m \times m$. Note that the case in which $k = 0$ is not allowed. Show that by choosing $A(k) = -\Theta(k + 1, k)$, the new measurement is

$$\zeta(k) = [H(k + 1)\Phi(k + 1, k) - \theta(k + 1, k)H(k)]x(k)$$
$$+ [H(k + 1)\Gamma(k + 1, k)w(k) + \xi(k)]$$

for $k = 1, 2, \ldots$, wherein it is obvious that the measurement error is a gaussian white sequence. This technique of differencing measurements to obtain a measurement in which the error is "white" is due to Bryson and Henrikson [10]. It obviously avoids the problem of augmenting the state vector. Finally, it is clear that the mean of the measurement error is

$$H(k + 1)\Gamma(k + 1, k)\bar{w}(k) + \bar{\xi}(k)$$

and its covariance matrix is

$$H(k + 1)\Gamma(k + 1, k)Q(k)\Gamma'(k + 1, k)H'(k + 1) + Z(k)\dagger$$

† For the continuous-time case, measurement differencing is replaced by measurement differentiation (see Prob. 8.17).

(c) Can the procedure in (b) be utilized if $x(0)$ and $v(0)$ are correlated? Explain.

4.14. For $t > \tau \geq t_0$, determine an expression for $P(t,\tau)$ for the Gauss-Markov process model of Eq. (4.52).

4.15. Consider the scalar Gauss-Markov process $\{x(t),\ t \geq 0\}$ which is defined by the relation $\dot{x} = -x + w(t)$, $t \geq 0$, where $x(0)$ is a zero mean gaussian random variable which is independent of the scalar zero mean gaussian white noise $\{w(t),\ t \geq 0\}$. Assume that $E[x^2(0)] = P(0)$ and $E[w(t)w(\tau)] = Q\delta(t - \tau)$ for all $t,\tau \geq 0$, where Q = positive constant. Under what relationship between $P(0)$ and Q is $\{x(t),\ t \geq 0\}$ covariance stationary? Is it also strict-sense stationary for the same relationship?

4.16. If $x(t_0)$ is not independent of $\{w(t),\ t \geq t_0\}$ in the model in Eq. (4.52), is $\{x(t),\ t \geq t_0\}$ still a Gauss-Markov process?

4.17. For the system model in Eq. (4.52), suppose that $x(t_0)$ is independent of $\{w(t),\ t \geq t_0\}$ but not of the measurement error process $\{v(t),\ t \geq t_0\}$. Suppose further that the latter two processes are independent of each other. Is $\{x(t),\ t \geq t_0\}$ still Gauss-Markov under these conditions?

4.18. Consider the model in Eq. (4.52) when a control input is present: $\dot{x} = F(t) + G(t)w(t) + C(t)u(t)$. Suppose that a feedback control of the form $u(t) = M(t)x(t) + e(t)$ is used in the system where $M(t)$ is a continuous $r \times n$ matrix and $\{e(t),\ t \geq t_0\}$ is an r-dimensional gaussian white noise which is independent of $x(t_0)$ but may be correlated with $\{w(t),\ t \geq t_0\}$. Is $\{x(t),\ t \geq t_0\}$ a Gauss-Markov process in the presence of this "noisy" feedback control?

4.19. (a) Suppose that a particle of unit mass, constrained to rectilinear motion, say along the x axis, leaves the origin at time $t = 0$ with a velocity which is a zero mean gaussian random variable with variance $\sigma_0^2 > 0$. Determine the means and variances of its position and velocity coordinates as functions of time for $t \geq 0$. What is the probability density function for the particle's position when $t = 10$? Is the two-dimensional stochastic process of the particle's position and velocity coordinates a Gauss-Markov process?

(b) Suppose that in addition to the conditions in (a), the particle is subject to an accelerating force which can be modeled as a zero mean gaussian white noise which is independent of the particle's initial position and velocity and has a constant variance $\sigma_w^2 > 0$. Show that the two-dimensional (position-velocity) stochastic process is Gauss-Markov and determine its covariance matrix as a function of time. How certain can one be of the particle's position and velocity for $t \gg 0$?

4.20. Consider the system

$$\dot{x} = \begin{bmatrix} 0 & 1 \\ -\omega_0^2 & 0 \end{bmatrix} x \qquad t \geq 0$$

where x is a 2 vector and ω_0 = constant. Assuming that $x(0)$ is gaussian with zero mean and covariance matrix equal to the 2×2 identity matrix, determine the covariance matrix $P(t) = E[x(t)x'(t)]$ for $t \geq t_0$.

REFERENCES

1. Parzen, E., "Stochastic Processes," Holden-Day, Inc., San Francisco, 1962.
2. Papoulis, A., "Probability, Random Variables, and Stochastic Processes," McGraw-Hill Book Company, New York, 1965.
3. Laning, J. H., and R. H. Battin, "Random Processes in Automatic Control," McGraw-Hill Book Company, New York, 1956.

4. Davenport, W. B., and W. L. Root, "An Introduction to the Theory of Random Signals and Noise," McGraw-Hill Book Company, New York, 1958.
5. Lee, Y. W., "Statistical Theory of Communication," John Wiley & Sons, Inc., New York, 1960.
6. James, H. M., N. B. Nichols, and R. S. Phillips, "Theory of Servomechanisms," McGraw-Hill Book Company, New York, 1947.
7. Kalman, R. E., New Methods in Wiener Filtering Theory, in J. L. Bogdanoff and F. Kozin, eds., "Proceedings of the First Symposium on Engineering Application of Random Function Theory and Probability," John Wiley & Sons, Inc., New York, 1963, pp. 270–388. Also, New Methods and Results in Linear Prediction and Filtering Theory, *Tech. Rept.* 61-1, Research Institute for Advanced Studies (RIAS), Martin Company, Baltimore, 1961.
8. Wiener, N., "Extrapolation, Interpolation, and Smoothing of Stationary Time Series," John Wiley & Sons, Inc., New York, 1949.
9. Kolomogorov, A. N., Interpolation and Extrapolation of Stationary Time Series, *Bull. Acad. Sci. USSR Math. Ser.*, vol. 5, p. 3, 1941.
10. Bryson, A. E., Jr., and L. J. Henrikson, Estimation Using Sampled-Data Containing Sequentially Correlated Noise, *Tech. Rept. No.* 533, Division of Engineering and Applied Physics, Harvard University, Cambridge, Mass., June, 1967.

5

Optimal Prediction and Filtering for Discrete Linear Systems

5.1 INTRODUCTION

We are now in a position to initiate our study of estimation theory. We shall consider first a rather general problem in estimation in which the state and measurement processes of a dynamic system are arbitrary discrete-time stochastic processes. We then present a solution to this problem in Theorem 5.1, which is of fundamental significance in estimation theory.

Following the theorem, we develop a corollary to it, as well as two additional theorems that deal with special cases of the original problem.

In Secs. 5.3 and 5.4, we apply the corollary to derive algorithms for optimal prediction and filtering, respectively, for the system model which was given in Sec. 4.3. These algorithms are the central results of this chapter.

Although only Theorem 5.1 and its corollary are germane to our work, we include the other theorems to give the student a broader foundation in estimation theory and to provide him with insight into the position that our specific results assume in a rather general framework of the theory.

5.2 OPTIMAL ESTIMATION FOR DISCRETE SYSTEMS

PROBLEM FORMULATION

Let us consider a dynamic system S whose state as a function of time is an n-dimensional discrete-time stochastic process $\{x(k), k \in I\}$, where either $I = \{k: k = 0, 1, \ldots, N\}$ or $I = \{k: k = 0, 1, \ldots\}$. Suppose that we are interested in knowing the value of $x(k)$ for some fixed k, but that $x(k)$ is not directly accessible to us for observation. Suppose further, then, that we have made a sequence of measurements $z(1), \ldots, z(j)$ which are causally related to $x(k)$ by means of some measurement system M as shown in Fig. 5.1 and that we wish to utilize these data to infer the value of $x(k)$. We assume that $\{z(i), i = 1, \ldots, j\}$ is an m-dimensional, discrete-time stochastic process.

Since only the measurements $z(1), \ldots, z(j)$ are available from which to estimate $x(k)$, we denote an estimate of $x(k)$ based on these measurements by $\hat{x}(k|j)$ and define it to be some n-dimensional, vector-valued function ϕ_k of the measurements, viz.,

$$\hat{x}(k|j) = \phi_k[z(i), i = 1, \ldots, j]$$

Very simply, the estimation problem is one of determining ϕ_k in some rational and meaningful manner. If $k > j$, the problem is one of *prediction;* if $k = j$, one of *filtering;* and, if $k < j$, one of *smoothing* or *interpolation.*

The approach to this problem that we take here is based on consideration of the *estimation error* which is defined by the relation

$$\tilde{x}(k|j) = x(k) - \hat{x}(k|j)$$

This definition has the geometric interpretation indicated in Fig. 5.2.

Ideally, $\tilde{x}(k|j) = 0$, and the estimate is exact. When $\tilde{x}(k|j) \neq 0$, we choose to assign a *penalty* for the incorrect estimate. We do this by specifying a *penalty* or *loss function* $L = L[\tilde{x}(k|j)]$ which has the following properties:

1. L is a scalar-valued function of n variables.
2. $L(0) = 0$, where the first 0 denotes the null n vector.

Fig. 5.1 Block diagram of dynamic system S with state $x(k)$ and measurement system M with measurements $z(i)$.

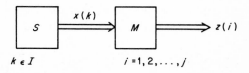

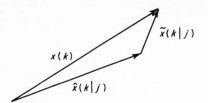

Fig. 5.2 Geometric interpretation of definition of estimation error.

3. $L[\tilde{x}^b(k|j)] \geq L[\tilde{x}^a(k|j)]$ whenever $\rho[\tilde{x}^b(k|j)] \geq \rho[\tilde{x}^a(k|j)]$, where ρ is a scalar-valued, nonnegative, convex function of n variables; that is, $\rho[\lambda x + (1 - \lambda)y] \leq \lambda\rho(x) + (1 - \lambda)\rho(y)$ for all n-vectors x and y, and $0 \leq \lambda \leq 1$.

4. $L[\tilde{x}(k|j)] = L[-\tilde{x}(k|j)]$

    The first property is one of convenience; that is, we choose to characterize the penalty as a scalar for analytical ease. The second property simply says that no penalty is assigned when the estimate is exact. In the third property, ρ is a measure of the distance of $\tilde{x}(k|j)$ from the origin in n-dimensional euclidean space, and L is specified to be a nondecreasing function of this distance. Thus, the "closer" $\tilde{x}(k|j)$ is to zero, the smaller the penalty. The fourth property means that $L[\cdot]$ is symmetric about the origin. A loss function that possesses the above four properties is termed an *admissible loss function*. It should be noted that L need not be a convex function.

    Typical examples of admissible loss functions are:

1. $L[\tilde{x}(k|j)] = \sum\limits_{i=1}^{n} \alpha_i|\tilde{x}_i(k|j)|$, where $\alpha_i \geq 0$, but not all identically zero

2. $L[\tilde{x}(k|j)] = \alpha\|\tilde{x}(k|j)\|^{2p}$, where $\alpha > 0$, $p =$ positive integer, and

$$\|x\| = \left(\sum_{i=1}^{n} x_i^2\right)^{\frac{1}{2}}$$

3. $L[\tilde{x}(k|j)] = \alpha_1\{1 - \exp[-\alpha_2\|\tilde{x}(k|j)\|^2]\}$, where α_1 and $\alpha_2 > 0$

4. $L[\tilde{x}(k|j)] = \begin{cases} 0 & \text{for } \|\tilde{x}(k|j)\|^4 < \alpha_1 \\ \mu & \text{for } \|\tilde{x}(k|j)\|^4 \geq \alpha_1 \end{cases}$
    where α_1 and $\mu > 0$

    Since $x(k)$ and $\hat{x}(k|j)$ are random vectors, it follows that $\tilde{x}(k|j)$ will also be a random vector and that L will be a random variable. In order to obtain a useful measure of the loss, we choose to define a *performance measure J* as the mean value of L, viz.,

$$J[\tilde{x}(k|j)] = E\{L[\tilde{x}(k|j)]\} \tag{5.1}$$

For an admissible loss function, it is clear that

$$J[\tilde{x}^b(k|j)] \geq J[\tilde{x}^a(k|j)]$$

whenever $L[\tilde{x}^b(k|j)] \geq L[\tilde{x}^a(k|j)]$. Hence, J is a nondecreasing function of the loss. As a result, we say that an estimate $\hat{x}(k|j)$ that minimizes $J[\tilde{x}(k|j)]$ is a "best" or *optimal estimate*.

We note carefully that an optimal estimate does *not* minimize the loss; rather, it minimizes the mean value of the loss. Hence, by optimal estimate, we mean an estimate which is "optimum-on-the-average."

The estimation problem as we have formulated it here can be summarized as follows.

PROBLEM STATEMENT

Given the measurements $z(1)$, . . . , $z(j)$, determine an estimate

$$\hat{x}(k|j) = \phi_k[z(i), i = 1, 2, \ldots, j]$$

of $x(k)$ such that the performance measure $J[\tilde{x}(k|j)] = E\{L[\tilde{x}(k|j)]\}$, where L is an admissible loss function, is minimized.

We remark that as a consequence of the properties of conditional expectation,

$$E_{\tilde{x}}\{L[\tilde{x}(k|j)]\} = E_{z^*}[E_{\tilde{x}}\{L[\tilde{x}(k|j)]|z(1), \ldots, z(j)\}]$$

where z^* denotes the jm vector

$$z^* = \begin{bmatrix} z(1) \\ \cdot \\ \cdot \\ \cdot \\ z(j) \end{bmatrix}$$

Since the first expectation on the right-hand side of this relation does not depend upon the choice of $\hat{x}(k|j)$, it follows that minimization of $E\{L[\tilde{x}(k|j)]\}$ is equivalent to minimizing $E\{L[\tilde{x}(k|j)]|z(1), z(2), \ldots, z(j)\}$. Therefore, Eq. (5.1) can also be written as

$$J[\tilde{x}(k|j)] = E_{\tilde{x}}\{L[\tilde{x}(k|j)]|z(1), \ldots, z(j)\} \qquad (5.2)$$

Hence, the value of the performance measure, as we might have expected intuitively, is conditioned on the actual values which the measurements assume. One could very well take Eq. (5.2) as the defining relation for the performance measure.

In any event, the notion of conditional expectation has been brought into the picture. This notion is fundamental in estimation, and we shall

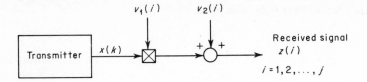

Fig. 5.3 Communication system block diagram.

examine it in some detail in the sequel. Before pursuing this point further, however, let us present an example to illustrate the problem formulation.

Example 5.1 Suppose that one component of a message that is to be transmitted is a random variable $x(k)$, k fixed, which is unknown to the receiver. Suppose further that in the process of transmission, the message is corrupted by a multiplicative noise $v_1(i)$ in the transmission channel and an additive noise $v_2(i)$ at the receiver, so that the received signal, assumed to be sampled j times, can be expressed in the form $z(i) = x(k)v_1(i) + v_2(i)$, $i = 1, 2, \ldots, j$, where $k = $ constant. The process is depicted in block diagram form in Fig. 5.3. The problem is that of inferring the value of $x(k)$ given the measurements $z(1)$, $z(2)$, . . . , $z(j)$. We formulate this problem as one of optimal estimation by saying that any estimate of $x(k)$ of the form $\hat{x}(k|j) = \phi[z(i), i = 1, 2, \ldots, j]$, where ϕ is a scalar-valued function of j variables, which minimizes $J[\tilde{x}(k|j)] = E[|x(k) - \hat{x}(kj)|]$ is an optimal estimate. In this example, the loss function is simply the magnitude of the estimation error.

FUNDAMENTAL THEORY OF ESTIMATION

Let the conditional probability distribution function of $x(k)$ given the values of $z(1)$, . . . , $z(j)$ be denoted by the expression

$$P[x(k) \leq \xi | z^*(j)] = F[\xi | z^*(j)]$$

for all n vectors ξ, where $z^*(j)$ is the jm vector

$$z^*(j) = \begin{bmatrix} z(1) \\ \cdot \\ \cdot \\ \cdot \\ z(j) \end{bmatrix}$$

Then, subject to a particular hypothesis on $F[\xi | z^*(j)]$, the solution of the estimation problem which was stated above is given in the following theorem.

Theorem 5.1 *If (1) L is an admissible loss function and (2) $F[\xi \mid z^*(j)]$ is:*
a. Symmetric about its mean $\bar{\xi}$; that is,

$$F[\xi - \bar{\xi}|z^*(j)] = 1 - F[\bar{\xi} - \xi|z^*(j)] \qquad \textit{for all } \xi$$

b. *Convex for all $\xi \leq \bar{\xi}$; that is,*

$$F[\lambda\xi^1 + (1 - \lambda)\xi^2 | z^*(j)] \leq \lambda F[\xi^1 | z^*(j)] + (1 - \lambda)F[\xi^2 | z^*(j)]$$

$$\text{for all } n \text{ vectors } \xi^1, \xi^2 \leq \bar{\xi} \text{ and } 0 \leq \lambda \leq 1$$

then the optimal estimate is

$$\hat{x}(k|j) = E[x(k)|z^*(j)] \tag{5.3}$$

This fundamental and powerful result is due to Sherman [1]. In words, the theorem says that, under the requirements of symmetry and convexity on the conditional probability distribution function, the optimal estimate is the conditional expectation. Obviously then, the optimal estimate is specified immediately if the conditional probability distribution function $F[x(k)|z^*(j)]$, or, equivalently, the conditional probability density function $f[x(k)|z^*(j)]$, if it exists, is known. More significantly, however, it should be noted that $E[x(k)|z^*(j)]$ is the optimal estimate for *all* combinations of admissible loss functions, and symmetric and convex conditional probability distribution functions.

An example of a probability distribution function which is symmetric about its mean and convex for all arguments less than or equal to its mean is illustrated in Fig. 5.4 for a random variable.

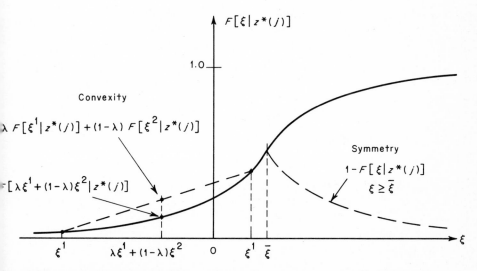

Fig. 5.4 Illustration of symmetric and convex conditional probability distribution function for a random variable.

One immediately obvious consequence of Theorem 5.1 is that if the stochastic processes $\{x(k), k \in I\}$ and $\{z(i), i = 1, \ldots, j\}$ are independent, then the optimal estimate is $\hat{x}(k|j) = \bar{x}(k)$ for all k and j.

The proof of Theorem 5.1 is based on the following lemma of probability theory [2,3].

Lemma 5.1 *Let L be an admissible loss function and y a zero mean random n vector whose probability density function is symmetric about its mean (zero) and convex for all $y \leq 0$. Then,*

$$E[L(y)] \leq E[L(y - a)]$$

for all n vectors a.

We leave as an exercise the task of applying this lemma to obtain Theorem 5.1.

Finally, in connection with Theorem 5.1, we remark that the optimal estimate is still $E[x(k)|z^*(j)]$ if the requirement that $F[\xi|z^*(j)]$ be convex is replaced by the condition that L be convex [4].

We now consider a corollary of Theorem 5.1 which deals with a special case of the more general problem posed earlier.

Corollary 5.1 *If (1) L is an admissible loss function, and (2) $\{x(k), k \in I\}$ and $\{z(i), i = 1, \ldots, j\}$ are discrete-time gaussian stochastic processes, then $\hat{x}(k|j) = E[x(k)|z^*(j)]$.*

Proof The proof follows from Theorem 5.1 if we can show that the gaussian conditional probability distribution function $F[x(k)|z^*(j)]$ is symmetric about its mean and convex for all $\xi \leq \bar{\xi}$.

Omitting the time arguments for simplicity, we have

$$F(\xi|z^*) = \int_{-\infty}^{\xi_n} \cdots \int_{-\infty}^{\xi_1} \alpha \exp\left[-\tfrac{1}{2}(x - \bar{\xi})'Q^{-1}(x - \bar{\xi})\right] dx \quad (5.4)$$

where $\bar{\xi} = \bar{x} + P_{xz^*}P_{z^*z^*}^{-1}(z^* - \bar{z}^*)$

$$Q = P_{xx} - P_{xz^*}P_{z^*z^*}^{-1}P_{z^*x} \quad (5.5)$$

$$\alpha = 1/\sqrt{(2\pi)^n|Q|}$$

under the assumption that the $jm \times jm$ matrix $P_{z^*z^*}$ and the $n \times n$ matrix Q are nonsingular.

We consider symmetry first. Replacing ξ by $\xi - \bar{\xi}$ in Eq. (5.4), we have

$$F(\xi - \bar{\xi}|z^*) = \int_{-\infty}^{\xi_n - \bar{\xi}_n} \cdots$$

$$\int_{-\infty}^{\xi_1 - \bar{\xi}_1} \alpha \exp\left[-\tfrac{1}{2}(x - \bar{\xi})'Q^{-1}(x - \bar{\xi})\right] dx \quad (5.6)$$

Similarly,

$$F(\bar{\xi} - \xi | z^*) = \int_{-\infty}^{\bar{\xi}_n - \xi_n} \cdots \int_{-\infty}^{\bar{\xi}_1 - \xi_1} \alpha \exp\left[-\tfrac{1}{2}(x - \bar{\xi})'Q^{-1}(x - \bar{\xi})\right] dx$$

from which it follows that

$$1 - F(\bar{\xi} - \xi | z^*) = \int_{\bar{\xi}_n - \xi_n}^{\infty} \cdots$$
$$\int_{\bar{\xi}_1 - \xi_1}^{\infty} \alpha \exp\left[-\tfrac{1}{2}(x - \bar{\xi})'Q^{-1}(x - \bar{\xi})\right] dx$$

Introducing the change of variable $y = -x$ in this last expression, it is seen that

$$1 - F(\bar{\xi} - \xi | z^*) = (-1)^n \int_{\xi_n - \bar{\xi}_n}^{-\infty} \cdots$$
$$\int_{\xi_1 - \bar{\xi}_1}^{-\infty} \alpha \exp\left[-\tfrac{1}{2}(y - \bar{\xi})'Q^{-1}(y - \bar{\xi})\right] dy$$

$$= \int_{-\infty}^{\xi_n - \bar{\xi}_n} \cdots$$
$$\int_{-\infty}^{\xi_1 - \bar{\xi}_1} \alpha \exp\left[-\tfrac{1}{2}(y - \bar{\xi})Q^{-1}(y - \bar{\xi})\right] dy \quad (5.7)$$

Comparing Eqs. (5.6) and (5.7), we observe that $F(\xi | z^*)$ is symmetric about $\bar{\xi}$.

Convexity for all $\xi \leq \bar{\xi}$ is established by showing that

$$\frac{\partial^2 F}{\partial \xi_i \, \partial \xi_j} \geq 0 \qquad i, j = 1, 2, \ldots, n$$

and is left as an exercise to complete the proof.

We remark that this corollary also holds for asymmetric loss functions [4]; i.e., the fourth property of admissible loss functions is not needed here.

As a consequence of this corollary, we have that the optimal estimate is

$$\hat{x}(k|j) = \bar{x}(k) + P_{x(k)z^*(j)} P_{z^*(j)z^*(j)}^{-1} [z^*(j) - \bar{z}^*(j)] \quad (5.8)$$

For simplicity, we assume hereafter that $\{x(k), k \in I\}$ and $\{z(i), i = 1, 2, \ldots, j\}$ have zero means, in which case the optimal estimate is

$$\hat{x} = P_{xz^*} P_{z^*z^*}^{-1} z^* \quad (5.9)$$

where $\hat{x} = \hat{x}(k|j)$, $x = x(k)$, and $z^* = z^*(j)$.

This estimate has the following important and useful properties, all of which follow from Theorem 3.2, the properties of conditional gaussian expectation which were developed in Sec. 3.6, the uniqueness of conditional expectation, and Eq. (5.9):

1. $\hat{x}(k|j)$ is a linear estimate, i.e., a linear combination of the available measurements.
2. $\hat{x}(k|j)$ and $\tilde{x}(k|j)$ are gaussian random n vectors.
3. $\tilde{x}(k|j)$ is independent of any linear combination of the available measurements. In particular, $\tilde{x}(k|j)$ is independent of $\hat{x}(k|j)$ which means that

$$E[\tilde{x}(k|j)\hat{x}'(k|j)] = 0 \qquad (5.10)$$

for all j and k.
4. $\hat{x}(k|j)$ is unique.

In the corollary, we restricted our attention to optimal estimates for gaussian processes, but required only that the loss functions considered be admissible. We now consider the problem of optimal estimation for a specific loss function where the probability distribution function $F[\xi|z^*(j)]$ need not be either symmetric or convex. We shall, however, require that the conditional probability density function $f[x(k)|z^*(j)]$ be defined and continuous at all but perhaps a finite number of points x_i in the range $-\infty < x_i < \infty$, $i = 1, 2, \ldots n$.

The loss function to which we restrict ourselves is $L[\tilde{x}(k|j)] = \tilde{x}'(k|j)\tilde{x}(k|j)$. The name commonly ascribed to the performance measure for this loss function is *mean square error*, which is motivated by the fact that $E[\tilde{x}'(k|j)\tilde{x}(k|j)]$ is the *mean* value of the *square* of the euclidean length of the *error* vector.

In this case, we have the following result (cf. Prob. 3.11).

Theorem 5.2 *If* $L[\tilde{x}(k|j)] = \tilde{x}'(k|j)\tilde{x}(k|j)$, *then* $\hat{x}(k|j) = E[x(k)|z^*(j)]$.

Proof We consider necessity first. For $L[\tilde{x}(k|j)] = \tilde{x}'(k|j)\tilde{x}(k|j)$, Eq. (5.2) can be written as

$$J[\tilde{x}(k|j)] = E\{[x(k) - \hat{x}(k|j)]'[x(k) - \hat{x}(k|j)]|z^*(j)\}$$

For convenience, we drop the time arguments and write

$$J(\tilde{x}) = E[(x - \hat{x})'(x - \hat{x})|z^*]$$

Then,

$$J(\tilde{x}) = \int_{-\infty}^{\infty} \cdots \int_{-\infty}^{\infty} (x - \hat{x})'(x - \hat{x})f(x|z^*)\, dx$$

where $f(x|z^*)$ is the conditional probability density function of $x(k)$ given $z(1)$, $z(2)$, $\ldots$, $z(j)$ and is assumed to be defined and continuous at all but perhaps a finite number of points in the range of integration.

Taking the gradient with respect to $\hat{x}$, we obtain

$$\nabla_{\hat{x}}[J(\hat{x})] = -2 \int_{-\infty}^{\infty} \cdots \int_{-\infty}^{\infty} (x - \hat{x})' f(x|z^*) \, dx \qquad (5.11)$$

Setting this gradient equal to zero, we see that

$$\hat{x}' \int_{-\infty}^{\infty} \cdots \int_{-\infty}^{\infty} f(x|z^*) \, dx = \int_{-\infty}^{\infty} \cdots \int_{-\infty}^{\infty} x' f(x|z^*) \, dx$$

Taking the transpose, evaluating the indicated integrals, and replacing the time arguments, we have immediately that

$$\hat{x}(k|j) = E[x(k)|z^*(j)]$$

The sufficiency follows trivially by considering the set of second partial derivatives $\partial^2 J / \partial \hat{x}_i \, \partial \hat{x}_j$ for $i, j = 1, 2, \ldots, n$. For $i = j$, it follows from Eq. (5.11) that

$$\frac{\partial^2 J}{\partial \hat{x}_i{}^2} = 2 \int_{-\infty}^{\infty} \cdots \int_{-\infty}^{\infty} f(x|z^*) \, dx = 2$$

For $i \neq j$,

$$\frac{\partial^2 J}{\partial \hat{x}_i \, \partial \hat{x}_j} = 0$$

From differential calculus, these two results are sufficient for $\hat{x}(k|j) = E[x(k)|z^*(j)]$ to be a relative minimum. However, since $J[\bar{x}(k|j)]$ is a convex, monotonically increasing function of $\bar{x}(k|j)$, the conditional expectation renders it an absolute minimum. This completes the proof.

From this result, we see that for at least one admissible loss function, viz., $L = \bar{x}'(k|j)\bar{x}(k|j)$, Theorem 5.1 is true without the restrictions of symmetry and convexity on the conditional probability distribution function. Whether or not this is true for any other admissible loss functions is not known.

Comparing Theorem 5.2 with Theorem 5.1, it is obvious that the former, although severely restrictive with respect to loss functions, is more general than the latter in the sense of the class of stochastic processes that can be handled. The interesting feature here is that if the assumptions made in formulating an estimation problem satisfy the hypotheses of Theorem 5.1, but not those of Theorem 5.2, or vice versa, the optimal estimate is the conditional expectation in either case. In this sense, the two theorems supplement each other.

We focus our attention next on optimal linear estimates. We let any linear estimate of $x(k)$ based on the measurements $z(1)$, $z(2)$, . . . , $z(j)$ be written as

$$\hat{x}(k|j) = \sum_{i=1}^{j} A(i)z(i) \tag{5.12}$$

where the $A(i)$ are $n \times m$ matrices.

We let $\{x(k),\ k \in I\}$ and $\{z(i),\ i = 1,\ .\ .\ .\ ,j\}$ be arbitrary stochastic processes and make the following assertion.

Theorem 5.3 *If only the first and second moments of the stochastic processes $\{x(k),\ k \in I\}$ and $\{z(i),\ i = 1,\ .\ .\ .\ ,j\}$ are known, then the estimate which is optimal for all admissible loss functions is the linear estimate*

$$\hat{x}(k|j) = \bar{x}(k) + P_{x(k)z^*(j)}P_{z^*(j)z^*(j)}^{-1}[z^*(j) - \bar{z}^*(j)] \tag{5.13}$$

Proof If the first and second moments, i.e., the means and covariance matrices, of the stochastic processes $\{x(k),\ k \in I\}$ and $\{z(i),\ i = 1,\ .\ .\ .\ ,j\}$ are given, we can specify unique gaussian processes $\{x^a(k),\ k \in I\}$ and $\{z^a(i),\ i = 1,\ .\ .\ .\ ,j\}$ having the same means and covariance matrices. This is true since gaussian processes are uniquely specified by their first and second moments.

However, in the gaussian case, it follows from Corollary 5.1 that the estimate which is optimal for all admissible loss functions is $E[x(k)|z^*(j)]$, as given by Eq. (5.13). The estimate is obviously linear, and our assertion is proved.

In our proof, we made use of the fact that since only the first and second moments of the stochastic processes $\{x(k),\ k \in I\}$ and $\{z(i),\ i = 1,\ .\ .\ .\ ,j\}$ are known, we could just as well model the two processes as gaussian with the same statistical parameters. In particular, we note that only $\bar{x}(k)$, $\bar{z}^*(j)$, $P_{x(k)z^*(j)}$, and $P_{z^*(k)z^*(k)}$ need be known.

Another way of viewing the above result is the following. If we restrict ourselves to estimates which are linear, i.e., estimates of the form in Eq. (5.12) (cf. Prob. 3.13), only the first and second moments are needed to characterize the estimates. Then, utilizing the fact that we can find unique gaussian processes with the same first and second moments, we apply Corollary 5.1 and obtain the result that the linear estimate which is optimal for one admissible loss function is optimal for all admissible loss functions.

On the other hand, if higher-order moments of the processes are known and the estimate is not necessarily restricted to be a linear one, Eq. (5.13) no longer gives the optimal estimate for all admissible loss functions, in general.

To summarize, Theorem 5.1 is the fundamental result and gives the solution for a very broad class of estimation problems.   Corollary 5.1 is a special case of the theorem for gaussian processes.   One principal result is that the estimate which is optimal for all admissible loss functions, namely, the conditional expectation, is a linear estimate as a consequence of the gaussian nature of the stochastic processes which are involved.

In Theorem 5.2, generality in the class of loss functions is sacrificed to gain generality in the class of stochastic processes which can be handled. For the mean square error performance measure, the optimal estimate is still the conditional expectation, but the conditional probability distribution function need be neither symmetric nor convex.

Theorem 5.3 pertains to situations in which only first and second moments of the stochastic processes involved are known, and it gives the estimate which is optimal for all admissible loss functions.   The simplicity of the estimate, namely its linear structure, makes the result attractive from a computational point of view.

Theorem 5.1 is due to Sherman [1], as noted earlier.   Corollary 5.1 and Theorem 5.2 are attributed to Doob [5], and Theorem 5.3 is an interpretation of some results due to Kalman [6].

5.3 OPTIMAL PREDICTION FOR DISCRETE LINEAR SYSTEMS

Although the results of the preceding section are fundamental ones, they are of limited practical utility.   This is seen by considering the gaussian case where

$$\hat{x}(k|j) = \bar{x}(k) + P_{x(k)z^*(j)}P_{z^*(j)z^*(j)}^{-1}[z^*(j) - \bar{z}^*(j)]$$

based entirely upon

For each set of measurements, it is necessary to compute the inverse of $P_{z^*(j)z^*(j)}$, a $jm \times jm$ matrix.   We recall that j is the number of measurements, while m is the number of elements in the measurement vector. If m is 1 or 2 and there are 40 measurements, this means that anywhere from a 40×40 to an 80×80 matrix must be inverted.

If both k and j are allowed to vary and it is desired to perform estimation "on-line," i.e., to process new measurements as they occur, application of the above expression to generate the optimal estimate becomes impractical.

What is desired from an engineering point of view are efficient and practical algorithms for processing the measurements sequentially, hopefully in real time, to obtain a current estimate.   This, it turns out, can be done quite readily for discrete-time Gauss-Markov processes which can be modeled as in Section 4.3.

In this section and the next, we shall develop such algorithms for optimal prediction and filtering, respectively, for this class of processes. The algorithms for optimal smoothing require consideration of three separate cases and are deferred to Chap. 6.

SYSTEM MODEL

Let us begin by reviewing the model for the relevant class of processes. We have the system

$$x(k + 1) = \Phi(k + 1, k)x(k) + \Gamma(k + 1, k)w(k) \tag{5.14}$$

$$z(k + 1) = H(k + 1)x(k + 1) + v(k + 1) \tag{5.15}$$

where x is an n vector, the state; w is a p vector, the disturbance; z is an m vector, the measurement (system output); v is an m vector, the measurement error; and $k = 0, 1, \ldots$, is the discrete-time index. In addition, Φ is an $n \times n$ matrix, the state transition matrix; Γ is an $n \times p$ matrix, the disturbance transition matrix; and H is an $m \times n$ matrix, the measurement matrix.

The process $\{w(k), k = 0, 1, \ldots\}$ is a p-dimensional gaussian white sequence for which

$$E[w(k)] = 0 \tag{5.16}$$

for all $k = 0, 1, \ldots$, and

$$E[w(j)w'(k)] = Q(k)\delta_{jk} \tag{5.17}$$

for all $j, k = 0, 1, \ldots$, where $Q(k)$ is a positive semidefinite $p \times p$ matrix. The process $\{v(k + 1), k = 0, 1, \ldots\}$ is an m-dimensional gaussian white sequence for which

$$E[v(k + 1)] = 0 \tag{5.18}$$

for all $k = 0, 1, \ldots$, and

$$E[v(j + 1)v'(k + 1)] = R(k + 1)\delta_{jk} \tag{5.19}$$

for all $j, k = 0, 1, \ldots$, where $R(k + 1)$ is a positive semidefinite $m \times m$ matrix. We consider here the case where these two stochastic processes are independent of each other, so that

$$E[v(j)w'(k)] = 0 \tag{5.20}$$

for all $j = 1, 2, \ldots$, and $k = 0, 1, \ldots$.

The initial state $x(0)$ is a gaussian random n vector with mean

$$E[x(0)] = 0 \tag{5.21}$$

and $n \times n$ positive semidefinite covariance matrix

$$E[x(0)x'(0)] = P(0) \tag{5.22}$$

It is assumed that $x(0)$ is independent of $\{w(k),\ k = 0,\ 1,\ \ldots\}$ and $\{v(k + 1),\ k = 0,\ 1,\ \ldots\}$, so that

$$E[x(0)w'(k)] = 0 \tag{5.23}$$

and

$$E[x(0)v'(k + 1)] = 0 \tag{5.24}$$

for all $k = 0,\ 1,\ \ldots\ .$

The model described by Eqs. (5.14) through (5.24) has the following properties:

1. The stochastic processes $\{x(k),\ k = 0,\ 1,\ \ldots\}$ and $\{z(i),\ i = 1,$ $\ldots,\ j\}$ are gaussian with identically zero means.
2. $E[x(j)w'(k)] = 0$ for all $k \geq j,\ j = 0,\ 1,\ \ldots$ (5.25)
3. $E[z(j)w'(k)] = 0$ for all $k \geq j,\ j = 1,\ 2,\ \ldots$ (5.26)
4. $E[x(j)v'(k)] = 0$ for all j and k (5.27)
    where $j = 0,\ 1,\ \ldots,$ and $k = 1,\ 2,\ \ldots\ .$
5. $E[z(j)v'(k)] = 0$ for all $k > j$ (5.28)
    where $j,\ k = 1,\ 2,\ \ldots\ .$

The first property follows from the results in Sec. 4.3 and Eqs. (5.16), (5.18), and (5.21).

We now verify the second and third properties, leaving the last two, which follow in a similar way, as an exercise.

From Eq. (4.31),

$$x(j) = \Phi(j,0)x(0) + \sum_{i=1}^{j} \Phi(j,i)\Gamma(i,\ i - 1)w(i - 1) \tag{5.29}$$

for $j = 1,\ 2,\ \ldots\ .$ Then,

$$E[x(j)w'(k)] = \Phi(j,0)E[x(0)w'(k)]$$
$$+ \sum_{i=1}^{j} \Phi(j,i)\Gamma(i,\ i - 1)E[w(i - 1)w'(k)]$$

From Eq. (5.23), the first term on the right-hand side of this expression vanishes for all $k = 0,\ 1,\ \ldots\ .$ Since $\{w(k),\ k = 0,\ 1,\ \ldots\}$ is a white sequence, it follows that $E[w(i - 1)w'(k)] = 0$ for all $k \neq i - 1$. Since $i = 1,\ 2,\ \ldots,\ j$, it is clear that the last term vanishes for all $k > j - 1$ or $k \geq j$. Hence, the second property is derived.

To verify the third property, we note from Eq. (5.15) that

$$z(j) = H(j)x(j) + v(j)$$

for $j = 1,\ 2,\ \ldots\ .$ It then follows that

$$E[z(j)w'(k)] = H(j)E[x(j)w'(k)] + E[v(j)w'(k)]$$

By virtue of the second property, Eq. (5.25), the first term on the right-hand side of this equation is zero for all $k \geq j$, $j = 0, 1, \ldots$. Since $\{w(k), k = 0, 1, \ldots\}$ and $\{v(k + 1), k = 0, 1, \ldots\}$ are independent, the second term is also zero for all $j = 1, 2, \ldots$, and $k = 0, 1, \ldots$, and the property is established.

OPTIMAL PREDICTION

We now develop the algorithm for the *optimal predicted estimate* $\hat{x}(k|j)$, $k > j$, $j = 0, 1, \ldots$, and establish some important properties of the corresponding *prediction error* $\tilde{x}(k|j) = x(k) - \hat{x}(k|j)$. In particular, we shall be concerned with the nature of the stochastic process $\{\tilde{x}(k|j), k = j + 1, j + 2, \ldots\}$ and the behavior of its corresponding covariance matrix $E[\tilde{x}(k|j)\tilde{x}'(k|j)] = P(k|j)$.

We assume that the *optimal filtered estimate* $\hat{x}(j|j)$ and the $n \times n$ covariance matrix $E[\tilde{x}(j|j)\tilde{x}'(j|j)] = P(j|j)$ of the corresponding *filtering error* $\tilde{x}(j|j) = x(j) - \hat{x}(j|j)$ are known for some $j = 0, 1, \ldots$. The procedure for obtaining $\hat{x}(j|j)$ and $P(j|j)$ for any j is given in the next section.

From the first property given above for the stochastic processes $\{x(k), k = 0, 1, \ldots\}$ and $\{z(i), i = 1, \ldots, j\}$, and Corollary 5.1, we have that

$$\hat{x}(j|j) = E[x(j)|z(1), \ldots, z(j)] \tag{5.30}$$

is the optimal filtered estimate of $x(j)$ for $j = 1, 2, \ldots$. For $j = 0$, we have no measurements, and it follows from Corollary 5.1 that

$$\hat{x}(0|0) = E[x(0)|\text{no measurements}]$$

or

$$\hat{x}(0|0) = E[x(0)] = 0 \tag{5.31}$$

It is clear that $\hat{x}(j|j)$ is gaussian with mean zero. Similarly, the filtering error

$$\tilde{x}(j|j) = x(j) - \hat{x}(j|j)$$

is a zero mean gaussian random n vector for which we have assumed that the covariance matrix $P(j|j)$ is given. For $j = 0$, we have

$$\tilde{x}(0|0) = x(0) - \hat{x}(0) = x(0)$$

so that

$$P(0|0) = E[\tilde{x}(0|0)\tilde{x}'(0|0)]$$
$$= E[x(0)x'(0)]$$

or

$$P(0|0) = P(0) \tag{5.32}$$

where the latter is assumed given in the system description.

We now can establish the following fundamental result for optimal prediction.

Theorem 5.4 *If the optimal filtered estimate $\hat{x}(j|j)$ and the covariance matrix $P(j|j)$ of the corresponding filtering error $\tilde{x}(j|j) = x(j) - \hat{x}(j|j)$ are known for some $j = 0, 1, \ldots$, then for all $k > j$:*

a. The optimal predicted estimate $\hat{x}(k|j)$, $k > j$, for all admissible loss functions is given by the expression

$$\hat{x}(k|j) = \Phi(k,j)\hat{x}(j|j) \tag{5.33}$$

b. The stochastic process $\{\tilde{x}(k|j), k = j + 1, j + 2, \ldots\}$ defined by the prediction error relation $\tilde{x}(k|j) = x(k) - \hat{x}(k|j)$ is a zero mean Gauss-Markov sequence whose covariance matrix is governed by the relation

$$P(k|j) = \Phi(k,j)P(j|j)\Phi'(k,j)$$
$$+ \sum_{i=j+1}^{k} \Phi(k,i)\Gamma(i, i - 1)Q(i - 1)\Gamma'(i, i - 1)\Phi'(k,i) \tag{5.34}$$

Proof

1. From Corollary 5.1,

$$\hat{x}(k|j) = E[x(k)|z(1), \ldots , z(j)] \tag{5.35}$$

However, from Eq. (4.30) we have

$$x(k) = \Phi(k,j)x(j) + \sum_{i=j+1}^{k} \Phi(k,i)\Gamma(i, i - 1)w(i - 1) \tag{5.36}$$

for $k \geq j + 1$.   Substituting Eq. (5.36) into Eq. (5.35) and utilizing basic properties of conditional expectation, we obtain the result

$$\hat{x}(k|j) = E\Big[\Phi(k,j)x(j)$$
$$+ \sum_{i=j+1}^{k} \Phi(k,i)\Gamma(i, i - 1)w(i - 1)|z(1), \ldots , z(j)\Big]$$
$$= E[\Phi(k,j)x(j)|z(1), \ldots , z(j)]$$
$$+ E\Big[\sum_{i=j+1}^{k} \Phi(k,i)\Gamma(i, i - 1)w(i - 1)|z(1), \ldots , z(j)\Big]$$
$$= \Phi(k,j)E[x(j)|z(1), \ldots , z(j)]$$
$$+ \sum_{i=j+1}^{k} \Phi(k,i)\Gamma(i, i - 1)E[w(i - 1)|z(1), \ldots , z(j)] \tag{5.37}$$

We note from Eq. (5.26) that the two sets of random vectors $\{w(i - 1), i = j + 1, j + 2, \ldots , k\}$ and $\{z(1), \ldots , z(j)\}$ are uncorrelated for all $k \geq j + 1$. Since each of the vectors is gaussian, this means that the two sets of random vectors are independent. Hence,

$$E[w(i - 1)|z(1), \ldots, z(j)] = E[w(i - 1)] = 0 \qquad (5.38)$$

for all $i = j + 1, j + 2, \ldots, k$, where we have made use of the fact that $\{w(k), k = 0, 1, \ldots\}$ has zero mean.

As a result of Eqs. (5.30) and (5.38), it is clear that Eq. (5.37) reduces to

$$\hat{x}(k|j) = \Phi(k,j)\hat{x}(j|j)$$

for all $k > j$.

2. For a given j and all $k > j$, we see from the definition of prediction error, Eqs. (5.33) and (5.36), and the definition of filtering error that

$$\tilde{x}(k|j) = x(k) - \hat{x}(k|j)$$

$$= \Phi(k,j)x(j) + \sum_{i=j+1}^{k} \Phi(k,i)\Gamma(i, i - 1)w(i - 1) - \Phi(k,j)\hat{x}(j|j)$$

$$= \Phi(k,j)\tilde{x}(j|j) + \sum_{i=j+1}^{k} \Phi(k,i)\Gamma(i, i - 1)w(i - 1) \qquad (5.39)$$

It is clear that $\{\tilde{x}(k|j), k = j + 1, j + 2, \ldots\}$ is a zero mean discrete-time gaussian process.

We demonstrate the Markov property next. Let I denote the time index set $\{j + 1, j + 2, \ldots\}$. Also, let $t_1 < t_2 < \cdots < t_m$ be any m time points in I where m is any integer. Then, from Eq. (5.39) and the properties of the state transition matrix, we have

$$\tilde{x}(m|j) = \Phi(m,j)\tilde{x}(j|j) + \sum_{i=j+1}^{m} \Phi(m,i)\Gamma(i, i - 1)w(i - 1)$$

$$= \Phi(m, m - 1)\Phi(m - 1, j)\tilde{x}(j|j)$$
$$\quad + \Phi(m,m)\Gamma(m, m - 1)w(m - 1)$$
$$\quad + \sum_{i=j+1}^{m-1} \Phi(m,i)\Gamma(i, i - 1)w(i - 1)$$

$$= \Phi(m, m - 1)\Phi(m - 1, j)\tilde{x}(j|j) + \Gamma(m, m - 1)w(m - 1)$$
$$\quad + \Phi(m, m - 1)\sum_{i=j+1}^{m-1} \Phi(m - 1, i)\Gamma(i, i - 1)w(i - 1)$$

$$= \Phi(m, m - 1)\left[\Phi(m - 1, j)\tilde{x}(j|j) \right.$$
$$\quad \left. + \sum_{i=j+1}^{m-1} \Phi(m - 1, i)\Gamma(i, i - 1)w(i - 1)\right]$$
$$\quad\quad\quad\quad + \Gamma(m, m - 1)w(m - 1)$$

Noting from Eq. (5.39) that the term in brackets is simply $\tilde{x}(m - 1|j)$, we have

$$\tilde{x}(m|j) = \Phi(m, m - 1)\tilde{x}(m - 1|j) + \Gamma(m, m - 1)w(m - 1)$$

from which the Markov property follows immediately. Hence $\{\tilde{x}(k|j),\ k = j + 1, j + 2, \ldots\}$ is Gauss-Markov.

We return to Eq. (5.39) to determine the expression for

$$P(k|j) = E[\tilde{x}(k|j)\tilde{x}'(k|j)]$$

From the definition of filtering error

$$
\begin{aligned}
E[\tilde{x}(j|j)w'(i - 1)] &= E\{[x(j) - \hat{x}(j|j)]w'(i - 1)\} \\
&= E[x(j)w'(i - 1)] - E[\hat{x}(j|j)w'(i - 1)]
\end{aligned}
$$

where $i = j + 1,\ j + 2, \ldots,\ k$. Examination of Eq. (5.25) reveals that the first term on the right-hand side of this expression is identically zero for all the relevant values of i. Further, since $\hat{x}(j|j)$ is a linear combination of the measurements, it can be written

$$\hat{x}(j|j) = \sum_{l=1}^{j} A(l)z(l)$$

as in Eq. (5.12). Hence,

$$E[\hat{x}(j|j)w'(i - 1)] = \sum_{l=1}^{j} A(l)E[z(l)w'(i - 1)]$$

which, in view of Eq. (5.26), vanishes for all $i = j + 1, j + 2, \ldots,$ k. Consequently,

$$E[\tilde{x}(j|j)w'(i - 1)] = 0$$

for all $i = j + 1, j + 2, \ldots, k$. This means that the cross terms which arise in the expected value of the product of the right-hand side of Eq. (5.39) with its transpose all vanish. Therefore,

$$
\begin{aligned}
P(k|j) = {}&\Phi(k,j)E[\tilde{x}(j|j)\tilde{x}'(j|j)]\Phi'(k,j) \\
&+ \sum_{i=j+1}^{k} \Phi(k,i)\Gamma(i, i - 1)E[w(i - 1)w'(i - 1)]\Gamma'(i, i - 1)\Phi'(k,i)
\end{aligned}
$$

where we have made use in the second term of the fact that

$$E[w(j)w'(k)] = 0$$

for $j \neq k$. Taking the indicated expected values, we have

$$P(k|j) = \Phi(k,j)P(j|j)\Phi'(k,j)$$
$$+ \sum_{i=j+1}^{k} \Phi(k,i)\Gamma(i, i-1)Q(i-1)\Gamma'(i, i-1)\Phi'(k,i)$$

for $k > j$, and our proof is complete.

As an alternative to Eq. (5.34) for the prediction error covariance matrix, we state the result

$$P(k|j) = \Phi(k, k-1)P(k-1|j)\Phi'(k, k-1)$$
$$+ \Gamma(k, k-1)Q(k-1)\Gamma'(k, k-1) \quad (5.40)$$

for $k = j + 1, j + 2, \ldots$. We leave the derivation of this result as an exercise.

We remark that $\hat{x}(k|j)$ is unique by virtue of the uniqueness of conditional expectation.

At this point, we note that Theorem 5.4 is of limited use as far as performing prediction is concerned, since the only value of j for which we know $\hat{x}(j|j)$ and $P(j|j)$ is $j = 0$. Specifically, we have from Eqs. (5.31) and (5.32) that $\hat{x}(0|0) = 0$ and $P(0|0) = P(0)$, respectively, so that

$$\hat{x}(k|0) = \Phi(k,0)\hat{x}(0|0) = 0$$

and

$$P(k|0) = \Phi(k,0)P(0)\Phi'(k,0)$$
$$+ \sum_{i=1}^{k} \Phi(k,i)\Gamma(i, i-1)Q(i-1)\Gamma'(i, i-1)\Phi'(k,i)$$

One special case in prediction which merits attention is that of *single-stage optimal prediction* in which we consider the optimal estimate $\hat{x}(k+1|k)$, $k = 0, 1, \ldots$. As we shall see, this result is extremely useful in our development of the equations for optimal filtering. The result is given in the following corollary to Theorem 5.4.

Corollary 5.2 *If the optimal filtered estimate $\hat{x}(k|k)$ and the covariance matrix $P(k|k)$ of the corresponding filtering error $\tilde{x}(k|k) = x(k) - \hat{x}(k|k)$ are known for some $k = 0, 1, \ldots$, then*

    a. The single-stage optimal predicted estimate for all admissible loss functions is given by the expression

$$\hat{x}(k+1|k) = \Phi(k+1, k)\hat{x}(k|k) \tag{5.41}$$

    b. The stochastic process $\{\tilde{x}(k+1|k), k = 0, 1, \ldots\}$ defined by the single-stage prediction error relation $\tilde{x}(k+1|k) = x(k+1) -$

$\hat{x}(k + 1|k)$ *is a zero mean Gauss-Markov sequence whose covariance matrix is given by the relation*

$$P(k + 1|k) = \Phi(k + 1, k)P(k|k)\Phi'(k + 1, k)$$
$$+ \Gamma(k + 1, k)Q(k)\Gamma'(k + 1, k) \quad (5.42)$$

Proof The corollary follows immediately from Theorem 5.4 by replacing k by $k + 1$ and j by k.

As in the case of Theorem 5.4, the corollary is of little immediate use for prediction, since only $\hat{x}(0|0)$ and $P(0|0)$ are known. We turn, therefore, to a study of the optimal filtering problem in which we obtain an algorithm which is very useful. We shall see that prediction and filtering are interdependent in terms of the determination of the filtered estimate given the predicted estimate and vice versa. The latter is already evident in both Eqs. (5.33) and (5.41).

Finally, we remark that the results in Theorem 5.4 and Corollary 5.2 are due to Kalman [6,7].

5.4 OPTIMAL FILTERING FOR DISCRETE LINEAR SYSTEMS

In developing the algorithm for optimal filtering for the system of Eqs. (5.14) and (5.15), we assume that only the initial estimate, $\hat{x}(0|0) = 0$, the covariance matrix of the filtering error at the initial time, $P(0|0) = E[\tilde{x}(0|0)\tilde{x}'(0|0)] = E[x(0)x'(0)] = P(0)$, and the set of measurements $\{z(1), \ldots, z(k), z(k + 1)\}$, $k =$ nonnegative integer, are given.

From Corollary 5.1, we know that the optimal filtered estimate $\hat{x}(k + 1|k + 1)$ is given by the relation

$$\hat{x}(k + 1|k + 1) = E[x(k + 1)|z(1), \ldots, z(k), z(k + 1)] \quad (5.43)$$

However, from Eqs. (3.50) and (3.51)

$$E[x(k + 1)|z(1), \ldots, z(k), z(k + 1)]$$
$$= E[x(k + 1)|z(1), \ldots, z(k), \tilde{z}(k + 1|k)]$$
$$= E[x(k + 1)|z(1), \ldots, z(k)] + E[x(k + 1)|\tilde{z}(k + 1|k)] \quad (5.44)$$

for $k = 0, 1, \ldots$, where

$$\tilde{z}(k + 1|k) = z(k + 1) - E[z(k + 1)|z(1), \ldots, z(k)]$$

In arriving at Eq. (5.44), we have utilized the fact that $\{x(k), k = 0, 1, \ldots\}$ has zero mean.

We observe that $E[z(k + 1)|z(1), \ldots, z(k)]$ is the optimal predicted estimate of $z(k + 1)$ given the measurements $\{z(1), \ldots, z(k)\}$, and we write

$$\hat{z}(k + 1|k) = E[z(k + 1)|z(1), \ldots, z(k)] \tag{5.45}$$

Then,

$$\tilde{z}(k + 1|k) = z(k + 1) - \hat{z}(k + 1|k) \tag{5.46}$$

is the difference between the actual and predicted measurements at $k + 1$. This difference is called the *measurement residual*.

Substituting Eq. (5.15) into Eq. (5.45), we see that

$$\begin{aligned}
\hat{z}(k + 1|k) &= E[H(k + 1)x(k + 1) + v(k + 1)|z(1), \ldots, z(k)] \\
&= H(k + 1)E[x(k + 1)|z(1), \ldots, z(k)] \\
&\qquad\qquad + E[v(k + 1)|z(1), \ldots, z(k)] \\
&= H(k + 1)\hat{x}(k + 1|k) + E[v(k + 1)|z(1), \ldots, z(k)]
\end{aligned}$$

From Eq. (5.28), $v(k + 1)$ is uncorrelated with each of the measurements $z(1), \ldots, z(k)$. Since all these vectors are gaussian, it follows from part 1 of Theorem 3.2 that $v(k + 1)$ is independent of the set of measurements $\{z(1), \ldots, z(k)\}$. Hence,

$$\begin{aligned}
\hat{z}(k + 1|k) &= H(k + 1)\hat{x}(k + 1|k) + E[v(k + 1)] \\
&= H(k + 1)\hat{x}(k + 1|k) \tag{5.47}
\end{aligned}$$

for $k = 0, 1, \ldots$, where we have made use of the fact that $\{v(k + 1), k = 0, 1, \ldots\}$ has zero mean.

With these preliminaries completed, we now state and prove the basic theorem of optimal filtering for discrete linear systems.

Theorem 5.5

   a. The optimal filtered estimate $\hat{x}(k + 1|k + 1)$ is given by the recursive relation

$$\begin{aligned}
\hat{x}(k + 1|k + 1) = \Phi(k + 1, k)\hat{x}(k|k) + K(k + 1)[z(k + 1) \\
- H(k + 1)\Phi(k + 1, k)\hat{x}(k|k)] \tag{5.48}
\end{aligned}$$

   for $k = 0, 1, \ldots$, where $\hat{x}(0|0) = 0$.

   b. $K(k + 1)$ is an $n \times m$ matrix which is specified by the set of relations

$$\begin{aligned}
K(k + 1) = P(k + 1|k)H'(k + 1)[H(k + 1)P(k + 1|k)H'(k + 1) \\
+ R(k + 1)]^{-1} \tag{5.49}
\end{aligned}$$

$$\begin{aligned}
P(k + 1|k) = \Phi(k + 1, k)P(k|k)\Phi'(k + 1, k) \\
+ \Gamma(k + 1, k)Q(k)\Gamma'(k + 1, k) \tag{5.50}
\end{aligned}$$

$$P(k + 1|k + 1) = [I - K(k + 1)H(k + 1)]P(k + 1|k) \tag{5.51}$$

for $k = 0, 1, \ldots$, *where I is the $n \times n$ identity matrix and $P(0|0) = P(0)$ is the initial condition for Eq.* (5.50).

c. *The stochastic process $\{\tilde{x}(k + 1|k + 1), k = 0, 1, \ldots\}$, which is defined by the filtering error relation*

$$\tilde{x}(k + 1|k + 1) = x(k + 1) - \hat{x}(k + 1|k + 1)$$

$k = 0, 1, \ldots$, *is a zero mean Gauss-Markov sequence whose covariance matrix is given by Eq.* (5.51).

Proof

1. From Eqs. (5.43) and (5.44),

$$\hat{x}(k + 1|k + 1) = \hat{x}(k + 1|k) + E[x(k + 1)|\tilde{z}(k + 1|k)] \qquad (5.52)$$

But from Eq. (5.41) of Corollary 5.2,

$$\hat{x}(k + 1|k) = \Phi(k + 1, k)\hat{x}(k|k) \qquad (5.53)$$

Further, since $x(k + 1)$ and $\tilde{z}(k + 1|k)$ are zero mean and gaussian, we have from Eq. (5.9) that

$$E[x(k + 1)|\tilde{z}(k + 1|k)] = P_{x\tilde{z}} P_{\tilde{z}\tilde{z}}^{-1}\tilde{z}$$

where the time arguments are omitted on the right-hand side for simplicity. We note, of course, that

$$P_{x\tilde{z}} = E[x(k + 1)\tilde{z}'(k + 1|k)]$$
$$P_{\tilde{z}\tilde{z}} = E[\tilde{z}(k + 1|k)\tilde{z}'(k + 1|k)]$$

Defining $K(k + 1) = P_{x\tilde{z}} P_{\tilde{z}\tilde{z}}^{-1}$, we write

$$E[x(k + 1)|\tilde{z}(k + 1|k)] = K(k + 1)\tilde{z}(k + 1|k)$$

However, it is clear from Eqs. (5.46), (5.47), and (5.53) that

$$\tilde{z}(k + 1|k) = z(k + 1) - \hat{z}(k + 1|k)$$
$$= z(k + 1) - H(k + 1)\hat{x}(k + 1|k)$$
$$= z(k + 1) - H(k + 1)\Phi(k + 1, k)\hat{x}(k|k) \qquad (5.54)$$

Hence,

$$E[x(k + 1)|\tilde{z}(k + 1|k)] = K(k + 1)[z(k + 1)$$
$$- H(k + 1)\Phi(k + 1, k)\hat{x}(k|k)]$$

Substituting this result and Eq. (5.53) into Eq. (5.52), we get

$$\hat{x}(k + 1|k + 1) = \Phi(k + 1, k)\hat{x}(k|k) + K(k + 1)[z(k + 1)$$
$$- H(k + 1)\Phi(k + 1, k)\hat{x}(k|k)]$$

for $k = 0, 1, \ldots$, which is identically Eq. (5.48).

From Eq. (5.31), the appropriate initial condition is obviously $\hat{x}(0|0) = 0$.

2. We evaluate $K(k + 1)$ next. From the second line in Eq. (5.54), Eq. (5.15) and the definition of prediction error we have

$$
\begin{aligned}
\tilde{z}(k + 1|k) &= z(k + 1) - H(k + 1)\hat{x}(k + 1|k) \\
&= H(k + 1)x(k + 1) + v(k + 1) - H(k + 1)\hat{x}(k + 1|k) \\
&= H(k + 1)\tilde{x}(k + 1|k) + v(k + 1) \qquad (5.55)
\end{aligned}
$$

Consequently,

$$
\begin{aligned}
P_{\tilde{z}\tilde{z}} &= E\{[H(k + 1)\tilde{x}(k + 1|k) + v(k + 1)] \\
&\qquad\qquad [H(k + 1)\tilde{x}(k + 1|k) + v(k + 1)]'\} \\
&= H(k + 1)E[\tilde{x}(k + 1|k)\tilde{x}'(k + 1|k)]H'(k + 1) \\
&\quad + H(k + 1)E[\tilde{x}(k + 1|k)v'(k + 1)] \\
&\quad + E[v(k + 1)\tilde{x}'(k + 1|k)]H'(k + 1) + E[v(k + 1)v'(k + 1)] \\
&= H(k + 1)P(k + 1|k)H'(k + 1) \\
&\quad + H(k + 1)E[\tilde{x}(k + 1|k)v'(k + 1)] \\
&\quad + E[v(k + 1)\tilde{x}'(k + 1|k)]H'(k + 1) + R(k + 1) \quad (5.56)
\end{aligned}
$$

Let us now show that the expected values in the middle two terms above vanish. Since one is merely the transpose of the other, it is sufficient to consider only the first one:

$$
\begin{aligned}
E[\tilde{x}(k + 1|k)v'(k + 1)] &= E[x(k + 1)v'(k + 1)] \\
&\qquad - E[\hat{x}(k + 1|k)v'(k + 1)]
\end{aligned}
$$

From Eq. (5.27), $E[x(k + 1)v'(k + 1)] = 0$ for all $k = 0, 1, \ldots$. Recalling that $\hat{x}(k + 1|k)$ can be written

$$
\hat{x}(k + 1|k) = \sum_{i=1}^{k} A(i)z(i)
$$

we see that, as a result of Eq. (5.28),

$$
E[\hat{x}(k + 1|k)v'(k + 1)] = \sum_{i=1}^{k} A(i)E[z(i)v'(k + 1)] = 0
$$

for all $k = 1, 2, \ldots$. For $k = 0$, we get

$$
\begin{aligned}
E[\hat{x}(1|0)v'(1)] &= E[\Phi(1,0)\hat{x}(0|0)v'(1)] \\
&= \Phi(1,0)E[\hat{x}(0|0)v'(1)] = 0
\end{aligned}
$$

since $\hat{x}(0|0) = 0$. Hence, we have that

$$E[\hat{x}(k + 1|k)v'(k + 1)] = 0 \tag{5.57}$$

for all $k = 0, 1, \ldots$.

It now follows immediately that

$$E[\tilde{x}(k + 1|k)v'(k + 1)] = 0 \tag{5.58}$$

for all $k = 0, 1, \ldots$, and Eq. (5.56) assumes the form

$$P_{\tilde{z}\tilde{z}} = H(k + 1)P(k + 1|k)H'(k + 1) + R(k + 1) \tag{5.59}$$

for all $k = 0, 1, \ldots$.

Turning next to $P_{x\tilde{z}}$, we see that

$$P_{x\tilde{z}} = E[x(k + 1)\tilde{z}'(k + 1|k)]$$
$$= E\{[\tilde{x}(k + 1|k) + \hat{x}(k + 1|k)]$$
$$[H(k + 1)\tilde{x}(k + 1|k) + v(k + 1)]'\}$$

where the second line follows from Eq. (5.55) and the fact that $\tilde{x}(k + 1|k) = x(k + 1) - \hat{x}(k + 1|k)$ can also be written as $x(k + 1) = \tilde{x}(k + 1|k) + \hat{x}(k + 1|k)$.

Expanding this result, we obtain

$$P_{x\tilde{z}} = E[\tilde{x}(k + 1|k)\tilde{x}'(k + 1|k)]H'(k + 1) + E[\tilde{x}(k + 1|k)v'(k + 1)]$$
$$+ E[\hat{x}(k + 1|k)\tilde{x}'(k + 1|k)]H'(k + 1) + E[\hat{x}(k + 1|k)v'(k + 1)]$$

From Eqs. (5.58), (5.10), and (5.57), we see that the second, third, and fourth terms, respectively, in this expression are identically zero. Hence,

$$P_{x\tilde{z}} = P(k + 1|k)H'(k + 1) \tag{5.60}$$

Substituting Eqs. (5.59) and (5.60) into the defining relation for K, we obtain the desired result:

$$K(k + 1) = P(k + 1|k)H'(k + 1)$$
$$[H(k + 1)P(k + 1|k)H'(k + 1) + R(k + 1)]^{-1}$$

for $k = 0, 1, \ldots$. If $R(k + 1)$ is assumed to be positive definite, it follows that the required inverse always exists.

From Eq. (5.42) in Corollary 5.2, we have that

$$P(k + 1|k) = \Phi(k + 1, k)P(k|k)\Phi'(k + 1, k)$$
$$+ \Gamma(k + 1, k)Q(k)\Gamma'(k + 1, k)$$

We seek now the expression for the covariance matrix of the filtering error

$$\tilde{x}(k + 1|k + 1) = x(k + 1) - \hat{x}(k + 1|k + 1)$$

Noting that $\hat{x}(k + 1|k + 1)$ can be written

$$\hat{x}(k + 1|k + 1) = \hat{x}(k + 1|k) + K(k + 1)\tilde{z}(k + 1|k)$$

we get, with the aid of Eq. (5.55), the result

$$\begin{aligned}
\tilde{x}(k + 1|k + 1) &= x(k + 1) - [\hat{x}(k + 1|k) + K(k + 1)\tilde{z}(k + 1|k)] \\
&= \tilde{x}(k + 1|k) \\
&\quad - K(k + 1)[H(k + 1)\tilde{x}(k + 1|k) + v(k + 1)] \\
&= [I - K(k + 1)H(k + 1)]\tilde{x}(k + 1|k) \\
&\qquad - K(k + 1)v(k + 1) \quad (5.61)
\end{aligned}$$

where I is the $n \times n$ identity matrix.

We note that evaluation of $P(k + 1|k + 1) = E[\tilde{x}(k + 1|k)\tilde{x}'(k + 1|k)]$ will involve the term $E[\tilde{x}(k + 1|k)v'(k + 1)]$ and its transpose, both of which are zero in view of Eq. (5.58). Therefore,

$$\begin{aligned}
P(k + 1|k + 1) &= [I - K(k + 1)H(k + 1)]E[\tilde{x}(k + 1|k)\tilde{x}'(k + 1|k)] \\
&\qquad [I - K(k + 1)H(k + 1)]' \\
&\quad + K(k + 1)E[v(k + 1)v'(k + 1)]K'(k + 1) \\
&= [I - K(k + 1)H(k + 1)]P(k + 1|k) \\
&\qquad [I - K(k + 1)H(k + 1)]' \\
&\quad + K(k + 1)R(k + 1)K'(k + 1) \quad (5.62)
\end{aligned}$$

Dropping the time indices on the right-hand side of Eq. (5.62) for convenience, expanding, and regrouping terms, we obtain

$$\begin{aligned}
P(k + 1|k + 1) &= (P - KHP)(I - KH)' + KRK' \\
&= P - KHP - PH'K' + KHPH'K' + KRK' \\
&= (I - KH)P - PH'K' + K(HPH' + R)K'
\end{aligned}$$

But we note from Eq. (5.49) that $K(HPH' + R) = PH'$. Hence,

$$\begin{aligned}
P(k + 1|k + 1) &= (I - KH)P - PH'K' + PH'K' \\
&= [I - K(k + 1)H(k + 1)]P(k + 1|k)
\end{aligned}$$

for $k = 0, 1, \ldots$ which gives us Eq. (5.51) of the theorem.

3. Replacing k by $k + 1$ and j by k in Eq. (5.39), we have

$$\tilde{x}(k + 1|k) = \Phi(k + 1, k)\tilde{x}(k|k) + \Gamma(k + 1, k)w(k) \qquad (5.63)$$

which we substitute into Eq. (5.61) to obtain

$$\tilde{x}(k + 1|k + 1) = [I - K(k + 1)H(k + 1)]\Phi(k + 1, k)\tilde{x}(k|k)$$
$$+ [I - K(k + 1)H(k + 1)]\Gamma(k + 1, k)w(k)$$
$$- K(k + 1)v(k + 1) \qquad (5.64)$$

Defining $A(k + 1) = [I - K(k + 1)H(k + 1)]$ and $\Phi^*(k + 1, k) = A(k + 1)\Phi(k + 1, k)$, we can write Eq. (5.64) as

$$\tilde{x}(k + 1|k + 1) = \Phi^*(k + 1, k)\tilde{x}(k|k)$$
$$+ [A(k + 1)\Gamma(k + 1, k) \mathbin{\vdots} - K(k + 1)] \begin{bmatrix} w(k) \\ \hline v(k + 1) \end{bmatrix}$$

Then, letting

$$\Gamma^*(k + 1, k) = [A(k + 1)\Gamma(k + 1, k) \mathbin{\vdots} - K(k + 1)]$$

and

$$w^*(k) = \begin{bmatrix} w(k) \\ \hline v(k + 1) \end{bmatrix}$$

where we note that Γ^* is an $n \times (p + m)$ matrix and w^* is a $(p + m)$ vector, we have

$$\tilde{x}(k + 1|k + 1) = \Phi^*(k + 1, k)\tilde{x}(k|k) + \Gamma^*(k + 1, k)w^*(k) \qquad (5.65)$$

From the definition of $w^*(k)$, it is clear that $\{w^*(k), k = 0, 1, \ldots\}$ is a zero mean gaussian white sequence whose covariance matrix is

$$E[w^*(j)w^{*\prime}(k)] = \begin{bmatrix} Q(k) & \mathbin{\vdots} & 0_1 \\ \hline 0_2 & \mathbin{\vdots} & R(k + 1) \end{bmatrix} \delta_{jk}$$

where 0_1 and 0_2 are $p \times m$ and $m \times p$ null matrices, respectively.

For $k = 0$, we recall that $\tilde{x}(0|0) = x(0) - \hat{x}(0|0) = x(0)$ since $\hat{x}(0|0) = 0$. Hence, $\tilde{x}(0|0)$ is a zero mean gaussian random n vector whose covariance matrix is $E[\tilde{x}(0|0)\tilde{x}'(0|0)] = E[x(0)x'(0)] = P(0)$. In addition, we have $E[\tilde{x}(0|0)w^{*\prime}(k)] = E[x(0)w^{*\prime}(k)] = 0$ for all $k = 0, 1, \ldots$, a result that follows immediately from the definition of $w^*(k)$, and Eqs. (5.23) and (5.24). Consequently, the stochastic process defined by Eq. (5.65) is of the same form and is subject to the same hypotheses as the process defined by Eq. (4.28) in Sec. 4.3. Hence, $\{\tilde{x}(k + 1|k + 1), k = 0, 1, \ldots\}$ is a Gauss-Markov sequence.

Since $\{\tilde{x}(k + 1|k + 1), \; k = 0, \; 1, \; \ldots\}$ is Gauss-Markov, it is completely described by its mean, $E[\tilde{x}(k + 1|k + 1)]$, and its covariance matrix $P(k + 1|k + 1)$. The latter is given in Eq. (5.51), while the former is zero for all k, since $\tilde{x}(0|0)$ and $\{w^*(k), \; k = 0, \; 1, \; \ldots\}$ each have zero mean. This completes the proof.

This theorem was first proved by Kalman [7] in 1960, and the algorithm for recursive filtering which is described by Eqs. (5.48) to (5.51) is called the *Kalman filter*.

COMPUTATIONAL ASPECTS

One of the most significant features of the Kalman filter is its recursive form, a property that makes it extremely useful in processing measurements to obtain the optimal filtered estimate utilizing a digital computer. The measurements can be processed as they occur, and there is no need to store any measurement data. In fact, so far as storage of the measurements and the state is concerned, only $\hat{x}(k|k)$ need be stored in proceeding from time k to time $k + 1$. However, the algorithm does require storage of the time histories of the matrices $\Phi(k + 1, k)$, $\Gamma(k + 1, k)$, $H(k + 1)$, $Q(k)$ and $R(k + 1)$ for all $k = 0, \; 1, \; \ldots \; .$

The information flow in the filter can be discussed very simply by considering the block diagram of Fig. 5.5, which is a representation of Eq. (5.48),

$$\hat{x}(k + 1|k + 1) = \Phi(k + 1, k)\hat{x}(k|k)$$
$$+ K(k + 1)[z(k + 1) - H(k + 1)\Phi(k + 1, k)\hat{x}(k|k)]$$

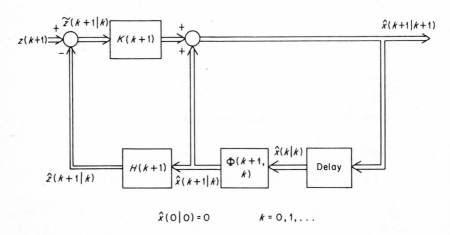

$$\hat{x}(0|0) = 0 \qquad k = 0, 1, \ldots$$

Fig. 5.5 Block diagram of Kalman filter.

Let us suppose that $\hat{x}(k|k)$ is known for some k and that we seek to determine $\hat{x}(k + 1|k + 1)$ given $z(k + 1)$. The computational cycle would proceed as follows:

1. The estimate $\hat{x}(k|k)$ is "propagated forward" by premultiplying it by the state transition matrix $\Phi(k + 1, k)$. This gives the predicted estimate $\hat{x}(k + 1|k)$. The step can be viewed as dynamic extrapolation of the preceding estimate.
2. $\hat{x}(k + 1|k)$ is premultiplied by $H(k + 1)$, giving $\hat{z}(k + 1|k)$ which is subtracted from the actual measurement $z(k + 1)$ to obtain the measurement residual $\tilde{z}(k + 1|k)$.
3. The residual is premultiplied by the matrix $K(k + 1)$, and the result is added to $\hat{x}(k + 1|k)$ to give $\hat{x}(k + 1|k + 1)$.
4. $\hat{x}(k + 1|k + 1)$ is stored until the next measurement is made at which time the cycle is repeated.

We observe that the filter operates in a "predict-correct" fashion. That is, the "correction" term $K(k + 1)\tilde{z}(k + 1|k)$ is added to the predicted estimate $\hat{x}(k + 1|k)$ to determine the filtered estimate. The correction term involves a weighting of the measurement residual by the matrix $K(k + 1)$. This latter matrix is commonly referred to as the *weighting matrix*, the *filter gain matrix*, or the *Kalman gain matrix*.

The interplay between prediction and filtering is evident at this point. We see that one of these estimates is obtained utilizing the other and vice versa; namely, we have

$$\hat{x}(k + 1|k) = \Phi(k + 1, k)\hat{x}(k|k)$$

and

$$\hat{x}(k + 1|k + 1) = \hat{x}(k + 1|k)$$
$$+ K(k + 1)[z(k + 1) - H(k + 1)\hat{x}(k + 1|k)]$$

In order to initiate filtering, we start with $\hat{x}(0|0) = 0$, and we see immediately that

$$\hat{x}(1|1) = K(1)z(1)$$

Then, $\hat{x}(2|2)$, $\hat{x}(3|3)$, . . . follow recursively, as indicated in the above four steps. If it is desired at any point to predict several stages ahead, say m stages, for example, then we have from Theorem 5.4 that

$$\hat{x}(k + m|k) = \Phi(k + m, k)\hat{x}(k|k)$$

Finally, we note in Fig. 5.5 that the optimal filter consists of the model of the dynamic process which performs the function of prediction and a feedback correction scheme in which the gain-times-residual term is applied to the model as a forcing function.

We consider next the computation of the filter gain matrix $K(k +$ and the two covariance matrices $P(k + 1|k)$ and $P(k + 1|k + 1)$. Th relevant equations, we recall, are

$$K(k + 1) = P(k + 1|k)H'(k + 1)$$
$$[H(k + 1)P(k + 1|k)H'(k + 1) + R(k + 1)]^{-1} \quad (5.49)$$

$$P(k + 1|k) = \Phi(k + 1, k)P(k|k)\Phi'(k + 1, k)$$
$$+ \Gamma(k + 1, k)Q(k)\Gamma'(k + 1, k) \quad (5.50)$$

$$P(k + 1|k + 1) = [I - K(k + 1)H(k + 1)]P(k + 1|k) \quad (5.51)$$

for $k = 0, 1, \ldots$, with $P(0|0) = E[x(0)x'(0)]$.

A typical computational cycle would proceed as follows:

1. Given $P(k|k)$, $Q(k)$, $\Phi(k + 1, k)$, and $\Gamma(k + 1, k)$, $P(k + 1|k)$ is com puted using Eq. (5.50).
2. $P(k + 1|k)$, $H(k + 1)$, and $R(k + 1)$ are substituted into Eq. (5.49 to obtain $K(k + 1)$, which is used in step 3 of the filter compu tations given earlier.
3. $P(k + 1|k)$, $K(k + 1)$, and $H(k + 1)$ are substituted into Eq. (5.51 to determine $P(k + 1|k + 1)$, which is stored until the time of th next measurement when the cycle is repeated.

The matrix inverse which must be computed in Eq. (5.49) generally poses no real problem. The matrix to be inverted is $m \times m$, where m is the number of elements in the measurement vector. In most systems m is kept small to avoid the high cost of complex instrumentation Consequently, it is not unusual to encounter systems with 12 to 15 state variables, but only 2 to 3 measurement variables.

Equations (5.49) through (5.51), of course, define the algorithm foi the recursive computation of the optimal filter gain matrix. However we know from Corollary 5.2 and Theorem 5.5 that $P(k + 1|k)$ anc $P(k + 1|k + 1)$ are the covariance matrices of the zero mean Gauss Markov processes $\{\tilde{x}(k + 1|k), k = 0, 1, \ldots\}$ and $\{\tilde{x}(k + 1|k + 1)$ $k = 0, 1, \ldots\}$, respectively. Therefore, in the process of computing the filter gain matrix, we obtain also the distribution of the prediction anc filtering errors. More specifically, the diagonal elements of $P(k + 1|k$ and $P(k + 1|k + 1)$ are the variances of the components of the predic tion and filtering error vectors, respectively.

For performance analysis purposes, it is obviously unnecessary t implement the filter. We need concern ourselves only with Eqs. (5.49 through (5.51) to assess filter performance. Specifically, we examine th diagonal elements of the two covariance matrices to obtain the tim histories of the error variances. Since the filter is optimal for all admi sible loss functions, we know that these variances are minimized. Hence, the Kalman filter is also the minimal variance filter.

In deriving the filter error covariance relation, Eq. (5.51), we first developed Eq. (5.62):

$$P(k + 1|k + 1) = [I - K(k + 1)H(k + 1)]P(k + 1|k)$$
$$[I - K(k + 1)H(k + 1)]' + K(k + 1)R(k + 1)K'(k + 1)$$

This expression is valid for any gain matrix $K(k + 1)$; it reduces to Eq. (5.51) only if we use the optimal gain matrix as given by Eq. (5.49). Consequently, Eq. (5.62) is a general expression for the filter error covariance matrix for any linear filter that has the structure given in Fig. 5.5.

EXAMPLES

In order to gain more of a working understanding of the Kalman filter, let us examine three rather simple examples. In order to avoid obscuring principles with algebraic details, we restrict our attention to scalar processes.

Example 5.2 Let us consider the stochastic process defined by the scalar relation

$$x(k + 1) = (-1)^{2k+1}x(k)$$

for $k = 0, 1, \ldots$, where $x(0)$ is a zero mean gaussian random variable with variance $P(0)$. The process is then one which takes on the same magnitude at each sampling instant with alternating sign. Two sample functions are indicated in Fig. 5.6.

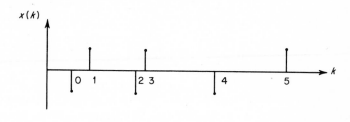

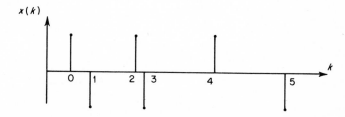

Fig. 5.6 Sample functions for discrete-time stochastic process.

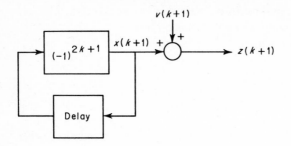

Fig. 5.7 System block diagram for Example 5.2.

Suppose that we observe this process in the presence of a zero mean gaussian white sequence $\{v(k + 1), k = 0, 1, . . .\}$ whose variance is $R(k + 1)$. Then the measurement equation is $z(k + 1) = x(k + 1) + v(k + 1), k = 0, 1, . . . ,$ and the model for both dynamics and measurements can be depicted as shown in Fig. 5.7. We see that $\Phi(k + 1, k) = (-1)^{2k+1}$ and $H(k + 1) = 1$.

From Eq. (5.49), we have the filter

$$\hat{x}(k + 1|k + 1) = (-1)^{2k+1}\hat{x}(k|k) + K(k + 1)[z(k + 1) - (-1)^{2k+1}\hat{x}(k|k)]$$

with $\hat{x}(0|0) = 0$. The corresponding block diagram is illustrated in Fig. 5.8.

Since there is no input disturbance to the dynamics, $Q(k) = 0$ for all $k = 0, 1, . . . ,$ and Eq. (5.50) is

$$P(k + 1|k) = (-1)^{2k+1}P(k|k)(-1)^{2k+1} = P(k|k) \tag{5.66}$$

for this example. The optimal filter gain is then seen to be

$$K(k + 1) = P(k|k)[P(k|k) + R(k + 1)]^{-1} = \frac{P(k|k)}{P(k|k) + R(k + 1)} \tag{5.67}$$

Substituting these two results into Eq. (5.51), we obtain

$$P(k + 1|k + 1) = \left[1 - \frac{P(k|k)}{P(k|k) + R(k + 1)}\right] P(k|k) = \frac{R(k + 1)P(k|k)}{P(k|k) + R(k + 1)} \tag{5.68}$$

for $k = 0, 1, . . . ,$ subject to the initial condition $P(0|0) = P(0)$.

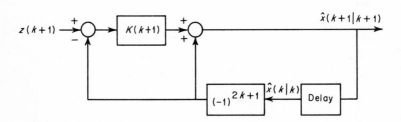

Fig. 5.8 Optimal filter block diagram for Example 5.2.

The computations here are exceedingly simple. Starting with $P(k|k)$, which is $P(0)$ for $k = 0$, we have immediately from Eq. (5.66) that the prediction error variance is $P(k|k)$ with no computation whatsoever. The filter gain is easily determined using Eq. (5.67), and it is obvious from an examination of Eqs. (5.67) and (5.68) that

$$P(k + 1|k + 1) = R(k + 1)K(k + 1) \tag{5.69}$$

For computational purposes, it is clear that we need consider only Eqs. (5.67) and (5.69).

The filtering error variance as given by Eq. (5.68) has the interesting property that

$$P(k + 1|k + 1) \leq \min [P(k|k), R(k + 1)]$$

This follows from the right-hand side of Eq. (5.68) and the fact that $R(k + 1)$ and $P(k|k)$ are variances, and, therefore, nonnegative for all $k = 0, 1, \ldots$. The above result implies that $P(k + 1|k + 1)$ is a monotone nonincreasing function of k. If $P(0|0) > 0$, and $R(k + 1) > 0$ and finite for all k, then

$$P(k + 1|k+1) < \min [P(k|k), R(k + 1)]$$

and $P(k + 1|k+1)$ is a monotone decreasing function of k.

If $R(k + 1)$ is infinite for any k, we see that $P(k + 1|k + 1) = P(k|k)$, $K(k + 1) = 0$, and

$$\hat{x}(k + 1|k + 1) = (-1)^{2k+1}\hat{x}(k|k)$$

This certainly satisfies our intuition, for if the measurement error variance is infinite, the corresponding measurement is clearly useless. On the other hand, if $R(k + 1)$ is zero for any k, we observe from Eq. (5.67) that $K(k + 1) = 1$, and the filter equation reduces to

$$\hat{x}(k + 1|k + 1) = z(k + 1)$$

This too satisfies our intuition, for if the measurement error variance is zero, the measurement is perfect and gives $x(k + 1)$ exactly. Equation (5.68), of course, reduces to $P(k + 1|k + 1) = 0$, and the filtering is exact thereafter. Moreover, if we note that $x(k + 1) = (-1)^{k+1}x(0)$, $k = 0, 1, \ldots$, we see that if $x(k + 1)$ is known exactly for any k, $x(0)$ can be determined exactly from it, and the state is known exactly for all $k = 0, 1, \ldots$.

Finally, let us consider the equilibrium or "steady-state" behavior of Eq. (5.68) for large values of k. Letting $P(k + 1|k + 1) = P(k|k) = \bar{P}$, we have in Eq. (5.68) that

$$\bar{P} = \frac{R(k + 1)\bar{P}}{\bar{P} + R(k + 1)}$$

The solution is obviously $\bar{P} = 0$. Since $P(k + 1|k + 1)$ is a monotone nonincreasing function of k, it follows that $\bar{P} = 0$ is a stable equilibrium point, and the "steady-state" solution of the variance equation for this example.

Example 5.3 In our second example, we consider the scalar system

$$x(k + 1) = \Phi x(k) + w(k)$$
$$z(k + 1) = x(k + 1) + v(k + 1)$$

for $k = 0, 1, \ldots$, where $\{w(k), k = 0, 1, \ldots\}$ is a zero mean gaussian white sequence with constant variance Q, $\{v(k + 1), k = 0, 1, \ldots\}$ is a zero mean gaussian white sequence with constant variance R, $x(0)$ is a zero mean gaussian random variable with variance $P(0)$, and Φ is a constant. We assume that the two gaussian white sequences and $x(0)$ are independent.

The optimal filter equation is

$$\hat{x}(k + 1|k + 1) = \Phi\hat{x}(k|k) + K(k + 1)[z(k + 1) - \Phi\hat{x}(k|k)] \tag{5.70}$$

From Eqs. (5.50) and (5.49), we obtain the results

$$P(k + 1|k) = \Phi^2 P(k|k) + Q \tag{5.71}$$

and

$$K(k + 1) = [\Phi^2 P(k|k) + Q][\Phi^2 P(k|k) + Q + R]^{-1} = \frac{\Phi^2 P(k|k) + Q}{\Phi^2 P(k|k) + Q + R} \tag{5.72}$$

respectively. The filtering error variance equation is then seen to be

$$\begin{aligned} P(k + 1|k+1) &= \left[1 - \frac{\Phi^2 P(k|k) + Q}{\Phi^2 P(k|k) + Q + R}\right][\Phi^2 P(k|k) + Q] \\ &= \frac{R[\Phi^2 P(k|k) + Q]}{\Phi^2 P(k|k) + Q + R} \end{aligned} \tag{5.73}$$

subject to the initial condition $P(0|0) = P(0)$.

From Eq. (5.71), we see that $P(k + 1|k) \geq Q$, since $P(k|k) \geq 0$. This means that the variance of the system disturbance sets the performance limit on prediction accuracy.

The range on the filter gain is seen to be $0 \leq K(k + 1) \leq 1$ from Eq. (5.72) except for the trivial case $P(k|k) = Q = R = 0$. Also, when Eqs. (5.72) and (5.73) are combined to give

$$P(k + 1|k + 1) = RK(k + 1)$$

it is obvious that $0 \leq P(k + 1|k + 1) \leq R$ for $k = 0, 1, \ldots$. As a result, if $P(0) \gg R$, the first measurement, that is, $z(1)$, will lead to a large reduction in the filtering error variance from $P(0)$ to $P(1|1) \leq R \ll P(0)$.

For the special case where $Q \gg R$, we observe from Eqs. (5.72) and (5.73) that $K(k + 1) \approx 1$ and $P(k + 1|k + 1) \approx R$ for all $k = 0, 1, \ldots$. In this case, the performance limit on filtering accuracy is obviously dictated by the measurement error variance.

Finally, let us examine the steady-state behavior of the variance equation, Eq. (5.73), under the assumption that $Q = 0$. Letting $P(k + 1|k + 1) = P(k|k) = \bar{P}$ with $Q = 0$, we have

$$\bar{P} = \frac{R\Phi^2\bar{P}}{\Phi^2\bar{P} + R}$$

which possesses two solutions,

$$\bar{P} = 0 \quad \text{and} \quad \bar{P} = \frac{(\Phi^2 - 1)R}{\Phi^2}$$

Since $\bar{P}$ is a variance, only the first solution is allowed when $\Phi^2 < 1$. However, both solutions must be considered whenever $\Phi^2 > 1$. Since Φ is a real constant, we know that $\Phi^2 \geq 0$.

Defining $\delta P(k + 1|k+1) = P(k + 1|k + 1) - \bar{P}$ and $\delta P(k|k) = P(k|k) - \bar{P}$, we see that

$$\delta P(k + 1|k + 1) = \frac{R\Phi^2 P(k|k)}{\Phi^2 P(k|k) + R} + \frac{R\Phi^2 \bar{P}}{\Phi^2 \bar{P} + R}$$

$$= \frac{R\Phi^2 P(k|k)(\Phi^2\bar{P} + R) - R\Phi^2\bar{P}[\Phi^2 P(k|k) + R]}{[\Phi^2 P(k|k) + R](\Phi^2\bar{P} + R)}$$

$$= \frac{\Phi^2 R}{\Phi^2\bar{P} + R} \frac{R}{\Phi^2 P(k|k) + R} \delta P(k|k)$$

For $\Phi^2 < 1$, we see that the coefficient on the right-hand side of this expression is less than one for all k, so that $\delta P(k + 1|k + 1) < \delta P(k|k)$ for all k. For $\bar{P} = 0$, we see further that $P(k + 1|k + 1) < P(k|k)$, so that $\bar{P} = 0$ is a stable equilibrium point for the filtering error variance equation whenever $\Phi^2 < 1$.

For $\Phi^2 > 1$, we must consider both $\bar{P} = 0$ and $\bar{P} = (\Phi^2 - 1)R/\Phi^2$. For $\bar{P} = 0$,

$$\delta P(k + 1|k + 1) = \frac{\Phi^2 R}{\Phi^2 P(k|k) + R} \delta P(k|k)$$

Taking $P(k|k) = 0$ for some k, we see that $\delta P(k + 1|k + 1) = \Phi^2 \delta P(k|k)$. Since $\Phi^2 > 1$, this means that even if the filtering error variance does become zero, it will not remain zero; that is, $\bar{P} = 0$ is an unstable equilibrium point of Eq. (5.73).

For the second value of $\bar{P}$,

$$\delta P(k + 1|k + 1) = \frac{R}{\Phi^2 P(k|k) + R} \delta P(k|k)$$

which means that $\delta P(k + 1|k + 1) < \delta P(k|k)$, since $\Phi^2 > 1$. Consequently, $\bar{P} = (\Phi^2 - 1)R/\Phi^2$ is a stable equilibrium point for $\Phi^2 > 1$.

In summary, the filtering error variance will converge to zero if $\Phi^2 < 1$ and to $(\Phi^2 - 1)R/\Phi^2$ if $\Phi^2 > 1$. Hence, for sufficiently long filtering times, the state can be determined exactly when $-1 < \Phi < 1$, but can only be specified to within an error variance of $(\Phi^2 - 1)R/\Phi^2$ when $Q = 0$.

Example 5.4   Let us examine a specific numerical example involving the system in the preceding example. In particular, we take $\Phi = 1$, $P(0) = 100$, $Q = 25$, and $R = 15$. Then, Eqs. (5.71) to (5.73) become

$$P(k + 1|k) = P(k|k) + 25$$

$$K(k + 1) = \frac{P(k|k) + 25}{P(k|k) + 40}$$

$$P(k + 1|k + 1) = \frac{15[P(k|k) + 25]}{P(k|k) + 40} = 15K(k + 1)$$

Starting with $P(0|0) = P(0) = 10$, it is a relatively simple matter to compute $P(k + 1|k)$, $K(k + 1)$, and $P(k + 1|k + 1)$ for $k = 0, 1, \ldots$. The results for the first few computational cycles are tabulated below.

k	$P(k|k - 1)$	$K(k)$	$P(k|k)$
0	. . . . .	. . . . .	100
1	125	0.893	13.40
2	38.4	0.720	10.80
3	35.8	0.704	10.57
4	35.6	0.703	10.55

The equilibrium value of $P(k|k)$ is obtained by setting $P(k + 1|k + 1) = P(k|k) = \bar{P}$ in the last of the above three relations to obtain

$$\bar{P}^2 + 25\bar{P} - 375 = 0$$

Since $\bar{P}$ is a variance, it must be nonnegative, which means that only the solution $\bar{P} = 10.55$ is valid. Comparing this result with $P(4|4)$, we see that the filter is in the steady state to within the indicated computational accuracy after processing four measurements. From Eq. (5.70), this means that the filter equation is

$$\hat{x}(k + 1|k + 1) = \hat{x}(k|k) + 0.703[z(k + 1) - \hat{x}(k|k)]$$
$$= 0.297\hat{x}(k|k) + 0.703z(k + 1)$$

in the steady-state, that is, for $k = 4, 5, \ldots$.

ALTERNATE GAIN EXPRESSION

By means of various vector-matrix manipulations, the Kalman filter equations can be put into a number of equivalent forms. However, experience has shown that the formulation given in Eqs. (5.48) through (5.51) is, in general, the most tractable one. Nevertheless, we shall give an illustration of an equivalent formulation by developing an alternate expression for the filter gain matrix, $K(k + 1)$. In order to do this, we shall need two particular matrix identities which we now develop.

Let P and R be nonsingular $n \times n$ and $m \times m$ matrices, respectively, and let H be an $m \times n$ matrix. The first matrix identity is

$$PH'(HPH' + R)^{-1} = (P^{-1} + H'R^{-1}H)^{-1}H'R^{-1} \tag{5.74}$$

and the second is

$$(P^{-1} + H'R^{-1}H)^{-1} = P - PH'(HPH' + R)^{-1}HP \tag{5.75}$$

We verify Eq. (5.74) by direct expansion:

$$PH'(HPH' + R)^{-1} = (P^{-1} + H'R^{-1}H)^{-1}(P^{-1} + H'R^{-1}H)$$
$$PH'(HPH' + R)^{-1}$$
$$= (P^{-1} + H'R^{-1}H)^{-1}(H' + H'R^{-1}HPH')$$
$$(HPH' + R)^{-1}$$
$$= (P^{-1} + H'R^{-1}H)^{-1}H'R^{-1}(R + HPH')$$
$$(HPH' + R)^{-1}$$
$$= (P^{-1} + H'R^{-1}H)^{-1}H'R^{-1}$$

To verify Eq. (5.75), we proceed in essentially the same manner:

$$(P^{-1} + H'R^{-1}H) = (P^{-1} + H'R^{-1}H)[P - PH'(HPH' + R)^{-1}HP]$$
$$[P - PH'(HPH' + R)^{-1}HP]^{-1}$$

Substituting into the second factor on the right from Eq. (5.74), we have

$$(P^{-1} + H'R^{-1}H)$$
$$= (P^{-1} + H'R^{-1}H)[P - (P^{-1} + H'R^{-1}H)^{-1}H'R^{-1}HP]$$
$$[P - PH'(HPH' + R)^{-1}HP]^{-1}$$
$$= (I + H'R^{-1}HP - H'R^{-1}HP)$$
$$[P - PH'(HPH' + R)^{-1}HP]^{-1}$$
$$= [P - PH'(HPH' + R)^{-1}HP]^{-1}$$

Taking the inverse of this result, we are led immediately to Eq. (5.75).

From Eqs. (5.49) and (5.74), we see that the optimal filter gain matrix can be expressed as

$$K(k + 1) = [P^{-1}(k + 1|k) + H'(k + 1)R^{-1}(k + 1)H(k + 1)]^{-1}$$
$$H'(k + 1)R^{-1}(k + 1) \quad (5.76)$$

However, from Eqs. (5.49), (5.51), and (5.75), we have

$$P(k + 1|k + 1)$$
$$= P(k + 1|k) - K(k + 1)H(k + 1)P(k + 1|k)$$
$$= P(k + 1|k) - P(k + 1|k)H'(k + 1)$$
$$[H(k + 1)P(k + 1|k)H'(k + 1) + R(k + 1)]^{-1}$$
$$H(k + 1)P(k + 1|k)$$
$$= [P^{-1}(k + 1|k) + H'(k + 1)R^{-1}(k + 1)H(k + 1)]^{-1} \quad (5.77)$$

Hence, substituting Eq. (5.77) into Eq. (5.76), we obtain the alternate gain expression

$$K(k + 1) = P(k + 1|k + 1)H'(k + 1)R^{-1}(k + 1) \quad (5.78)$$

Whereas the sequence of computations is $P(k + 1|k) \to K(k + 1) \to$ $P(k + 1|k + 1)$ for the original formulation of the Kalman filter in Theorem 5.5, it is clear that the sequence is $P(k + 1|k) \to P(k + 1|k + 1)$ $\to K(k + 1)$ if Eq. (5.78) is utilized. In this latter formulation, Eq. (5.51) must be put into the form

$$P(k + 1|k + 1) = P(k + 1|k) - P(k + 1|k)H'(k + 1)$$
$$[H(k + 1)P(k + 1|k)H'(k + 1) + R(k + 1)]^{-1}$$
$$H(k + 1)P(k + 1|k) \quad (5.79)$$

by substituting into it from Eq. (5.49). The relevant equations for this formulation are then Eqs. (5.50), (5.79), and (5.78), with computations proceeding in the order indicated.

This formulation offers no computational advantage over the original one. Indeed, it suffers from the disadvantage that $K(k + 1)$ must be computed twice, once in the process of computing $P(k + 1|k + 1)$ in Eq. (5.79) and again in Eq. (5.78). On the other hand, if only an error covariance analysis is desired, with no regard for the time history of $K(k + 1)$, one can use Eqs. (5.50) and (5.79) directly and alternate between them to obtain $P(k + 1|k)$ and $P(k + 1|k + 1)$, $k = 0, 1, \ldots$. An additional simplification is possible if there is interest only in the time history of the filtering error covariance matrix. Namely, Eq. (5.50) can be substituted into Eq. (5.79) to obtain a first-order matrix relation between $P(k + 1|k + 1)$ and $P(k|k)$.

Finally, we remark that Eq. (5.77), which is an alternate expression for Eq. (5.51) for determining $P(k + 1|k + 1)$, is not especially attractive for computational purposes because of the matrix inversions required, specifically $P^{-1}(k + 1|k)$ and the inverse of $[P^{-1}(k + 1|k) + H'(k + 1)R^{-1}(k + 1)H(k + 1)]$. Both are inverses of $n \times n$ matrices, where we recall that n is the number of state variables. Superficially, it may appear that Eq. (5.77) is tractable if n is 2, 3, or 4. Still, three matrix inversions are required in contrast to only one if Eqs. (5.49), (5.50), and (5.51) are used. Since m is generally less than n, the choice is obvious.

5.5 OPTIMAL FILTERING IN THE PRESENCE OF TIME–CORRELATED DISTURBANCES AND MEASUREMENT ERRORS

In considering systems of the type described by Eqs. (5.14) and (5.15),

$$x(k + 1) = \Phi(k + 1, k)x(k) + \Gamma(k + 1, k)w(k)$$
$$z(k + 1) = H(k + 1)x(k + 1) + v(k + 1)$$

it is not always appropriate to assume that the stochastic processes $\{w(k), \ k = 0, \ 1, \ . \ . \ .\}$ and $\{v(k + 1), \ k = 0, \ 1, \ . \ . \ .\}$ are each white processes.

For example, let us consider the tracking of a satellite as it passes over the United States, first by a radar tracking station in California and then by one in Florida. As the satellite passes over each station, let us suppose that a number of range measurements (distance from the tracking radar to the satellite) are taken. Unless the two radars are identical to the last detail, we would expect the errors in range measurements made at the California station to be independent of those at the Florida station. On the other hand, it is not likely that the error made in any one of the range measurements at either station is independent of the error in the other range measurements taken at the same station. Thus, it is evident that there exist situations in which measurement errors are correlated in time.

In a similar vein, we would not expect the random wind gusts experienced by an aircraft passing through a storm front to be independent of one another.

The most straightforward approach to follow when $\{w(k), \ k = 0,$ $1, \ . \ . \ .\}$ and $\{v(k + 1), \ k = 0, 1, \ . \ . \ .\}$ are time-correlated processes is to model them as Gauss-Markov sequences. To illustrate the approach, let us consider two examples.

Example 5.5 Suppose that a technician makes an alignment error in the process of mounting an angle sensing device to its base plate. This will cause the device to read consistently "high" or "low" depending on the sign of the mounting alignment error. A constant error such as this is termed a *bias error*.

Assuming that the measurement is a scalar and ignoring all other error sources, the error model is

$$v(k + 1) = v(k)$$

for $k = 0, 1, \ . \ . \ .$, where v is a scalar. If we consider an ensemble of such devices, we could attach a probabilistic description to $v(0)$ and assume that it is a zero mean gaussian random variable with a known variance σ^2. The process $\{v(k + 1), \ k = 0, 1, \ . \ . \ .\}$ is then obviously a zero mean Gauss-Markov sequence with variance $P(k + 1) = \sigma^2, k = 0, 1, \ . \ . \ .$.

Finally, we could go one step further and consider the model where the measurement error $v(k + 1)$ is composed of two components, a bias term and a purely random term. In this case, we model the error as

$$v(k + 1) = v(k) + \eta(k)$$

for $k = 0, 1, \ . \ . \ .$, where we now assume additionally that $\{\eta(k), \ k = 0, 1, \ . \ . \ .\}$ is a scalar zero mean gaussian white sequence which is independent of $v(0)$ for all k and has variance $\sigma_\eta^2(k)$. The process $\{v(k + 1), \ k = 0, 1, \ . \ . \ .\}$

is clearly Gauss-Markov with mean zero and variance

$$P(k + 1) = P(k) + \sigma_\eta{}^2(k)$$

for $k = 0, 1, \ldots$, where the initial condition is $P(0) = \sigma^2$.

Example 5.6 Let us consider a scalar process $\{w(k), k = 0, 1, \ldots\}$, which has zero mean and whose covariance kernel is $P(j,k) = \alpha e^{-|j-k|}$, where $\alpha =$ constant > 0. Clearly, the process is covariance stationary. A sketch of the covariance kernel as a function of $j - k$ is given in Fig. 5.9. Since only the first and second moments of the process are given, we attempt to determine a gaussian sequence with the same properties.

In particular, let us attempt to find a scalar Gauss-Markov sequence of the form

$$w(k + 1) = \Phi(k + 1, k)w(k) + \eta(k) \tag{5.80}$$

with $k = 0, 1, \ldots$, that has the same first and second moments as the given process. First, we choose $w(0)$ to be a zero mean gaussian random variable whose variance is to be determined. Second, we let $\{\eta(k), k = 0, 1, \ldots\}$ be a zero mean gaussian white sequence whose variance is also to be determined.

From Eq. (5.80) and the properties of Gauss-Markov sequences, we have

$$E[w(k + 1)w(k)] = \Phi(k + 1, k)E[w^2(k)] + E[\eta(k)w(k)]$$
$$= \Phi(k + 1, k)E[w^2(k)]$$

From the given data, $E[w(k + 1)w(k)] = P(k + 1, k) = \alpha e^{-1}$ and $E[w^2(k)] = P(k,k) = \alpha$. Hence

$$\alpha e^{-1} = \Phi(k + 1, k)\alpha$$

or

$$\Phi(k + 1, k) = e^{-1}$$

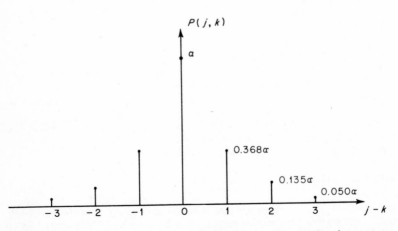

Fig. 5.9 Covariance kernel for time-correlated discrete-time stochastic process.

and our model becomes

$$w(k + 1) = e^{-1}w(k) + \eta(k) \tag{5.81}$$

The corresponding covariance equation is

$$P(k + 1) = e^{-2}P(k) + Q(k)$$

where $Q(k)$ is the variance of $\{\eta(k),\ k = 0,\ 1,\ \ldots\}$. However, $P(k + 1) = P(k + 1, k + 1) = \alpha$ and $P(k) = P(k,k) = \alpha$. Consequently, we see that

$$\alpha = e^{-2}\alpha + Q(k)$$

which means that $\{\eta(k),\ k = 0,\ 1,\ \ldots\}$ is to be modeled as a scalar gaussian white sequence with mean zero and variance $Q(k) = \alpha(1 - e^{-2})$ for all k.

Clearly, $P(0,0) = \alpha$. Hence, we let $w(0)$ be a zero mean gaussian random variable whose variance is α. This completes the specification of the model for the correlated disturbance process.

As a check on our model, let us verify that the Gauss-Markov model has a covariance kernel which is identical to the one given. Since the process must be covariance stationary, we compute $E[w(j)w(0)] = P(j,0)$ for $j = 1, 2, \ldots$. As in Eq. (4.31),

$$w(j) = e^{-j}w(0) + \sum_{i=1}^{j} e^{-(j-i)}\eta(i - 1)$$

and we obtain

$$E[w(j)w(0)] = e^{-j}E[w(0)w(0)] + \sum_{i=1}^{j} e^{-(j-i)}E[\eta(i - 1)w(0)] = \alpha e^{-j}$$

The last term vanishes, since $\eta(i - 1)$, $i = 1, 2, \ldots$, is independent of $w(0)$ by hypothesis.

The correlation coefficient for $w(k + 1)$ and $w(k)$ is obviously

$$\rho = \frac{P(k + 1, k)}{\sqrt{P(k + 1, k + 1)}\ \sqrt{P(k,k)}} = \frac{0.368\alpha}{\sqrt{\alpha}\ \sqrt{\alpha}} = 0.368$$

We note that the process has a constant variance $P(k) = P(k,k) = \alpha$ for all $k = 0, 1, \ldots$. This means that the process is in the steady state.

Assuming now that disturbance and measurement errors can be modeled as Gauss-Markov sequences, let us show how to utilize the results of Theorem 5.5 to perform optimal filtering in such cases.

In order to do this, we make use of the augmented state vector approach [6] which was introduced in Sec. 4.3.

Let us suppose that we have the model

$$y(k + 1) = \Phi_1(k + 1, k)y(k) + \Gamma_1(k + 1, k)w_1(k) \tag{5.82}$$

$$w_1(k + 1) = \Phi_2(k + 1, k)w_1(k) + \Gamma_2(k + 1, k)\eta_1(k) \tag{5.83}$$

$$z(k + 1) = H_1(k + 1)y(k + 1) + v(k + 1) \tag{5.84}$$

$$v(k + 1) = \Phi_3(k + 1, k)v(k) + \Gamma_3(k + 1, k)\eta_2(k) \tag{5.85}$$

for $k = 0, 1, \ldots,$ where y is an n vector, w_1 a p vector, η_1 a q vector, z and v m vectors, and η_2 an r vector. In addition, Φ_1 is $n \times n$, Γ_1 $n \times p$, Φ_2 $p \times p$, Γ_2 $p \times q$, H_1 $m \times n$, Φ_3 $m \times m$, and Γ_3 $m \times r$. We assume that $\{\eta_1(k), k = 0, 1, \ldots\}$ and $\{\eta_2(k), k = 0, 1, \ldots\}$ are zero mean gaussian white sequences with covariance matrices

$$E[\eta_1(j)\eta_1'(k)] = N_{11}(k)\delta_{jk}$$

and

$$E[\eta_2(j)\eta_2'(k)] = N_{22}(k)\delta_{jk}$$

respectively. We allow the possibility that the two processes may be correlated with each other so that

$$E[\eta_1(j)\eta_2'(k)] = N_{12}(k)\delta_{jk}$$

The latter three matrices are $q \times q$, $r \times r$, and $q \times r$, respectively.

We assume further that $y(0)$, $w_1(0)$, and $v(0)$ are each zero mean gaussian random vectors with covariance matrices $P_{yy}(0)$, $P_{ww}(0)$, and $P_{vv}(0)$, respectively. We also allow the possibility that these three vectors may be correlated with cross-covariance matrices

$$E[y(0)w_1'(0)] = P_{yw}(0)$$

$$E[y(0)v'(0)] = P_{yv}(0)$$

$$E[w_1(0)v'(0)] = P_{wv}(0)$$

We observe that $\{w_1(k + 1), k = 0, 1, \ldots\}$ and $\{v(k + 1), k = 0, 1, \ldots\}$ are both Gauss-Markov sequences which "appear" in the measurement $z(k + 1)$, the former indirectly as the forcing function in the equation for $y(k + 1)$, and the latter directly, as is obvious from Eq. (5.84). Hence, we are motivated to make w_1 and v part of the state vector that is to be estimated. We do this by defining $x(k + 1)$ to be the $(n + p + m)$ vector

$$x(k + 1) = \begin{bmatrix} y(k + 1) \\ w_1(k + 1) \\ v(k + 1) \end{bmatrix}$$

Then, Eqs. (5.82), (5.83), and (5.85) can be combined and expressed in the form

$$x(k + 1) = \begin{bmatrix} \Phi_1(k + 1, k) & \Gamma_1(k + 1, k) & 0 \\ 0 & \Phi_2(k + 1, k) & 0 \\ 0 & 0 & \Phi_3(k + 1, k) \end{bmatrix} x(k)$$

$$+ \begin{bmatrix} 0 & 0 \\ \Gamma_2(k + 1, k) & 0 \\ 0 & \Gamma_3(k + 1, k) \end{bmatrix} \begin{bmatrix} \eta_1(k) \\ \eta_2(k) \end{bmatrix} \quad (5.86)$$

where the 0's are null matrices which have the appropriate number of rows and columns so that the indicated multiplications are conformable.

We next let $\Phi(k + 1, k)$ denote the $(n + p + m) \times (n + p + m)$ matrix

$$\Phi(k + 1, k) = \begin{bmatrix} \Phi_1(k + 1, k) & \Gamma_1(k + 1, k) & 0 \\ 0 & \Phi_2(k + 1, k) & 0 \\ 0 & 0 & \Phi_3(k + 1, k) \end{bmatrix}$$

and $\Gamma(k + 1, k)$ the $(n + p + m) \times (q + r)$ matrix

$$\Gamma(k + 1, k) = \begin{bmatrix} 0 & 0 \\ \Gamma_2(k + 1, k) & 0 \\ 0 & \Gamma_3(k + 1, k) \end{bmatrix}$$

Additionally, we let $w(k)$ denote the $(q + r)$ vector

$$w(k) = \begin{bmatrix} \eta_1(k) \\ \eta_2(k) \end{bmatrix}$$

Utilizing the above three definitions, we observe immediately that Eq. (5.86) can be written simply as

$$x(k + 1) = \Phi(k + 1, k)x(k) + \Gamma(k + 1, k)w(k) \tag{5.87}$$

for $k = 0, 1, \ldots$.

From its definition, we see that $x(0)$ is a zero mean gaussian random $(n + p + m)$ vector whose covariance matrix is

$$P(0) = \begin{bmatrix} P_{yy}(0) & P_{yw}(0) & P_{yv}(0) \\ P'_{yw}(0) & P_{ww}(0) & P_{wv}(0) \\ P'_{yv}(0) & P'_{wv}(0) & P_{vv}(0) \end{bmatrix}$$

In addition, it is clear that $\{w(k), k = 0, 1, \ldots\}$ is a $(q + r)$-dimensional gaussian white sequence with mean zero and covariance matrix

$$E[w(j)w'(k)] = Q(k)\delta_{jk} = \begin{bmatrix} N_{11}(k) & N_{12}(k) \\ N'_{12}(k) & N_{22}(k) \end{bmatrix} \delta_{jk}$$

Consequently, the stochastic process $\{x(k + 1), k = 0, 1, \ldots\}$ which is defined by Eq. (5.87) is a zero mean Gauss-Markov sequence.

In terms of the definition of $x(k + 1)$, it is evident that the measurement equation, Eq. (5.84), can be written as

$$z(k + 1) = [H_1(k + 1) \quad I \quad 0]x(k + 1)$$

where I is the $m \times m$ identity matrix and 0 is the $m \times p$ null matrix.

Letting $H(k + 1)$ denote the $m \times (n + m + p)$ matrix

$$H(k + 1) = [H_1(k + 1) \quad I \quad 0]$$

we have

$$z(k + 1) = H(k + 1)x(k + 1) \tag{5.88}$$

We observe now that the system model described by Eqs. (5.87) and (5.88) is the one to which Theorem 5.5 pertains with one exception. There is no additive gaussian white sequence in the measurement equation. Hence, there will be no $R(k + 1)$ matrix in the filter gain and covariance relations.

Applying Theorem 5.5 to the problem at hand, we have

$$\hat{x}(k + 1|k + 1) = \Phi(k + 1, k)\hat{x}(k|k)$$
$$+ K(k + 1)[z(k + 1) - H(k + 1)\Phi(k + 1, k)\hat{x}(k|k)] \tag{5.89}$$

$$K(k + 1) = P(k + 1|k)H'(k + 1)[H(k + 1)P(k + 1|k)H'(k + 1)]^{-1} \tag{5.90}$$

$$P(k + 1|k) = \Phi(k + 1, k)P(k|k)\Phi'(k + 1, k)$$
$$+ \Gamma(k + 1, k)Q(k)\Gamma'(k + 1, k) \tag{5.91}$$

$$P(k + 1|k + 1) = [I - K(k + 1)H(k + 1)]P(k + 1|k) \tag{5.92}$$

where the initial condition on Eq. (5.89) is $\hat{x}(0|0) = 0$, and the initial condition on Eq. (5.91) is $P(0|0) = P(0)$.

Equations (5.89) through (5.92) specify the Kalman filter for problems involving correlated disturbances and measurement errors which can be modeled as Gauss-Markov sequences. There are two important points that should be noted in connection with the result. First, in order to apply Theorem 5.5, it has been necessary to increase the number of elements in the state vector from n to $n + p + m$ even though one may be interested primarily in estimating only the state vector y in Eq. (5.82). This means an increase in the number of computations in both the filter and the gain-covariance relations. For example, suppose that the state vector y is a 10 vector, the disturbance vector is a 3 vector, and the measurement and measurement error vectors are 2 vectors; that is, $n = 10$, $p = 3$, and $m = 2$. If the disturbance and measurement error processes were each white, the gain matrix would be 10×2 and would require 20 elements to specify it, and the two covariance matrices would each be 10×10 and would require $10(11)/2 = 55$ elements each to specify them. In the present case, however, K is $(n + j + m) \times m = 15 \times 2$ with 30 elements, and the covariance matrices are each 15×15 with $15(16)/2 = 120$ elements each. The number of elements in each of the covariance matrices has more than doubled!

Second, the matrix that must be inverted in computing the gain matrix, Eq. (5.90), is $m \times m$, just as in the original formulation where $\{w(k), k = 0, 1, \ldots\}$ and $\{v(k + 1), k = 0, 1, \ldots\}$ are white. However, one difficulty may arise here which is not inherent in the original formulation. Namely, in the latter, the matrix to be inverted is $H(k + 1) \cdot P(k + 1|K)H'(k + 1) + R(k + 1)$. If $R(k + 1)$ is assumed to be positive definite for all k, the inverse always exists. On the other hand, the additive $R(k + 1)$ is not present in Eq. (5.90), and the matrix to be inverted may be singular.

It is clear that this difficulty will not exist if the measurement equation can be expressed as

$$z(k + 1) = H_1(k + 1)y(k + 1) + v(k + 1) + \eta_3(k + 1)$$

where $v(k + 1)$ is the Gauss-Markov process defined by Eq. (5.85), and $\{\eta_3(k + 1), k = 0, 1, \ldots\}$ is a zero mean gaussian white sequence with a covariance matrix $R(k + 1)$ which is positive definite for all k.

If the presence of the gaussian white sequence component $\eta_3(k + 1)$ in the measurement error is not justified, it is necessary to employ the measurement differencing scheme of Bryson and Henrikson [8] which was introduced in Prob. 4.13b. This point is pursued further in Prob. 6.19. In this case, if the system disturbance is a Gauss-Markov sequence, the augmented state vector approach must still be used to define the state vector:

$$x(k + 1) = \begin{bmatrix} y(k + 1) \\ w(k + 1) \end{bmatrix}$$

PROBLEMS

5.1. Specify three admissible loss functions for estimation other than the examples given in Sec. 5.2. Include at least one which is not convex.

5.2. Use Lemma 5.1 to obtain Theorem 5.1.

5.3. Establish the convexity of $F(\xi|z^*)$ which is required in the proof of Corollary 5.1.

5.4. The joint probability density function of two gaussian distributed random variables x_1 and x_2 is

$$f(x_1, x_2) = \frac{1}{2\pi\sigma_1\sigma_2(1 - \rho^2)^{1/2}}$$

$$\exp\left\{ -\frac{\sigma_2^2(x_1 - \bar{x}_1)^2 - 2\rho\sigma_1\sigma_2(x_1 - \bar{x}_1)(x_2 - \bar{x}_2) + \sigma_1^2(x_2 - \bar{x}_2)^2}{2\sigma_1^2\sigma_2^2(1 - \rho)^2} \right\}$$

where $\bar{x}_1 = E(x_1)$, $\bar{x}_2 = E(x_2)$, $\sigma_1^2 = E[(x_1 - \bar{x}_1)^2] > 0$, $\sigma_2^2 = E[(x_2 - \bar{x}_2)^2] > 0$, and $\rho = E[(x_1 - \bar{x}_1)(x_2 - \bar{x}_2)]/\sigma_1\sigma_2$ is the correlation coefficient, $-1 \le \rho \le 1$.

(a) If x_2 is measured with no error, what is the estimate of x_1 that is optimal for all admissible loss functions where the estimation error is $\tilde{x}_1 = x_1 - \hat{x}_1$?

(b) What is the probability density function of the estimation error?

5.5. Derive Eq. (5.40).

5.6. A scalar discrete-time stochastic process $\{x(k),\ k = 0, 1, \ldots\}$ is known to have mean zero and variance $\sigma^2 = $ constant for all k. The process is observed directly in the presence of an additive measurement error v, so that the measurement equation is $z(k + 1) = x(k + 1) + v(k + 1),\ k = 0, 1, \ldots$. The measurement errors are assumed to be independent of each other and of $x(k)$ for all k and to have mean zero and variance $\sigma_v^2 = $ constant. Note that only the first and second moments of the two processes $\{x(k),\ k = 0, 1, \ldots\}$ and $\{v(k + 1),\ k = 0, 1, \ldots\}$ are known.

(a) Devise a Gauss-Markov model of the type described by Eqs. (5.14) and (5.15) in Sec. 5.3 for the above processes.

(b) Develop the Kalman filter equations for the model of part a.

5.7. Consider the system described by the equations

$$x(k + 1) = x(k)$$
$$z(k + 1) = x(k + 1) + v(k + 1)$$

$k = 0, 1, \ldots$, where x is an n vector and $\{v(k + 1),\ k = 0, 1, \ldots\}$ is an n-dimensional zero mean gaussian white sequence with a positive definite covariance matrix $R(k + 1)$ which is finite for all k. Assume that $x(0)$ is a zero mean gaussian random n vector which is independent of $\{v(k + 1),\ k = 0, 1, \ldots\}$. Assume further that there is so much uncertainty associated with $x(0)$ that its covariance matrix $P(0) = E[x(0)x'(0)]$ can be considered as diagonal with arbitrarily large terms; that is, $\sigma_{ii}^2(0) = E[x_i^2(0)] \to \infty$ for $i = 1, 2, \ldots, n$. Note that x is an unknown constant. Apply the Kalman filter algorithm to show that x can be estimated from the measurements z with

$$\hat{x}(1|1) = z(1) \quad \text{— measurement}$$
$$P(1|1) = R(1) \quad \text{— meas. variance}$$

in particular.

5.8. Assume that the system of Eq. (5.14) is subject to a known control input sequence $\{u(k),\ k = 0, 1, \ldots\}$, so that the dynamics are modeled by the relation

$$x(k + 1) = \Phi(k + 1, k)x(k) + \Gamma(k + 1, k)w(k) + \Psi(k + 1, k)u(k)$$

where u is an r vector and Ψ is an $n \times r$ matrix, the control transition matrix. Assume that, aside from the introduction of a known control sequence, the system model is the same as that described in Sec. 5.3, with the measurement system governed by the equation $z(k + 1) = H(k + 1)x(k + 1) + v(k + 1)$.

(a) Show that the optimal predictor for all admissible loss functions is

$$\hat{x}(k|j) = \Phi(k,j)\hat{x}(j|j) + \sum_{i=j+1}^{k} \Phi(k,i)\Psi(i, i - 1)u(i - 1)$$

for all $k > j$ assuming $\hat{x}(j|j)$ is given along with $u(i - 1)$ for $i = j + 1, j + 2, \ldots, k$. Show also that the covariance matrix of the corresponding prediction error $\tilde{x}(k|j) = x(k) - \hat{x}(k|j)$ is given by Eq. (5.34) of Theorem 5.4.

(b) Show that the optimal filter for all admissible loss functions is governed by the relations

$$\hat{x}(k + 1|k + 1) = \hat{x}(k + 1|k) + K(k + 1)[z(k + 1) - H(k + 1)\hat{x}(k + 1|k)]$$
$$\hat{x}(k + 1) = \Phi(k + 1, k)\hat{x}(k|k) + \Psi(k + 1, k)u(k)$$

for $k = 0, 1, \ldots$, where $\hat{x}(0|0) = 0$, and the gain matrix $K(k + 1)$, prediction error covariance matrix $P(k + 1|k)$, and filtering error covariance matrix $P(k + 1|k + 1)$ are governed by Eqs. (5.49) to (5.51), respectively, of Theorem 5.5, with $P(0|0) = P(0)$.

(c) Draw the block diagram of the filter of part b.

5.9. Show how the results of Corollary 5.2 and Theorem 5.5 must be modified to perform discrete optimal linear prediction and filtering in cases where $x(0)$, $\{w(k), k = 0, 1, \ldots\}$, and $\{v(k + 1), k = 0, 1, \ldots\}$ have arbitrary nonzero mean values.

5.10. Show how the initial computational cycle for the Kalman filter and the corresponding gain and covariance relations must be modified if the first measurement is $z(0)$ rather than $z(1)$.

5.11. Suppose that for the system model of Eq. (5.14) the corresponding measurement model is $z(k) = H(k)x(k) + v(k)$ where $k = 0, 1, \ldots$. Take $\{w(k), k = 0, 1, \ldots\}$ and $\{v(k), k = 0, 1, \ldots\}$ to be zero mean gaussian white sequences which are independent of $x(0)$ and have covariance matrices $Q(k)$ and $R(k)$, respectively. Assume that these two sequences are correlated with each other with $p \times m$ cross-covariance matrix $E[w(j)v'(k)] = S(k)\delta_{jk}$. Show that one possible estimation algorithm for this case is

$$\hat{x}(k + 1|k) = \Phi(k + 1, k)\hat{x}(k|k - 1) + K(k)[z(k) - H(k)\hat{x}(k|k - 1)]$$

$$K(k) = [\Phi(k + 1, k)P(k|k - 1)H'(k)$$
$$+ \Gamma(k + 1, k)S(k)][H(k)P(k|k - 1)H'(k) + R(k)]^{-1}$$

$$P(k + 1|k) = \Phi(k + 1, k)\{P(k|k - 1) - [P(k|k - 1)H'(k)$$
$$+ \Gamma(k + 1, k)S(k)][H(k)P(k|k - 1)H'(k) + R(k)]^{-1} \cdot [H(k)P(k|k - 1)$$
$$+ S'(k)\Gamma'(k + 1, k)]\}\Phi'(k + 1, k) + \Gamma(k + 1, k)Q(k)\Gamma'(k + 1, k)$$

for $k = 0, 1, \ldots$, where $\hat{x}(0| - 1) = 0$ and $P(0|-1) = E[x(0)x'(0)]$, [6].

5.12. Consider the nonlinear system

$$x(k + 1) = f[x(k),k] + G[x(k),k]w(k)$$
$$z(k + 1) = h[x(k + 1), k + 1] + v(k + 1)$$

for $k = 0, 1, \ldots$, where x is an n vector, w a p vector, and z and v are m vectors. Let f be an n-dimensional vector-valued function of $x(k)$ and k, h an m-dimensional vector-valued function of $x(k + 1)$ and $k + 1$, and G an $m \times p$ matrix function of $x(k)$ and k. Assume that $\{w(k), k = 0, 1, \ldots\}$ and $\{v(k + 1), k = 0, 1, \ldots\}$ are zero mean gaussian white sequences which are independent of each other and of $x(0)$, and whose covariance matrices are $Q(k)$ and $R(k + 1)$, respectively. Take $x(0)$ to be a zero mean gaussian random n vector with covariance matrix $P(0)$.

(a) Linearize the above system about an assumed nominal $\{x^\circ(k), k = 0, 1, \ldots\}$ and determine the corresponding Kalman filter, including the gain and covariance calculations.

(b) The problem of storing or computing the nominal in part a can be obviated by linearizing at each measurement point about the then available estimate of x. Develop the filter, gain, and covariance equations for this case. These results are an approximate solution of the problem of optimal filtering for the above nonlinear system. The exact solution is unknown; other approximations may be found, for example, in Refs. [1–3] and [8] of Chap. 1.

REFERENCES

1. Sherman, S., Non-Mean-Square Error Criteria, *IRE Trans. Inform. Theory,* vol. IT-4, p. 125, 1958.
2. ———, A Theorem on Convex Sets with Applications, *Ann. Math. Statist.,* vol. 26, p. 763, 1955.
3. Anderson, T. W., The Integral of a Symmetric Unimodal Function Over a Symmetric Convex Set and Some Probability Inequalities, *Proc. Am. Math. Soc.,* vol. 6, p. 170, 1955.
4. Deutsch, R., "Estimation Theory," Prentice-Hall, Inc., Englewood Cliffs, N.J., 1965.
5. Doob, J. L., "Stochastic Processes," John Wiley & Sons, Inc., New York, 1953.
6. Kalman, R. E., New Methods in Wiener Filtering Theory, in J. L. Bogdanoff and F. Kozin, ed., "Proceedings of the First Symposium on Engineering Applications of Random Function Theory and Probability," John Wiley & Sons, Inc., New York, 1963, pp. 270–388. Also, New Methods and Results in Linear Prediction and Filtering Theory, *Tech. Rept.* 61-1, Research Institute for Advanced Studies (RIAS), Martin Company, Baltimore, 1961.
7. ———, A New Approach to Linear Filtering and Prediction Problems, *J. Basic Eng.,* vol. 82, p. 35, 1960.
8. Bryson, A. E., Jr., and L. J. Henrikson, Estimation Using Sampled-Data Containing Sequentially Correlated Noise, *Tech. Rept.* 533, Division of Engineering and Applied Physics, Harvard University, Cambridge, Mass., June, 1967.

6
Optimal Smoothing for Discrete Linear Systems

6.1 INTRODUCTION

We continue our study of optimal estimation for discrete linear systems in this chapter with an examination of the smoothing problem. We recall from Sec. 5.2 that this problem deals with estimates of the system's state which are of the form

$$\hat{x}(k|j) = \phi_k[z(i), \; i = 1, \; . \; . \; . \; , j]$$

where $j > k$; that is, the time at which it is desired to estimate the state lies to the left of the time of the last measurement $z(j)$ on the time scale.

From Corollary 5.1, we know that the smoothed estimate which is optimal for all admissible loss functions is

$$\hat{x}(k|j) = E[x(k)|z(1), \; . \; . \; . \; , z(j)]$$

In addition, since the conditional expectation is unique, we know that all optimal smoothed estimates are also unique.

Our primary interest here is to exploit the above relationship to obtain data smoothing algorithms for the system model in Eqs. (5.14) and (5.15). In particular, we wish to develop algorithms which are recursive in time, thereby permitting us to perform smoothing efficiently with a digital computer. Before attempting this, however, it will prove expedient first to classify smoothed estimates according to possible relationships between the two time indices, k and j. The need for this classification arises because both indices may be variable, or one may be fixed and the other variable. For example, we may seek a smoothing algorithm which is recursive in j for a fixed k, or vice versa. Although numerous classifications are possible, we give only three in Sec. 6.2, motivating each with a discussion of the type of practical problem to which it pertains.

We begin our development of the algorithms for optimal smoothing in Sec. 6.3 by examining the simple problems of single- and double-stage optimal smoothing where we obtain algorithms for $\hat{x}(k|k + 1)$ and $\hat{x}(k|k + 2)$, respectively, for $k = 0, 1, \ldots$. This work forms the basis for our derivations in the sequel, and we proceed, utilizing finite induction, to develop the optimal smoothing algorithms for the first two classifications in Secs. 6.4 and 6.5, respectively.

The optimal smoothing algorithm for the third classification of smoothed estimates is developed in Sec. 6.6 by algebraic manipulation of the optimal algorithm for the other two classifications.

The three optimal smoothing algorithms which are given in this chapter are due to Rauch [1], Carlton [2], and Rauch, et al. [3]. However, the development here is different, and more closely parallels that given by Meditch [4].

It is pointed out in the three theorems here that the smoothing error stochastic processes for all three classifications of smoothed estimates are Gauss-Markov-2 rather than simply Gauss-Markov as in the prediction and filtering problems.

6.2 CLASSIFICATION OF SMOOTHED ESTIMATES

The three classes of smoothed estimates of the state of the system of Eq. (5.14) which we consider in this chapter are:

1. $\hat{x}(k|N)$, $k = 0, 1, \ldots, N - 1$, N = fixed positive integer, termed the fixed-interval smoothed estimate.
2. $\hat{x}(k|j)$, $j = k + 1, k + 2, \ldots$, k = fixed integer, termed the fixed-point or single-point smoothed estimate.
3. $\hat{x}(k|k + N)$, $k = 0, 1, \ldots$, N = fixed positive integer, termed the fixed-lag smoothed estimate.

Let us examine each of these separately in terms of some physical problems.

FIXED-INTERVAL SMOOTHING

Consider an experiment in which a satellite is to be launched from earth and injected into an orbit about the earth. Since it is impossible to fabricate perfect guidance and propulsion systems, we cannot expect the satellite to follow exactly the desired flight path from liftoff through boost and finally into orbit injection. Nevertheless, the resulting orbit is usually sufficiently close for the satellite's intended purpose; if not, it is sometimes possible to make small corrections later in the flight to adjust the orbit. In either event, it is desirable to determine the performance of the guidance and propulsion systems during the initial phases of the flight: liftoff, boost, and orbit injection. The answers to such questions as: (1) "How closely did the satellite follow the desired flight path?", (2) "What error sources within the guidance system contributed significantly to deviations from the desired flight conditions?", and (3) "Over what time intervals did the propulsion system provide too much or too little thrust, and what was the extent of these thrust deviations?" provide not only an assessment of the overall mission, but also point out weaknesses which may be corrected by suitable redesign.

To answer these questions and others like them, telemetry and tracking data, taken during the flight, are recorded and processed. Since the data are subject to errors, it is usually necessary to use some estimation scheme to reduce the effects of these errors.

By linearizing the model for the flight, including trajectory, guidance system, and propulsion system dynamics, and the model for the telemetry and tracking system measurements, about some appropriate set of nominal conditions, it is usually possible to obtain a linearized representation of the form in Eqs. (5.14) and (5.15). Then it is clear that the optimal filtering algorithm of Theorem 5.5 can be used to obtain filtered estimates of the parameters of interest.

Let us suppose that during liftoff, boost, and orbit injection, we obtain measurements at the N time points t_k, $k = 1, \ldots, N$, which are not necessarily equally spaced in time. Utilizing optimal filtering, we then determine $\hat{x}(k|k)$, $k = 0, 1, \ldots, N$, the optimal estimate of $x(k)$ based on k measurements.

Assume now that we have recorded the set of measurements $\{z(j), j = 1, \ldots, N\}$. With the experiment completed and the optimal filtered estimates computed, we note that for $k = 0, 1, \ldots, N - 1$, we have measurement data related to $x(k)$, not only up to time k but also beyond. We thus wonder whether or not it is possible to "refine" the set of filtered estimates $\{\hat{x}(k|k), k = 0, 1, \ldots, N\}$ utilizing the additionally available measurement data. For example, with $N = 200$ and $k = 51$, we pose the question, "Is the optimal smoothed estimate of $x(51)$, based on the 200 measurements which are available, a 'better' estimate of the state at that time than the filtered estimate $\hat{x}(51|51)$?" Intuitively, we feel that it should be, since one of the estimates is based on only 51 pieces of data, whereas the other is based on 200. Although there is no absolute a priori guarantee that $\hat{x}(51|200)$ is a "better" estimate of $x(51)$ than is $\hat{x}(51|51)$, we are nonetheless motivated by our intuition and curiosity to examine the question which we posed above in a general way.

Basically, our situation is the following. With the experiment completed, we have measurement data over the *fixed interval* $[0,N]$. For each time point k within the interval, we wish to obtain the optimal estimate of the state $x(k)$ which is based on *all* the available measurement data $\{z(j), j = 1, 2, \ldots, N\}$. From Corollary 5.1, we know that the estimate which we seek is

$$\hat{x}(k|N) = E[x(k)|z(1), \ldots, z(N)] \tag{6.1}$$

where $k = 0, 1, \ldots, N$. This estimate is termed the optimal fixed-interval smoothed estimate of $x(k)$. Our problem is to obtain an efficient procedure for computing it.

Of all the estimates which are defined by Eq. (6.1), we recognize only $\hat{x}(N|N)$ as having appeared in our previous work. We shall find that this estimate constitutes the "starting point" for optimal fixed-interval smoothing, and that the appropriate algorithm is "backward" recursive in time.

On two occasions above, we have used the phrase "with the experiment completed." This points out the postexperimental nature of fixed-interval smoothing. As we might expect, this means that such smoothing cannot be carried out on-line during an experiment, as can optimal filtering. Very simply then, fixed-interval smoothing is of use, following an experiment, in an attempt to refine the filtered estimates of the state. The reader can readily visualize numerous instances in which such data smoothing would be desirable, e.g., assessing the performance of aircraft following various flight tests, or of chemical processes during development of prototype systems, or, in the postexperimental processing of seismic data for geologic studies.

FIXED-POINT SMOOTHING

Returning to the satellite launching example which we gave above, we remark that one very important and critical point in the mission is the time at which the satellite is injected into orbit after which it is in a state of "free fall." If, as is common, the boost and orbit injection phases of the flight are executed with a multistage rocket, the time points at which one stage burns out and another ignites are also critical times.

Let us suppose now that after orbit injection the satellite is tracked by a ground-based radar tracking network, say for 10 to 15 orbits, and that we wish to use these additional data to obtain, and hopefully improve, the estimate of $x(N)$, that is, the state at the time of injection. Up to the injection point, we have $\hat{x}(N|N)$ from the measurements $\{z(j), j = 1, \ldots, N\}$. What we wish to do now is to determine $\hat{x}(N|N + 1)$, $\hat{x}(N|N + 2)$, etc. We see here that our concern is with an estimate of the system's state at a fixed point in time which is based not only on measurements up to that time, but additionally, on measurements taken beyond it.

Speaking more generally, we wish to determine

$$\hat{x}(k|j) = E[x(k)|z(1), \ldots, z(j)] \tag{6.2}$$

for some fixed k with $j > k$. This equation defines what we call the optimal fixed-point smoothed estimate. Since the time of the last measurement to be taken may not be known a priori, we allow it to be free in Eq. (6.2).

As in our previous work, we seek a recursive algorithm for implementing the computation which is defined by Eq. (6.2). In Sec. 6.5, we shall develop such an algorithm which can also be used on-line if desired.

Such algorithms are useful in numerous situations. In addition to the example above, we include the problem of determining the initial concentrations of the reacting substances in a chemical process from, say, temperature, pressure, and concentration measurements taken during the reaction.

FIXED–LAG SMOOTHING

As the third and final classification of optimal smoothed estimates, we consider estimates which are given by the relation

$$\hat{x}(k|k + N) = E[x(k)|z(1), \ldots, z(k), z(k + 1), \ldots, z(k + N)] \tag{6.3}$$

where $k = 0, 1, \ldots$, and N is some positive integer. Equation (6.3) defines what is called the optimal fixed-lag smoothed estimate, the name arising from the fact that the time point at which we seek the estimate of the system's state lags the time point of the most recent measurement by a fixed interval of time N; that is, $t_{k+N} - t_k = \text{constant} > 0$ for all $k = 0, 1, \ldots$.

This type of estimate is a natural extension of the optimal filtered estimate $\hat{x}(k|k)$, where the time at which the estimate of the state is desired and the time of the most recent measurement are coincident; i.e., estimate and measurement "run along together." Alternately, the optimal fixed-lag smoothed estimate is the antithesis of what we call the optimal fixed-lead predicted estimate,

$$\hat{x}(k + N|k) = \Phi(k + N, k)\hat{x}(k|k)$$

where $k = 0, 1, \ldots$, and N = positive integer, wherein the time of the optimal estimate leads the time of the most recent measurement by a fixed amount.

The intuitive justification for examining this class of smoothed estimates is the same as for the other two classifications. Namely, we feel that a smoothed estimate should be "better" than a predicted or filtered one.

In Sec. 6.6, we shall obtain a forward-time recursive algorithm for optimal fixed-lag smoothing. This algorithm can be used on-line in estimation problems where a lag between the measurements and the estimates is permissible. For example, let us consider a communication system in which the transmitted signals are sample functions of a stochastic process which is defined by Eq. (5.14) and in which the received signal can be modeled by Eq. (5.15) with the measurement error due to atmospheric and antenna noise. Logically, we would use optimal filtering in an attempt to extract the transmitted signal from the received one. In principle, we could do this on-line, and would have a current estimate of the transmitted information. On the other hand, if we could tolerate a fixed delay between the received signal and the estimated state, we could utilize fixed-lag smoothing to determine $\hat{x}(k|k + N)$, which we intuitively feel is "better" than $\hat{x}(k|k)$.

In general, optimal fixed-lag smoothing appears attractive primarily in communication and telemetry data reduction problems, its analogy to optimal filtering being obvious.

6.3 SINGLE- AND DOUBLE-STAGE OPTIMAL SMOOTHING

SINGLE-STAGE OPTIMAL SMOOTHING

We examine first the problem of obtaining an algorithm for computing $\hat{x}(k|k + 1)$, the optimal single-stage smoothed estimate for a given $k = 0$, $1, \ldots$. We proceed in a fashion essentially identical to our development of Theorem 5.5 for optimal filtering.

As in the proof of Theorem 5.5, we take $\{z(1), \ldots , z(k), \tilde{z}(k + 1|k)\}$ as our set of measurements, recalling from Eqs. (5.54) and

(5.55) that

$$\tilde{z}(k + 1|k) = z(k + 1) - \hat{z}(k + 1|k)$$
$$= z(k + 1) - H(k + 1)\hat{x}(k + 1|k)$$
$$= H(k + 1)\tilde{x}(k + 1|k) + v(k + 1) \qquad (6.4)$$

From Corollary 5.1 and the fact that $\tilde{z}(k + 1|k)$ is independent of the set of measurements $\{z(1), \ldots, z(k)\}$, it follows that

$$\hat{x}(k|k + 1) = E[x(k)|z(1), \ldots, z(k), \tilde{z}(k + 1|k)]$$
$$= E[x(k)|z(1), \ldots, z(k)] + E[x(k)|\tilde{z}(k + 1|k)]$$

The first term on the right-hand side of this equation is obviously $\hat{x}(k|k)$, while the second term, since $x(k)$ and $\tilde{z}(k + 1|k)$ are jointly gaussian, each with zero mean, is

$$P_{x\tilde{z}}P_{\tilde{z}\tilde{z}}^{-1}\tilde{z}(k + 1|k)$$

where $P_{x\tilde{z}} = E[x(k)\tilde{z}'(k + 1|k)]$ and $P_{\tilde{z}\tilde{z}} = E[\tilde{z}(k + 1|k)\tilde{z}'(k + 1|k)]$. Hence,

$$\hat{x}(k|k + 1) = \hat{x}(k|k) + P_{x\tilde{z}}P_{\tilde{z}\tilde{z}}^{-1}\tilde{z}(k + 1|k) \qquad (6.5)$$

Let us now evaluate the $n \times m$ "gain" matrix in Eq. (6.5). Utilizing Eq. (6.4), we have

$$P_{x\tilde{z}} = E[x(k)\tilde{x}'(k + 1|k)]H'(k + 1) + E[x(k)v'(k + 1)]$$

From Eq. (5.27), the second term is zero for all k. Then, substituting

$$\tilde{x}(k + 1|k) = \Phi(k + 1, k)\tilde{x}(k|k) + \Gamma(k + 1, k)w(k) \qquad (6.6)$$

into this expression and recalling from Eq. (5.25) that $E[x(k)w'(k)] = 0$ for all k, we have

$$P_{x\tilde{z}} = E[x(k)\tilde{x}'(k|k)]\Phi'(k + 1, k)H'(k + 1) \qquad (6.7)$$

From the definition of filtering error, we have $x(k) = \tilde{x}(k|k) + \hat{x}(k|k)$. However, from Eq. (5.10), $E[\hat{x}(k|k)\tilde{x}'(k|k)] = 0$, and Eq. (6.7) reduces to

$$P_{x\tilde{z}} = E[\tilde{x}(k|k)\tilde{x}'(k|k)]\Phi'(k + 1, k)H'(k + 1)$$
$$= P(k|k)\Phi'(k + 1, k)H'(k + 1) \qquad (6.8)$$

where $P(k|k)$ is the optimal filtering error covariance matrix.

Parenthetically, we remark that, as part of our development here, we have shown that

$$E[x(k)\tilde{x}'(k + 1|k)] = P(k|k)\Phi'(k + 1, k) \tag{6.9}$$

a result which will prove quite useful in the sequel.

From Eq. (5.59), we recall that

$$P_{\tilde{z}\tilde{z}} = E[\tilde{z}(k + 1|k)\tilde{z}'(k + 1|k)]$$
$$= H(k + 1)P(k + 1|k)H'(k + 1) + R(k + 1) \tag{6.10}$$

where $P(k + 1|k)$ is the optimal prediction error covariance matrix.

Now letting $M(k|k + 1)$ denote the smoothing filter gain matrix $P_{x\tilde{z}}P_{\tilde{z}\tilde{z}}^{-1}$ where the two time arguments refer to the estimate and measurement time points, respectively, we have from Eqs. (6.8) and (6.10) that

$$M(k|k + 1) = P(k|k)\Phi'(k + 1, k)H'(k + 1)$$
$$[H(k + 1)P(k + 1|k)H'(k + 1) + R(k + 1)]^{-1} \tag{6.11}$$

Equation (6.5) can now be written

$$\hat{x}(k|k + 1) = \hat{x}(k|k)$$
$$+ M(k|k + 1)[z(k + 1) - H(k + 1)\Phi(k + 1, k)\hat{x}(k|k)] \tag{6.12}$$

for $k = 0, 1, \ldots$, with $\hat{x}(0|0) = \bar{x}(0) = 0$ as the initial condition.

Equation (6.12) gives the desired algorithm for optimal single-stage smoothing. We observe that its structure is the same as that of the Kalman filter, a result which is not surprising. Here, the correction (gain-times-residual) term is added to the optimal filtered estimate instead of to the optimal predicted estimate as in the Kalman filter. The gain matrix $M(k|k + 1)$ is, of course, different from $K(k + 1)$. We shall see below that one particular interpretation of Eq. (6.11) shows that $K(k + 1)$ is a factor in $M(k|k + 1)$.

We note from Eq. (6.11) that in contrast to the optimal filter computations where the gain matrix $K(k + 1)$ is a function of the prediction or filtering error covariance matrices—see Eqs. (5.49) and (5.78)—$M(k|k + 1)$ is independent of the single-stage smoothing error covariance matrix $P(k|k + 1) = E[\tilde{x}(k|k + 1)\tilde{x}'(k|k + 1)]$. Apparently then, determination of $P(k|k + 1)$ is a separate computation. We shall not establish here an expression for $P(k|k + 1)$, since it will be a special case of the general results which we develop later.

It is clear that the smoothing filter requires $\hat{x}(k|k)$ as an input for each k. This means that optimal single-stage smoothing cannot be carried out unless accompanied by optimal filtering. Additionally, it is

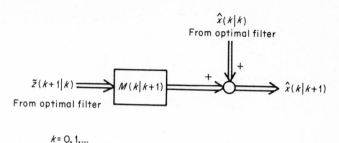

Fig. 6.1 Block diagram of optimal single-stage smoothing filter.

seen that computation of $M(k|k+1)$ requires $P(k|k)$ and $P(k+1|k)$ from the filtering solution.   Hence, the optimal single-stage smoothing filter is heavily dependent upon the optimal filter.

A block diagram for the smoothing filter of Eq. (6.12) is shown in Fig. 6.1.

With respect to the three smoothing estimate classifications in Sec. 6.2, we see that Eq. (6.12) gives the following:

1. Only one of the fixed-interval smoothed estimates.   That is, for $k = N - 1$, it gives only $\hat{x}(N - 1|N)$.
2. The fixed-point smoothed estimate for only one measurement beyond the fixed point.   That is, it gives $\hat{x}(k|j)$ only for $j = k + 1$.
3. All fixed-lag smoothed estimates for unit delay.   That is, it gives all $\hat{x}(k|k + N)$ where $N = 1$ and $k = 0, 1, \ldots$.

For our purposes in the sequel, it is necessary that we express Eqs. (6.11) and (6.12) in a different way.   To do this, we note that under the assumption that $P(k+1|k)$ is nonsingular, we have the result

$$H'(k+1)[H(k+1)P(k+1|k)H'(k+1) + R(k+1)]^{-1}$$
$$= P^{-1}(k+1|k)K(k+1) \quad (6.13)$$

from Eq. (5.49).   It then follows that Eq. (6.11) can be expressed as

$$M(k|k+1) = A(k)K(k+1) \tag{6.14}$$

where $A(k)$ is defined by

$$A(k) = P(k|k)\Phi'(k+1, k)P^{-1}(k+1|k)$$

The fact that $K(k+1)$ is a factor in $M(k|k+1)$ is obvious from Eq. (6.14).

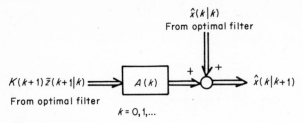

Fig. 6.2 Alternate block diagram for optimal single-stage smoothing filter.

Substituting into Eq. (6.12) from Eq. (6.14), we have

$$\hat{x}(k|k + 1) = \hat{x}(k|k) + A(k)K(k + 1)[z(k + 1)$$
$$- H(k + 1)\Phi(k + 1, k)\hat{x}(k|k)] \quad (6.15)$$

Since

$$K(k + 1)[z(k + 1) - H(k + 1)\Phi(k + 1, k)\hat{x}(k|k)]$$
$$= K(k + 1)\tilde{z}(k + 1|k)$$

is directly available from the optimal filter, the block diagram for the optimal single-stage smoothing algorithm of Eq. (6.15) can be represented as in Fig. 6.2.

Going one step further, we have from Eq. (5.48) of Theorem 5.5 that

$$K(k + 1)[z(k + 1) - H(k + 1)\Phi(k + 1, k)\hat{x}(k|k)]$$
$$= \hat{x}(k + 1|k + 1) - \Phi(k + 1, k)\hat{x}(k|k)$$
$$= \hat{x}(k + 1|k + 1) - \hat{x}(k + 1|k) \quad (6.16)$$

so that Eq. (6.15) can also be written

$$\hat{x}(k|k + 1) = \hat{x}(k|k) + A(k)[\hat{x}(k + 1|k + 1) - \hat{x}(k + 1|k)] \quad (6.17)$$

for $k = 0, 1, \ldots$, where

$$A(k) = P(k|k)\Phi'(k + 1, k)P^{-1}(k + 1|k) \quad (6.18)$$

For computational purposes, it is clear that the formulation of Eqs. (6.11) and (6.12) is preferable to the one of Eqs. (6.17) and (6.18), since the former avoids the computation of $P^{-1}(k + 1|k)$, which is inherent in the latter formulation. However, the latter is of primary utility in the sequel.

DOUBLE-STAGE OPTIMAL SMOOTHING

Continuing now to the determination of $\hat{x}(k|k + 2)$, we take $\{z(1), \ldots, z(k + 1), \tilde{z}(k + 2|k + 1)\}$ as the set of measurements, and by a direct application of Corollary 5.1 have

$$\hat{x}(k|k+2) = E[x(k)|z(1), \ldots, z(k+1), \tilde{z}(k+2|k+1)]$$

Since $\tilde{z}(k+2|k+1)$ is independent of the set of measurements $\{z(1), \ldots, z(k+1)\}$, we can write

$$\hat{x}(k|k+2) = E[x(k)|z(1), \ldots, z(k+1)]$$
$$+ E[x(k)|\tilde{z}(k+2|k+1)]$$

which reduces to

$$\hat{x}(k|k+2) = \hat{x}(k|k+1) + P_{x\tilde{z}}P_{\tilde{z}\tilde{z}}^{-1}\tilde{z}(k+2|k+1) \tag{6.19}$$

where now

$$P_{x\tilde{z}} = E[x(k)\tilde{z}'(k+2|k+1)]$$

and

$$P_{\tilde{z}\tilde{z}} = E[\tilde{z}(k+2|k+1)\tilde{z}'(k+2|k+1)]$$

Evaluation of $P_{\tilde{z}\tilde{z}}$ is trivial, requiring only that we increase each time index in Eq. (6.10) by one to obtain

$$P_{\tilde{z}\tilde{z}} = H(k+2)P(k+2|k+1)H'(k+2) + R(k+2) \tag{6.20}$$

The evaluation of $P_{x\tilde{z}}$, although not trivial, is relatively straightforward. From Eq. (6.4),

$$\tilde{z}(k+2|k+1) = H(k+2)\tilde{x}(k+2|k+1) + v(k+2)$$

so that

$$P_{x\tilde{z}} = E[x(k)\tilde{x}'(k+2|k+1)]H'(k+2) + E[x(k)v'(k+2)]$$

The second term on the right-hand side of this equation is zero by virtue of Eq. (5.27), and we have

$$P_{x\tilde{z}} = E[x(k)\tilde{x}'(k+2|k+1)]H'(k+2) \tag{6.21}$$

Increasing all the time indices in Eq. (6.6) by one, we obtain

$$\tilde{x}(k+2|k+1) = \Phi(k+2, k+1)\tilde{x}(k+1|k+1)$$
$$+ \Gamma(k+2, k+1)w(k+1) \tag{6.22}$$

It now follows from Eqs. (6.21) and (6.22), and the fact that

$$E[x(k)w'(k+1)] = 0$$

for all k [see Eq. (5.25)], that

$$P_{x\tilde{z}} = E[x(k)\tilde{x}'(k+1|k+1)]\Phi'(k+2, k+1)H'(k+2) \tag{6.23}$$

for $k = 0, 1, \ldots$.

But from Eq. (5.61),

$$\tilde{x}(k + 1|k + 1) = [I - K(k + 1)H(k + 1)]\tilde{x}(k + 1|k) \\ - K(k + 1)v(k + 1)$$

Substituting this result into Eq. (6.23) and taking note again from Eq. (5.27) that $E[x(k)v'(k + 1)] = 0$ for all k, we see that

$$P_{x\tilde{z}} = E[x(k)\tilde{x}'(k + 1|k)][I - K(k + 1)H(k + 1)]'\Phi'(k + 2, k + 1) \\ H'(k + 2) \quad (6.24)$$

Substituting into Eq. (6.24) for $E[x(k)\tilde{x}'(k + 1|k)]$ from Eq. (6.9), we have

$$P_{x\tilde{z}} = P(k|k)\Phi'(k + 1, k)[I - K(k + 1)H(k + 1)]'\Phi'(k + 2, k + 1) \\ H'(k + 2) \quad (6.25)$$

It now follows from the definition of M and Eqs. (6.20) and (6.25) that

$$M(k|k + 2) = P_{x\tilde{z}}P_{\tilde{z}\tilde{z}}^{-1} \\ = P(k|k)\Phi'(k + 1, k)[I - K(k + 1)H(k + 1)]' \\ \Phi'(k + 2, k + 1)H'(k + 2) \\ [H(k + 2)P(k + 2|k + 1)H'(k + 2) + R(k + 2)]^{-1} \\ (6.26)$$

Assuming that $P^{-1}(k + 1|k)$ exists, we obtain from Eq. (5.51) the result

$$[I - K(k + 1)H(k + 1)] = P(k + 1|k + 1)P^{-1}(k + 1|k)$$

and subsequently the result

$$[I - K(k + 1)H(k + 1)]' = P^{-1}(k + 1|k)P(k + 1|k + 1) \quad (6.27)$$

where we have made use of the fact that the two covariance matrices are symmetric.

Increasing each of the time indices in Eq. (6.13) by one, we have

$$H'(k + 2)[H(k + 2)P(k + 2|k + 1)H'(k + 2) + R(k + 2)]^{-1} \\ = P^{-1}(k + 2|k + 1)K(k + 2) \quad (6.28)$$

under the assumption that $P(k + 2|k + 1)$ is nonsingular.

Substituting Eqs. (6.27) and (6.28) into Eq. (6.26), we see that

$$M(k|k + 2) = P(k|k)\Phi'(k + 1, k)P^{-1}(k + 1|k)P(k + 1|k + 1) \\ \Phi'(k + 2, k + 1)P^{-1}(k + 2|k + 1)K(k + 2)$$

From Eq. (6.18), it is clear that the first three factors in this expression are equal to $A(k)$ and the second three are equal to $A(k + 1)$. Consequently,

$$M(k|k + 2) = A(k)A(k + 1)K(k + 2) \tag{6.29}$$

Returning now to Eq. (6.19) and expanding it, we have

$$\hat{x}(k|k + 2) = \hat{x}(k|k + 1) + M(k|k + 2)[z(k + 2)$$
$$- H(k + 2)\Phi(k + 2, k + 1)\hat{x}(k + 1|k + 1)] \tag{6.30}$$

As expected, $\hat{x}(k|k + 2)$ is obtained by adding a correction term to $\hat{x}(k|k + 1)$.

Proceeding as in the single-stage problem, we note that

$$K(k + 2)[z(k + 2) - H(k + 2)\Phi(k + 2, k + 1)\hat{x}(k + 1|k + 1)]$$
$$= \hat{x}(k + 2|k + 2) - \hat{x}(k + 2|k + 1)$$

Hence, Eq. (6.30) can be rewritten in the form

$$\hat{x}(k|k + 2) = \hat{x}(k|k + 1)$$
$$+ A(k)A(k + 1)[\hat{x}(k + 2|k + 2) - \hat{x}(k + 2|k + 1)] \tag{6.31}$$

Substituting into Eq. (6.31) for $\hat{x}(k|k + 1)$ from Eq. (6.17), and noting also from Eq. (6.17) with k replaced by $k + 1$ that

$$A(k + 1)[\hat{x}(k + 2|k + 2) - \hat{x}(k + 2|k + 1)]$$
$$= \hat{x}(k + 1|k + 2) - \hat{x}(k + 1|k + 1)$$

we have

$$\hat{x}(k|k + 2) = \hat{x}(k|k) + A(k)[\hat{x}(k + 1|k + 1) - \hat{x}(k + 1|k)]$$
$$+ A(k)[\hat{x}(k + 1|k + 2) - \hat{x}(k + 1|k + 1)]$$

which simplifies to

$$\hat{x}(k|k + 2) = \hat{x}(k|k) + A(k)[\hat{x}(k + 1|k + 2) - \hat{x}(k + 1|k)] \tag{6.32}$$

for $k = 0, 1, \ldots$, where $A(k)$ is given by Eq. (6.18).

We have already shown how the single-stage smoothing results which we obtained could be used to perform data smoothing for the three classifications in Sec. 6.2. Let us now do the same for the double-stage results.

1. Setting $k = N - 2$ in Eq. (6.32), we have

$$\hat{x}(N - 2|N) = \hat{x}(N - 2|N - 2)$$
$$+ A(N - 2)[\hat{x}(N - 1|N) - \hat{x}(N - 1|N - 2)]$$

which in conjunction with single-stage computation of $\hat{x}(N - 1|N)$ gives the optimal fixed-interval smoothed estimates $\hat{x}(k|N)$ for $k = N - 1$ and $N - 2$. Since $\hat{x}(N - 1|N)$ must be determined before $\hat{x}(N - 2|N)$ can be, we see the reverse-time recursive nature of fixed-interval smoothing.

2. For optimal fixed-point smoothing, we see that $\hat{x}(k|k + 2)$ can be obtained using $\hat{x}(k|k + 1)$, the single-stage result, and Eq. (6.31). Hence, at this point we have $\hat{x}(k|j)$ for $j = k + 1$ and $k + 2$ with the computations proceeding recursively forward in time.

3. Equation (6.32) gives the optimal fixed-lag smoothed estimate for two units of delay assuming that the optimal fixed-lag smoothed estimate for one unit of delay, $\hat{x}(k + 1|k + 2)$, has previously been determined for $k = 0, 1, \ldots$. We observe that the computations here are forward-time recursive.

Before proceeding to the general results, there is one important point which should be made. Namely, the algorithms which we are discussing here in terms of the three classifications of smoothed estimates are equivalent. That is, one can be derived from either of the other two by vector-matrix manipulations and suitable interpretation of the relations which are developed. That this is so should be no surprise since we are merely dealing with three particular classifications within the framework of a general problem of optimal smoothing.

6.4 OPTIMAL FIXED-INTERVAL SMOOTHING

OPTIMAL SMOOTHING FILTER

Proceeding now by finite induction, we have for any integer $N - 1 \geq k + 2$ from Eq. (6.32) that

$$\hat{x}(k|N - 1) = \hat{x}(k|k) + A(k)[\hat{x}(k + 1|N - 1) - \hat{x}(k + 1|k)] \quad (6.33)$$

where $A(k)$ is given by Eq. (6.18). We seek an expression for $\hat{x}(k|N)$.

Taking $\{z(1), \ldots, z(N - 1), \tilde{z}(N|N - 1)\}$ as the set of measurements, applying Corollary 5.1, and noting that $\tilde{z}(N|N - 1)$ is independent of all the other measurements, we have

$$\hat{x}(k|N) = E[x(k)|z(1), \ldots, z(N - 1), \tilde{z}(N|N - 1)]$$
$$= E[x(k)|z(1), \ldots, z(N - 1)] + E[x(k)|\tilde{z}(N|N - 1)]$$
$$= \hat{x}(k|N - 1) + P_{x\tilde{z}}P_{\tilde{z}\tilde{z}}^{-1}\tilde{z}(N|N - 1) \quad (6.34)$$

where

$$P_{x\tilde{z}} = E[x(k)\tilde{z}'(N|N - 1)]$$

and

$$P_{\tilde{z}\tilde{z}} = E[\tilde{z}(N|N - 1)\tilde{z}'(N|N - 1)]$$

Replacing k by $N - 1$ in Eq. (6.10), we have immediately that

$$P_{\tilde{z}\tilde{z}} = H(N)P(N|N - 1)H'(N) + R(N) \tag{6.35}$$

Evaluation of $P_{x\tilde{z}} = E[x(k)\tilde{z}'(N|N - 1)]$ simply requires a systematic repetition of the steps used to evaluate

$$P_{x\tilde{z}} = E[x(k)\tilde{z}'(k + 2|k + 1)]$$

in the double-stage smoothing problem. We first note that

$$\tilde{z}(N|N - 1) = H(N)\tilde{x}(N|N - 1) + v(N)$$

and, since $x(k)$ and $v(N)$ are, by Eq. (5.27), uncorrelated for all k and N, we obtain the result

$$P_{x\tilde{z}} = E[x(k)\tilde{x}'(N|N - 1)]H'(N) \tag{6.36}$$

Further, replacing k by $N - 1$ in Eq. (6.6), noting from Eq. (5.25) that $x(k)$ and $w(N - 1)$ are uncorrelated for all $N - 1 \geq k$, we have

$$P_{x\tilde{z}} = E[x(k)\tilde{x}'(N - 1|N - 1)]\Phi'(N, N - 1)H'(N) \tag{6.37}$$

With k replaced by $N - 2$, Eq. (5.61) becomes

$$\tilde{x}(N - 1|N - 1) = [I - K(N - 1)H(N - 1)]\tilde{x}(N - 1|N - 2) \\ - K(N - 1)v(N - 1) \tag{6.38}$$

However, $x(k)$ and $v(N - 1)$ are uncorrelated for all k and $N - 1$ as a consequence of Eq. (5.27). Hence, substitution of Eq. (6.38) into Eq. (6.37) yields

$$P_{x\tilde{z}} = E[x(k)\tilde{x}'(N - 1|N - 2)] \\ [I - K(N - 1)H(N - 1)]'\Phi'(N, N - 1)H'(N) \tag{6.39}$$

From Eq. (6.27), we have

$$[I - K(N - 1)H(N - 1)]' = P^{-1}(N - 1|N - 2)P(N - 1|N - 1)$$

under the assumption that $P(N - 1|N - 2)$ is nonsingular. Hence,

$$P_{x\tilde{z}} = E[x(k)\tilde{x}'(N - 1|N - 2)] \\ P^{-1}(N - 1|N - 2)P(N - 1|N - 1)\Phi'(N, N - 1)H'(N) \tag{6.40}$$

A reexamination of the steps in Eqs. (6.36) through (6.40) and a comparison of Eqs. (6.36) and (6.40) reveals that

$$E[x(k)\tilde{x}'(N|N - 1)] = E[x(k)\tilde{x}'(N - 1|N - 2)]$$
$$P^{-1}(N - 1|N - 2)P(N - 1|N - 1)\Phi'(N, N - 1) \quad (6.41)$$

Combining Eqs. (6.35) and (6.40) to obtain the smoothing filter gain matrix in Eq. (6.34), we have

$$P_{x\tilde{z}}P_{\tilde{z}\tilde{z}}^{-1} = M(k|N)$$
$$= E[x(k)\tilde{x}'(N - 1|N - 2)]P^{-1}(N - 1|N - 2)P(N - 1|N - 1)$$
$$\Phi'(N, N - 1)H'(N)[H(N)P(N|N - 1)H'(N) + R(N)]^{-1} \quad (6.42)$$

However, for $k = N - 1$, Eq. (6.13) is

$$H'(N)[H(N)P(N|N - 1)H'(N) + R(N)]^{-1} = P^{-1}(N|N - 1)K(N)$$

assuming that $P(N|N - 1)$ is nonsingular. It then follows that Eq. (6.42) reduces to

$$M(k|N) = E[x(k)\tilde{x}'(N - 1|N - 2)]$$
$$P^{-1}(N - 1|N - 2)P(N - 1|N - 1)$$
$$\Phi'(N, N - 1)P^{-1}(N|N - 1)K(N)$$
$$= E[x(k)\tilde{x}'(N - 1|N - 2)]$$
$$P^{-1}(N - 1|N - 2)A(N - 1)K(N) \quad (6.43)$$

where we have made use of the definition of the gain matrix A in Eq. (6.18).

If the steps which led to Eq. (6.41) are repeated for $E[x(k)\tilde{x}'(N - 1|N - 2)]$, there results

$$E[x(k)\tilde{x}'(N - 1|N - 2)] = E[x(k)\tilde{x}'(N - 2|N - 3)]$$
$$P^{-1}(N - 2|N - 3)P(N - 2|N - 2)\Phi'(N - 1, N - 2)$$

Substitution into Eq. (6.43) gives

$$M(k|N) = E[x(k)\tilde{x}'(N - 2|N - 3)]$$
$$P^{-1}(N - 2|N - 3)P(N - 2|N - 2)\Phi'(N - 1, N - 2)$$
$$P^{-1}(N - 1|N - 2)A(N - 1)K(N)$$
$$= E[\tilde{x}(k)\tilde{x}'(N - 2|N - 3)]$$
$$P^{-1}(N - 2|N - 3)A(N - 2)A(N - 1)K(N)$$

Continuing in this way, we arrive at the point

$$M(k|N) = E[x(k)\tilde{x}'(k+1|k)]$$
$$P^{-1}(k+1|k)A(k+1) \cdots A(N-2)A(N-1)K(N)$$

Substituting into this result from Eq. (6.9) and recalling Eq. (6.18) once more, we have

$$M(k|N) = P(k|k)\Phi'(k+1,k)$$
$$P^{-1}(k+1|k)A(k+1) \cdots A(N-1)K(N)$$
$$= A(k)A(k+1) \cdots A(N-1)K(N) \qquad (6.44)$$

Returning to the smoothing filter equation, Eq. (6.34), and substituting into it from Eqs. (6.33) and (6.44), the latter for the gain matrix $P_{x\tilde{z}}P_{\tilde{z}\tilde{z}}^{-1}$, we obtain

$$\hat{x}(k|N) = \hat{x}(k|k) + A(k)[\hat{x}(k+1|N-1) - \hat{x}(k+1|k)]$$
$$+ A(k)A(k+1) \cdots A(N-1)K(N)\tilde{z}(N|N-1)$$
$$= \hat{x}(k|k) - A(k)\hat{x}(k+1|k) + A(k)[\hat{x}(k+1|N-1)$$
$$+ A(k+1) \cdots A(N-1)K(N)\tilde{z}(N|N-1)] \qquad (6.45)$$

We assert that the term in brackets on the right-hand side of Eq. (6.45) is simply $\hat{x}(k+1|N)$. To show this, we replace k by $k+1$ in Eq. (6.34) and obtain

$$\hat{x}(k+1|N) = \hat{x}(k+1|N-1) + P_{x\tilde{z}}P_{\tilde{z}\tilde{z}}^{-1}\tilde{z}(N|N-1) \qquad (6.46)$$

where

$$P_{x\tilde{z}} = E[x(k+1)\tilde{z}'(N|N-1)]$$

and

$$P_{\tilde{z}\tilde{z}} = E[\tilde{z}(N|N-1)\tilde{z}'(N|N-1)]$$

Noting that now $P_{x\tilde{z}}P_{\tilde{z}\tilde{z}}^{-1} = M(k+1|N)$ and utilizing Eq. (6.44) with k replaced by $k+1$, we have

$$P_{x\tilde{z}}P_{\tilde{z}\tilde{z}}^{-1} = M(k+1|N)$$
$$= A(k+1)A(k+2) \cdots A(N-1)K(N)$$

in Eq. (6.46) and our assertion is proved.

Consequently, we have from Eq. (6.45) that, for a fixed value of N and for $k = 0, 1, \ldots, N-1$,

$$\hat{x}(k|N) = \hat{x}(k|k) - A(k)\hat{x}(k+1|k) + A(k)\hat{x}(k+1|N)$$
$$= \hat{x}(k|k) + A(k)[\hat{x}(k+1|N) - \hat{x}(k+1|k)] \qquad (6.47)$$

We recall again from Eq. (6.18) that

$$A(k) = P(k|k)\Phi'(k + 1, k)P^{-1}(k + 1|k)$$

Equation (6.47) specifies the algorithm for optimal fixed-interval smoothing. At each k, it requires the optimal filtered and predicted estimates $\hat{x}(k|k)$ and $\hat{x}(k + 1|k)$, respectively, as input.

We note that the algorithm is a system of n first-order difference equations in $\hat{x}(k|N)$ and $\hat{x}(k + 1|N)$.

The fact that computations must be carried out in reverse time becomes apparent when we attempt to find an appropriate "initial" condition for Eq. (6.47). We see that only for $k = N - 1$ within the interval $[0,N]$ do we have all of the information which we need. Specifically, for $k = N - 1$, Eq. (6.47) becomes

$$\hat{x}(N - 1|N) = \hat{x}(N - 1|N - 1) + A(N - 1)[\hat{x}(N|N) - \hat{x}(N|N - 1)]$$

Having so determined $\hat{x}(N - 1|N)$, we then proceed successively to determine $\hat{x}(N - 2|N)$, $\hat{x}(N - 3|N)$, . . . , $\hat{x}(0|N)$, in that order. Hence, the indexing on k in Eq. (6.47) is $k = N - 1, N - 2, . . . , 0$, and the boundary condition is the optimal filtered estimate $\hat{x}(N|N)$. The sequence of computations is indicated schematically on the time scale in Fig. 6.3. It is apparent that optimal fixed-interval smoothing does not have the potential for on-line data processing that optimal filtering does.

The block diagram for the optimal fixed-interval smoothing filter is given in Fig. 6.4. We see that the "residual" is the difference between the optimal fixed-interval smoothed estimate $\hat{x}(k + 1|N)$ and the optimal predicted estimate $\hat{x}(k + 1|k)$. At each time point k, we observe that the correction term $A(k)[\hat{x}(k + 1|N) - \hat{x}(k + 1|k)]$ is added to the optimal filtered estimate $\hat{x}(k|k)$ to obtain $\hat{x}(k|N)$. Since $A(k)$ is the only gain matrix in Eq. (6.47), we hereafter refer to it as the optimal fixed-interval smoothing filter gain matrix.

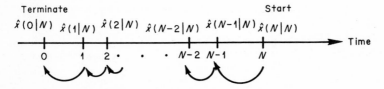

Fig. 6.3 Computational sequence for optimal fixed-interval smoothing.

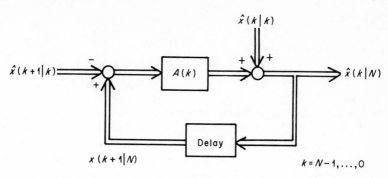

Fig. 6.4 Block diagram of optimal fixed-interval smoothing filter.

SMOOTHING ERROR PROCESS AND COVARIANCE MATRIX

From the definition of estimation error, we have.that the optimal fixed-interval smoothing error is

$$\tilde{x}(k|N) = x(k) - \hat{x}(k|N)$$

Substituting into this definition from Eq. (6.47), we obtain

$$\tilde{x}(k|N) = \tilde{x}(k|k) - A(k)[\hat{x}(k+1)|N) - \hat{x}(k+1|k)] \qquad (6.48)$$

for $k = N - 1, \ldots, 0$.

Also, from the definition of estimation error, we have

$$\hat{x}(k+1|N) = x(k+1) - \tilde{x}(k+1|N)$$

and

$$\hat{x}(k+1|k) = x(k+1) - \tilde{x}(k+1|k)$$

from which it is obvious that

$$\hat{x}(k+1|N) - \hat{x}(k+1|k) = \tilde{x}(k+1|k) - \tilde{x}(k+1|N)$$

Substituting this result into Eq. (6.48), we see that

$$\tilde{x}(k|N) = \tilde{x}(k|k) + A(k)[\tilde{x}(k+1|N) - \tilde{x}(k+1|k)] \qquad (6.49)$$

Since

$$\tilde{x}(k+1|k) = \Phi(k+1, k)\tilde{x}(k|k) + \Gamma(k+1, k)w(k)$$

it is clear that

$$
\begin{aligned}
\tilde{x}(k|N) &= A(k)\tilde{x}(k+1|N) + \tilde{x}(k|k) - A(k)\tilde{x}(k+1|k) \\
&= A(k)\tilde{x}(k+1|N) + [I - A(k)\Phi(k+1, k)]\tilde{x}(k|k) \\
&\qquad\qquad\qquad - A(k)\Gamma(k+1, k)w(k) \quad (6.50)
\end{aligned}
$$

where I is the $n \times n$ identity matrix.

Let us examine the stochastic process $\{\tilde{x}(k|N), \ k = N, \ N - 1, \ \ldots, 0\}$ which is defined by Eq. (6.50). We note from the equation that $\tilde{x}(k|N)$ can be viewed as the output of a linear first-order system operating in reverse time and that the system is subject to two forcing functions. The first of these is the filtering error process $\{\tilde{x}(k|k), \ k = 0, 1, \ \ldots\}$ which we know is a zero mean Gauss-Markov sequence. The second forcing function is the zero mean gaussian white sequence $\{w(k), \ k = 0, 1, \ \ldots\}$ which is independent of the filtering error process. Since $\{\tilde{x}(k|k), \ k = 0, 1, \ \ldots\}$ is Gauss-Markov, it follows that the reverse-time process $\{\tilde{x}(k|k), \ k = N - 1, N - 2, \ \ldots, 0\}$ is also.

The boundary condition for the system in Eq. (6.50) is $\tilde{x}(N|N)$, which is a zero mean gaussian random n vector. The block diagram for the system is given in Fig. 6.5.

For such a system, it was shown in Sec. 4.3 that the output, $\tilde{x}(k|N)$ in this case, is a zero mean Gauss-Markov-2 sequence.

Since $\{\tilde{x}(k|N), \ k = N, \ N - 1, \ \ldots, 0\}$ is not Gauss-Markov, it is not completely characterized by its mean (which is zero) and its covariance matrix $P(k|N) = E[\tilde{x}(k|N)\tilde{x}'(k|N)]$. It is clear from Sec. 4.3 that a complete description of the process requires second-order probability functions.

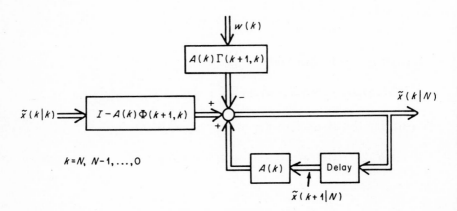

Fig. 6.5 Block diagram representation of optimal fixed-interval smoothing error process.

Despite the fact that $P(k|N)$ is only "part of the story" in the complete description of the stochastic process $\{\tilde{x}(k|N), k = N, N - 1, \ldots, 0\}$, it is of considerable use. In particular, it completely describes the probability distribution of the optimal fixed-interval smoothing error for each $k \in [0,N]$. If, for example, $P(k|N)$ is nonsingular, we have the probability density function

$$f[\tilde{x}(k|N)] = \frac{1}{\sqrt{(2\pi)^n|P(k|N)|}} \exp\left[-\tfrac{1}{2}\tilde{x}'(k|N)P^{-1}(k|N)\tilde{x}(k|N)\right]$$

of the smoothing error. More simply, the diagonal elements of $P(k|N)$ are the variances of the elements of the smoothing error vector $\tilde{x}(k|N)$. These can be compared with the respective elements of $P(k|k)$ to obtain, easily and quickly, a picture of the accuracy of optimal fixed-interval smoothing relative to optimal filtering. Hence, rather than obtaining a complete description of the stochastic process $\{\tilde{x}(k|N), k = N, N - 1, \ldots, 0\}$, we shall obtain only an expression for its covariance matrix.

To do this, we first write Eq. (6.48) in the form

$$\tilde{x}(k|N) + A(k)\hat{x}(k + 1|N) = \tilde{x}(k|k) + A(k)\hat{x}(k + 1|k) \tag{6.51}$$

From Eq. (5.10), we have

$$E[\tilde{x}(k|N)\hat{x}'(k + 1|N)] = 0$$

and

$$E[\tilde{x}(k|k)\hat{x}'(k + 1|k)] = 0$$

for all $k = 0, 1, \ldots, N$. Hence, postmultiplying each side of Eq. (6.51) by its transpose and taking the expected value of the result, we obtain

$$P(k|N) + A(k)P_{\hat{x}\hat{x}}(k + 1|N)A'(k) = P(k|k) + A(k)P_{\hat{x}\hat{x}}(k + 1|k)A'(k)$$

or equivalently

$$P(k|N) = P(k|k) + A(k)[P_{\hat{x}\hat{x}}(k + 1|k) - P_{\hat{x}\hat{x}}(k + 1|N)]A'(k) \tag{6.52}$$

where we have defined

$$P_{\hat{x}\hat{x}}(k + 1|k) = E[\hat{x}(k + 1|k)\hat{x}'(k + 1|k)]$$

and

$$P_{\hat{x}\hat{x}}(k + 1|N) = E[\hat{x}(k + 1|N)\hat{x}'(k + 1|N)]$$

Since $x(k + 1) = \hat{x}(k + 1|k) + \tilde{x}(k + 1|k)$, and since

$$E[\hat{x}(k + 1|k)\tilde{x}'(k + 1|k)] = 0$$

for all $k = 0, 1, \ldots$ by Eq. (5.10), it follows that

$$P(k + 1) = P_{\hat{x}\hat{x}}(k + 1|k) + P(k + 1|k) \tag{6.53}$$

where $P(k + 1) = E[x(k + 1)x'(k + 1)]$.

Similarly, since $x(k + 1) = \hat{x}(k + 1|N) + \tilde{x}(k + 1|N)$, it is clear that

$$P(k + 1) = P_{\hat{x}\hat{x}}(k + 1|N) + P(k + 1|N) \tag{6.54}$$

From Eqs. (6.53) and (6.54), we have

$$P_{\hat{x}\hat{x}}(k + 1|k) + P(k + 1|k) = P_{\hat{x}\hat{x}}(k + 1|N) + P(k + 1|N)$$

which means that

$$P_{\hat{x}\hat{x}}(k + 1|k) - P_{\hat{x}\hat{x}}(k + 1|N) = P(k + 1|N) - P(k + 1|k) \tag{6.55}$$

Substituting this result into Eq. (6.52), we obtain the result

$$P(k|N) = P(k|k) + A(k)[P(k + 1|N) - P(k + 1|k)]A'(k) \tag{6.56}$$

Equation (6.56) is a first-order $n \times n$ matrix difference equation for the optimal fixed-interval smoothing error covariance matrix. Its recursive nature as well as its dependence on input data, viz., $P(k|k)$ and $P(k + 1|k)$, from the optimal filtering solution are apparent.

The appropriate indexing for Eq. (6.56) is obviously $k = N - 1$, $N - 2, \ldots, 0$, and the boundary condition is $P(k + 1|N) = P(N|N)$ for $k = N - 1$.

This completes our development for optimal fixed-interval smoothing, and we summarize our results.

Theorem 6.1

a. *The optimal fixed-interval smoothed estimate $\hat{x}(k|N)$ is governed by the system of equations*

$$\hat{x}(k|N) = \hat{x}(k|k) + A(k)[\hat{x}(k + 1|N) - \hat{x}(k + 1|k)] \tag{6.57}$$

for $k = N - 1, N - 2, \ldots, 0$ where $\hat{x}(N|N)$ is the boundary condition for $k = N - 1$.

b. *$A(k)$ is the $n \times n$ smoothing filter gain matrix, which is given by the relation*

$$A(k) = P(k|k)\Phi'(k + 1, k)P^{-1}(k + 1|k) \tag{6.58}$$

c. *The optimal fixed-interval smoothing error stochastic process $\{\tilde{x}(k|N), k = N, N - 1, \ldots, 0\}$ is a zero mean Gauss-Markov-2 sequence whose covariance matrix is given by the system of equations*

$$P(k|N) = P(k|k) + A(k)[P(k + 1|N) - P(k + 1|k)]A'(k) \tag{6.59}$$

for $k = N - 1, N - 2, \ldots, 0$, *subject to the boundary condition* $P(N|N)$ *for* $k = N - 1$.

Example 6.1   In order to illustrate optimal fixed-interval smoothing in a very simple way and obtain at the same time a comparison of the relative accuracies of smoothing and filtering, let us consider the problem of optimal fixed-interval smoothing for the system in Example 5.4 with $N = 4$.

To review briefly, we have the scalar system $x(k + 1) = x(k) + w(k)$ with the scalar measurement $z(k + 1) = x(k + 1) + v(k + 1)$ and $P(0) = 100$, $Q(k) = 25$, and $R(k + 1) = 15$.

Since $\Phi(k + 1, k) = 1$, it follows that Eq. (6.57) becomes

$$\hat{x}(k|4) = \hat{x}(k|k) + A(k)[\hat{x}(k + 1|4) - \hat{x}(k|k)]$$

for $k = 3, 2, 1, 0$ in this example.   Also, from Example 5.4, $P(k + 1|k) = P(k|k) + 25$, so that Eq. (6.58) is

$$A(k) = P(k|k)P^{-1}(k + 1|k) = \frac{P(k|k)}{P(k + 1|k)} = \frac{P(k|k)}{P(k|k) + 25}$$

here.

By direct substitution into Eq. (6.59), we have

$$P(k|4) = P(k|k) + \left[\frac{P(k|k)}{P(k + 1|k)} \right]^2 [P(k + 1|4) - P(k + 1|k)]$$

for $k = 3, 2, 1, 0$.

Utilizing these last two expressions, we compute $A(k)$ and $P(k|N)$ for $k = 3, 2, 1, 0$ and present them in the accompanying table, together with the data which were computed in Example 5.4.   The three estimation error variances

k	$P(k|k - 1)$	$P(k|k)$	$P(k|4)$	$K(k)$	$A(k)$
0	. . . . .	100	26.25	. . . . .	0.800
1	125	13.40	9.84	0.893	0.349
2	38.4	10.80	8.30	0.720	0.302
3	35.8	10.57	8.37	0.704	0.324
4	35.6	10.55	10.55	0.703	. . . . .

are given in adjacent columns for easy comparison.

Figure 6.6 is a block diagram for the optimal smoothing filter.

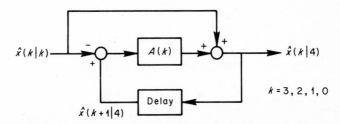

Fig. 6.6   Optimal fixed-interval smoothing filter for Example 6.1.

Example 6.2 Consider the problem of optimal fixed-interval smoothing for those systems in which there is no system disturbance, that is $Q(k) = 0$ for all $k = 0, 1, \ldots$. Then, $P(k + 1|k) = \Phi(k + 1, k)P(k|k)\Phi'(k + 1, k)$ and Eq. (6.58) becomes

$$A(k) = P(k|k)\Phi'(k + 1, k)[\Phi'(k + 1, k)]^{-1}P^{-1}(k|k)\Phi^{-1}(k + 1, k) = \Phi(k, k + 1)$$

Substituting this result and $\hat{x}(k + 1|k) = \Phi(k + 1, k)\hat{x}(k|k)$ into Eq. (6.57), and simplifying, we obtain

$$\hat{x}(k|N) = \hat{x}(k|k) + \Phi(k, k + 1)[\hat{x}(k + 1|N) - \Phi(k + 1, k)\hat{x}(k|k)]$$
$$= \Phi(k, k + 1)\hat{x}(k + 1|N)$$

from which it follows that

$$\hat{x}(k|N) = \Phi(k,N)\hat{x}(N|N)$$

In this case, it is clear that the smoothed estimate is obtained from the optimal filtered estimate at the end of the interval $[0,N]$ by a "backward" extrapolation of the latter estimate through the state transition matrix.

The corresponding estimation error covariance matrix is seen to be

$$P(k|N) = P(k|k) + \Phi(k, k + 1)[P(k + 1|N)$$
$$- \Phi(k + 1, k)P(k|k)\Phi'(k + 1, k)]\Phi'(k, k + 1)$$
$$= \Phi(k, k + 1)P(k + 1|N)\Phi'(k, k + 1)$$

or equivalently

$$P(k|N) = \Phi(k,N)P(N|N)\Phi'(k,N)$$

6.5 OPTIMAL FIXED-POINT SMOOTHING

OPTIMAL SMOOTHING FILTER

Replacing N by j in Eq. (6.34) and recalling that the gain matrix in that equation is $M(k|N)$, we have

$$\hat{x}(k|j) = \hat{x}(k|j - 1) + M(k|j)\tilde{z}(j|j - 1) \tag{6.60}$$

However, from Eq. (6.44),

$$M(k|j) = \left[\prod_{i=k}^{j-1} A(i)\right] K(j) \tag{6.61}$$

and from Eq. (6.16),

$$K(j)\tilde{z}(j|j - 1) = \hat{x}(j|j) - \hat{x}(j|j - 1)$$

Substituting these two results into Eq. (6.60) and defining

$$B(j) = \prod_{i=k}^{j-1} A(i) \tag{6.62}$$

we obtain the algorithm

$$\hat{x}(k|j) = \hat{x}(k|j-1) + B(j)[\hat{x}(j|j) - \hat{x}(j|j-1)] \tag{6.63}$$

for optimal fixed-point smoothing where $j = k+1, k+2, \ldots$. The $n \times n$ matrix $B(j)$ which is defined in Eq. (6.62) is called the optimal fixed-point smoothing filter gain matrix. The algorithm is obviously forward-time recursive in j. Additionally, the gain matrix can be determined recursively by noting that Eq. (6.62) can be written

$$B(j) = B(j-1)A(j-1) \tag{6.64}$$

for $j = k+2, k+3, \ldots$, where

$$A(j-1) = P(j-1|j-1)\Phi'(j, j-1)P^{-1}(j|j-1)$$

and

$$B(k+1) = A(k)$$

is the initial condition.

Computations are initiated at $j = k+1$, where we obtain

$$\hat{x}(k|k+1) = \hat{x}(k|k) + B(k+1)[\hat{x}(k+1|k+1) - \hat{x}(k+1|k)]$$

from Eq. (6.63). The appropriate initial condition for the computations is obviously $\hat{x}(k|j-1) = \hat{x}(k|k)$ for $j = k+1$.

The block diagram for the optimal fixed-point smoothing filter of Eq. (6.63) is shown in Fig. 6.7. The dependence on optimal filtering and prediction for input data is quite evident.

An alternate form of the algorithm which takes advantage of the fact that at each j the quantity $K(j)\tilde{z}(j|j-1)$ is directly available from the optimal filter follows immediately from Eqs. (6.60), (6.61), and (6.62).

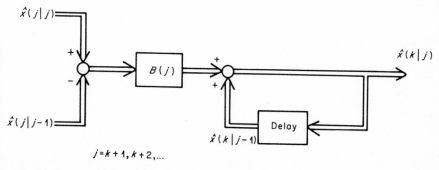

Fig. 6.7 Block diagram of optimal fixed-point smoothing filter.

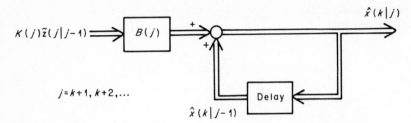

Fig. 6.8 Block diagram of alternate optimal fixed-point smoothing filter.

We obtain the relation

$$\hat{x}(k|j) = \hat{x}(k|j-1) + B(j)K(j)\tilde{z}(j|j-1) \tag{6.65}$$

for $j = k+1, k+2, \ldots$. The block diagram is given in Fig. 6.8.

We shall have occasion later to reexamine this alternate algorithm at length.

SMOOTHING ERROR PROCESS AND COVARIANCE MATRIX

Analysis of the optimal fixed-point smoothing error proceeds in essentially the same way as that for the optimal fixed-interval smoothing error, which was presented in Sec. 6.4.

From the definition of estimation error and Eq. (6.63), we have

$$
\begin{aligned}
\tilde{x}(k|j) &= x(k) - \hat{x}(k|j) \\
&= x(k) - \hat{x}(k|j-1) - B(j)[\hat{x}(j|j) - \hat{x}(j|j-1)] \\
&= \tilde{x}(k|j-1) - B(j)[\hat{x}(j|j) - \hat{x}(j|j-1)] \tag{6.66}
\end{aligned}
$$

As in Sec. 6.4,

$$\hat{x}(j|j) = x(j) - \tilde{x}(j|j)$$

and

$$\hat{x}(j|j-1) = x(j) - \tilde{x}(j|j-1)$$

so that

$$\hat{x}(j|j) - \hat{x}(j|j-1) = \tilde{x}(j|j-1) - \tilde{x}(j|j) \tag{6.67}$$

From Eq. (5.61),

$$\tilde{x}(j|j) = [I - K(j)H(j)]\tilde{x}(j|j-1) - K(j)v(j) \tag{6.68}$$

and it follows that

$$\tilde{x}(j|j-1) - \tilde{x}(j|j) = K(j)H(j)\tilde{x}(j|j-1) + K(j)v(j) \tag{6.69}$$

As a consequence of Eqs. (6.67) and (6.69), it is clear that Eq. (6.66) can be written

$$\tilde{x}(k|j) = \tilde{x}(k|j-1) - B(j)K(j)H(j)\tilde{x}(j|j-1) - B(j)K(j)v(j) \tag{6.70}$$

for $j = k + 1, k + 2, \dots$.

We observe that Eq. (6.70) is the model for a linear first-order system which has the zero mean, Gauss-Markov sequence $\{\tilde{x}(j|j-1),$ $j = k + 1, k + 2, \dots\}$ and the zero mean gaussian white sequence $\{v(j), j = k + 1, k + 2, \dots\}$ as inputs. Since the initial condition $\tilde{x}(k|k)$ is a zero mean gaussian random n vector, it follows that $\{\tilde{x}(k|j),$ $j = k, k + 1, \dots\}$ is a zero mean Gauss-Markov-2 sequence.

An alternate way of exhibiting this result is to note that

$$\tilde{x}(j|j-1) = \Phi(j,j-1)\tilde{x}(j-1|j-1) + \Gamma(j,j-1)w(j-1) \tag{6.71}$$

Reducing the time index in Eq. (6.68) by one and substituting the result in Eq. (6.71), we obtain

$$\tilde{x}(j|j-1) = \Phi(j,j-1)[I - K(j-1)H(j-1)]\tilde{x}(j-1|j-2)$$
$$- \Phi(j,j-1)K(j-1)v(j-1) + \Gamma(j,j-1)w(j-1) \tag{6.72}$$

for $j = k + 1, k + 2, \dots$. The initial condition for Eq. (6.72) is the zero mean gaussian random n vector $\tilde{x}(k|k-1)$.

The block diagram for the system which is represented by Eqs. (6.70) and (6.72) is shown in Fig. 6.9. The system output is obviously $\tilde{x}(k|j)$ and the inputs are the zero mean gaussian white sequences $\{w(j-1), j = k + 1, k + 2, \dots\}, \{v(j-1), j = k + 1, k + 2, \dots\},$

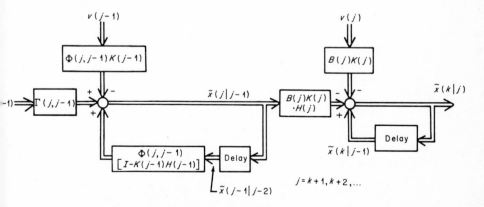

Fig. 6.9 Block diagram representation of optimal fixed-point smoothing error process.

and $\{v(j),\ j = k + 1,\ k + 2,\ \ldots\}$. The presence of the two delay operations shows that the system is second-order, and since the initial condition vectors $\tilde{x}(k|k - 1)$ and $\tilde{x}(k|k)$ are zero mean, gaussian, and independent of the inputs, we have the desired result; i.e., the process is Gauss-Markov-2.

As in the fixed-interval smoothing problem, we shall concern ourselves here only with determining a relation for the estimation error covariance matrix.

To do this, we proceed as before by first rewriting Eq. (6.66) as

$$\tilde{x}(k|j) + B(j)\hat{x}(j|j) = \tilde{x}(k|j - 1) + B(j)\hat{x}(j|j - 1)$$

Then, since $\tilde{x}(k|j)$ and $\hat{x}(j|j)$, and $\tilde{x}(k|j - 1)$ and $\hat{x}(j|j - 1)$, are, respectively, independent as a consequence of Eq. (5.10), we have

$$P(k|j) + B(j)P_{\hat{x}\hat{x}}(j|j)B'(j) = P(k|j - 1)$$
$$+ B(j)P_{\hat{x}\hat{x}}(j|j - 1)B'(j) \quad (6.73)$$

where $P(k|j) = E[\tilde{x}(k|j)\tilde{x}'(k|j)]$ and $P(k|j - 1) = E[\tilde{x}(k|j - 1)\tilde{x}'(k|j - 1)]$ are the fixed-point smoothing error covariance matrices, and $P_{\hat{x}\hat{x}}(j|j) = E[\hat{x}(j|j)\hat{x}'(j|j)]$ and $P_{\hat{x}\hat{x}}(j|j - 1) = E[\hat{x}(j|j - 1)\hat{x}(j|j - 1)]$.

Rearranging the terms in Eq. (6.73), we write

$$P(k|j) = P(k|j - 1) + B(j)[P_{\hat{x}\hat{x}}(j|j - 1) - P_{\hat{x}\hat{x}}(j|j)]B'(j) \quad (6.74)$$

Replacing k by $j - 1$ and N by j in Eq. (6.55), we see that

$$P_{\hat{x}\hat{x}}(j|j - 1) - P_{\hat{x}\hat{x}}(j|j) = P(j|j) - P(j|j - 1)$$

Consequently, Eq. (6.74) becomes

$$P(k|j) = P(k|j - 1) + B(j)[P(j|j) - P(j|j - 1)]B'(j) \quad (6.75)$$

for $j = k + 1, k + 2, \ldots$, where $B(j)$ is given by Eq. (6.62), and $P(j|j)$ and $P(j|j - 1)$ are the error covariance matrices for optimal filtering and prediction, respectively. The initial condition for Eq. (6.75) is obviously $P(k|k)$.

Noting from Eq. (5.51) that

$$P(j|j) - P(j|j - 1) = -K(j)H(j)P(j|j - 1)$$

we can also express Eq. (6.75) as

$$P(k|j) = P(k|j - 1) - B(j)K(j)H(j)P(j|j - 1)B'(j)$$

At this point, it is convenient for us to summarize the above results.

Theorem 6.2

    a. *The optimal fixed-point smoothed estimate $\hat{x}(k|j)$ is given by the system of equations*

$$\hat{x}(k|j) = \hat{x}(k|j-1) + B(j)[\hat{x}(j|j) - \hat{x}(j|j-1)] \tag{6.76}$$

    for a fixed k and $j = k+1,\ k+2,\ \ldots,$ where the initial condition is $\hat{x}(k|k)$.

    b. *$B(j)$ is the $n \times n$ smoothing filter gain matrix, and is given by the relations*

$$B(j) = \prod_{i=k}^{j-1} A(i) \tag{6.77}$$

    and

$$A(i) = P(i|i)\Phi'(i+1,\ i)P^{-1}(i+1|i) \tag{6.78}$$

    for $j = k+1,\ k+2,\ \ldots$.

    c. *The optimal fixed-point smoothing error stochastic process $\{\tilde{x}(k|j),\ j = k,\ k+1,\ \ldots\}$ is a zero mean Gauss-Markov-2 sequence whose covariance matrix is governed by the systems of equations*

$$P(k|j) = P(k|j-1) + B(j)[P(j|j) - P(j|j-1)]B'(j) \tag{6.79}$$

    or

$$P(k|j) = P(k|j-1) - B(j)K(j)H(j)P(j|j-1)B'(j) \tag{6.80}$$

    for $j = k+1,\ k+2,\ \ldots,$ subject to the initial condition $P(k|k)$.

An alternate formulation is, of course, also possible by replacing Eq. (6.76) by Eq. (6.65) and Eq. (6.77) by Eq. (6.64). In either case, the algorithm possesses the undesirable attribute of requiring the computation of the inverse of the prediction error covariance matrix as one of the steps in determining the smoothing filter gain matrix. This feature was also present in the optimal fixed-interval smoothing filter algorithm— see Eq. (6.58). In the latter algorithm, errors made in computing the matrix inverse at any time point affect the gain matrix computation only at that time point. However, in the former case, the error effect is cumulative, since the gain matrix is a continuing product. For example, if an error in computing $P^{-1}(k+1|k)$ were made at only one value of $k \in [0,N]$, the fixed-interval smoothing filter gain matrix would be incorrect only at that k. On the other hand, if the same error were made at one value of $i > k$, the value of $B(j)$ would be in error for all $j > k$, as is clear from Eqs. (6.77) and (6.78). Fortunately, it is possible to develop an algorithm for optimal fixed-point smoothing which avoids computation of $P^{-1}(i + 1|i)$. We present this result which is due to Fraser [5] as a corollary to Theorem 6.2.

Corollary 6.1

a. *The optimal fixed-point smoothed estimate is governed by the system of equations*

$$\hat{x}(k|j) = \hat{x}(k|j-1) + W(j)H'(j)R^{-1}(j)[z(j)$$
$$- H(j)\Phi(j,\, j-1)\hat{x}(j-1|j-1)] \quad (6.81)$$

where $j = k + 1,\, k + 2,\, \ldots$ and the initial condition is $\hat{x}(k|k)$.

b. *$W(j)$ is the $n \times n$ smoothing filter gain matrix which is determined recursively from the relation*

$$W(j) = W(j-1)\Phi'(j,\, j-1)[I - S(j)P(j|j)] \quad (6.82)$$

for $j = k + 1,\, k + 2,\, \ldots,$ where the initial condition is

$$W(k) = P(k|k)$$

and $S(j) = H'(j)R^{-1}(j)H(j)$.

c. *The stochastic process $\{\tilde{x}(k|j),\, j = k,\, k + 1,\, \ldots\}$ is a zero mean Gauss-Markov-2 process whose covariance matrix satisfies*

$$P(k|j) = P(k|j-1) - W(j)[S(j)P(j|j-1)S(j) + S(j)]W'(j)$$
$$(6.83)$$

for $j = k + 1,\, k + 2,\, \ldots,$ where the initial condition is $P(j|j)$.

Proof The proof, which follows readily from Theorem 6.2. is left as an exercise.

Essentially, the requirement of computing $P^{-1}(j + 1|j)$ has been replaced by that of computing $R^{-1}(j)$. Since the number of measurement variables is usually less than the number of state variables, the desirabil-

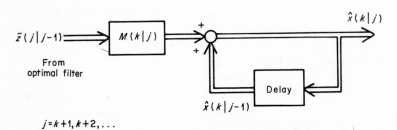

$j = k + 1,\, k + 2, \ldots$

Fig. 6.10 Block diagram for optimal fixed-point smoothing algorithm of Eq. (6.81).

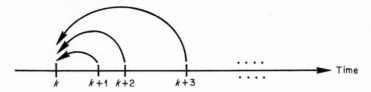

Fig. 6.11   Notion of "reflection" of measurement data in optimal fixed-point smoothing.

ity of this algorithm is evident.   However, it is now necessary that $R(j)$ be positive definite for all j.

Noting from Eqs. (6.60) and (6.81) that $M(k|j) = W(j)H'(j)R^{-1}(j)$, the block diagram for the algorithm in Eq. (6.81) assumes the form which is shown in Fig. 6.10.

Discussion   We observe that optimal fixed-point smoothing consists, at each time point, of adding a correction term to the estimate $\hat{x}(k|j-1)$ which was obtained at the preceding measurement time.   The correction term is the familiar product of a gain matrix $M(k|j)$ and the measurement residual $\tilde{z}(j|j-1)$.   The effect of this term is to "reflect" the "new" information about $x(k)$ back to the time point k.   The basic idea is illustrated in Fig. 6.11.

It is clear that optimal fixed-point smoothing can be used on-line in conjunction with optimal filtering and prediction.   Thus, fixed-point smoothing is not restricted to postexperimental data processing as is fixed-interval smoothing.   If desired, however, fixed-point smoothing could obviously be done postexperimentally.   In either event, it is not necessary to know the time point of the final measurement a priori, since the algorithm proceeds recursively forward in time and can be stopped whenever desired.

Example 6.3   Let us consider the problem of optimal fixed-point smoothing to obtain a "refined" estimate of the initial condition for the system in Example 6.1. In that example, we recall that $P(0|0) = 100$ and that by fixed-interval smoothing, we had obtained the result $P(0|4) = 26.25$, an obviously significant reduction in the uncertainty associated with the initial condition.   We now determine, in addition, $P(0|1)$, $P(0|2)$, and $P(0|3)$.

For this purpose, we shall use the algorithm in Corollary 6.1 and obtain also $W(j)$ and $M(0|j) = W(j)H'(j)R^{-1}(j)$ for $j = 1, 2, 3, 4$.

Recalling from Example 6.1 that $\Phi(j+1, j) = 1$, $H(j) = 1$ and $R(j) = 15$, we have first that $S(j) = H'(j)R^{-1}(j)H(j) = \frac{1}{15}$ and Eqs. (6.82) and (6.83) become

$$W(j) = [1 - \tfrac{1}{15}P(j|j)]W(j-1)$$

and

$$P(0|j) = P(0|j-1) - \tfrac{1}{15}W^2(j)[\tfrac{1}{15}P(j|j-1) + 1]$$

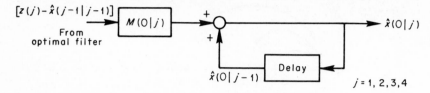

Fig. 6.12 Optimal fixed-point smoothing filter for Example 6.3.

for $j = 1, 2, 3, 4$, where $W(0) = P(0|0) = 100$. It is clear that $k = 0$ here since we are seeking an estimate of $x(0)$.

We have also that $M(0|j) = \frac{1}{15}W(j)$, and Eq. (6.81) is

$$\hat{x}(0|j) = \hat{x}(0|j - 1) + \frac{1}{15}W(j)[z(j) - \hat{x}(j - 1|j - 1)]$$

for $j = 1, 2, 3, 4$ with $\hat{x}(0|0) = 0$.

Utilizing the above relations, we obtain the set of values which are tabulated below. It is seen that a major reduction in the smoothing error variance occurs as soon as the first measurement is incorporated, and that the improvement in accuracy thereafter is relatively modest.

j	$W(j)$	$M(0|j)$	$P(0|j)$
0	......	......	100
1	10.70	0.711	28.60
2	2.98	0.199	26.45
3	0.883	0.0588	26.27
4	0.262	0.0175	26.25

The block diagram for the smoothing filter is indicated in Fig. 6.12.

6.6 OPTIMAL FIXED–LAG SMOOTHING

The algorithm for optimal fixed-lag smoothing for a lag of N units of time is given in the following theorem which we now state and prove.

Theorem 6.3

    a. The optimal fixed-lag smoothed estimate $\hat{x}(k + 1|k + 1 + N)$ is described by the system of equations

$$\hat{x}(k + 1|k + 1 + N)$$
$$= \Phi(k + 1, k)\hat{x}(k|k + N)$$
$$+ C(k + 1 + N)K(k + 1 + N)\tilde{z}(k + 1 + N|k + N)$$
$$+ U(k + 1)[\hat{x}(k|k + N) - \hat{x}(k|k)] \quad (6.84)$$

for $k = 0, 1, \ldots$, where the initial condition is $\hat{x}(0|N)$.

b. $C(k + N)$ *and* $U(k + 1)$ *are the* $n \times n$ *smoothing filter gain matrices, which are given by the relations*

$$C(k + 1 + N) = \prod_{i=k+1}^{k+N} A(i) \tag{6.85}$$

$$A(i) = P(i|i)\Phi'(i + 1, i)P^{-1}(i + 1|i)$$

$$U(k + 1) = \Gamma(k + 1, k)Q(k)\Gamma'(k + 1, k)\Phi'(k, k + 1)P^{-1}(k|k) \tag{6.86}$$

for $k = 0, 1, \ldots .$

c. The optimal fixed-lag smoothing error stochastic process $\{\tilde{x}(k|k + N), k = 0, 1, \ldots\}$ *is a zero mean Gauss-Markov-2 sequence whose covariance matrix is governed by the equation*

$$P(k + 1|k + 1 + N) = P(k + 1|k) - C(k + 1 + N)$$
$$K(k + 1 + N)H(k + 1 + N)$$
$$P(k + 1 + N|k + N)C'(k + 1 + N)$$
$$- A^{-1}(k)[P(k|k)$$
$$- P(k|k + N)][A'(k)]^{-1} \tag{6.87}$$

for $k = 0, 1, \ldots ,$ *where the initial condition is* $P(0|N)$.

Proof As remarked in Sec. 6.1, the proof of the above results will be carried out utilizing the results for the other two classifications of smoothed estimates. This is in contrast to the development which begins with the fact that $\hat{x}(k + 1|k + 1 + N) = E[x(k + 1)|z(1), z(2), \ldots , z(k + N), \tilde{z}(k + 1 + N|k + N)]$. The latter approach is left as an exercise.

We begin by recalling Eq. (6.57) from Theorem 6.1 for optimal fixed-interval smoothing,

$$\hat{x}(k|N) = \hat{x}(k|k) + A(k)[\hat{x}(k + 1|N) - \hat{x}(k + 1|k)]$$

which we rewrite as

$$\hat{x}(k + 1|N) = \Phi(k + 1, k)\hat{x}(k|k) + A^{-1}(k)[\hat{x}(k|N) - \hat{x}(k|k)] \tag{6.88}$$

utilizing the fact that $\hat{x}(k + 1|k) = \Phi(k + 1, k)\hat{x}(k|k)$ and assuming that $A(k)$ is nonsingular.

Adding and subtracting $\Phi(k + 1, k)\hat{x}(k|N)$ on the right-hand side of Eq. (6.88) and regrouping terms, we have

$$
\begin{aligned}
\hat{x}(k + 1|N) &= \Phi(k + 1, k)\hat{x}(k|N) + \Phi(k + 1, k)\hat{x}(k|k) \\
&\quad - \Phi(k + 1, k)\hat{x}(k|N) + A^{-1}(k)[\hat{x}(k|N) - \hat{x}(k|k)] \\
&= \Phi(k + 1, k)\hat{x}(k|N) - \Phi(k + 1, k)[\hat{x}(k|N) \\
&\quad - \hat{x}(k|k)] + A^{-1}(k)[\hat{x}(k|N) - \hat{x}(k|k)] \\
&= \Phi(k + 1, k)\hat{x}(k|N) \\
&\quad + [A^{-1}(k) - \Phi(k + 1, k)][\hat{x}(k|N) - \hat{x}(k|k)] \quad (6.89)
\end{aligned}
$$

From Eq. (6.58) of Theorem 6.1,

$$
\begin{aligned}
A^{-1}(k) &= P(k + 1|k)[\Phi'(k + 1, k)]^{-1}P^{-1}(k|k) \\
&= P(k + 1|k)\Phi'(k, k + 1)P^{-1}(k|k)
\end{aligned}
$$

Since

$$
\begin{aligned}
P(k + 1|k) &= \Phi(k + 1, k)P(k|k)\Phi'(k + 1, k) \\
&\quad + \Gamma(k + 1, k)Q(k)\Gamma'(k + 1, k)
\end{aligned}
$$

it follows that

$$
\begin{aligned}
A^{-1}(k) - \Phi(k + 1, k) &= P(k + 1|k)\Phi'(k, k + 1)P^{-1}(k|k) \\
&\quad - \Phi(k + 1, k) \\
&= [\Phi(k + 1, k)P(k|k)\Phi'(k + 1, k) \\
&\quad + \Gamma(k + 1, k)Q(k)\Gamma'(k + 1, k)] \\
&\quad \Phi'(k, k + 1)P^{-1}(k|k) - \Phi(k + 1, k) \\
&= \Phi(k + 1, k) + \Gamma(k + 1, k)Q(k)\Gamma'(k + 1, k) \\
&\quad \Phi'(k, k + 1)P^{-1}(k|k) - \Phi(k + 1, k) \\
&= \Gamma(k + 1, k)Q(k)\Gamma'(k + 1, k)\Phi'(k, k + 1) \\
&\quad P^{-1}(k|k) \quad (6.90)
\end{aligned}
$$

Defining $U(k) = A^{-1}(k) - \Phi(k + 1, k)$ as given by Eq. (6.90), we express Eq. (6.89) as

$$
\hat{x}(k + 1|N) = \Phi(k + 1, k)\hat{x}(k|N) + U(k)[\hat{x}(k|N) - \hat{x}(k|k)]
$$

This result is simply an alternate relation for optimal fixed-interval smoothing. Allowing the right endpoint of the interval to be variable by replacing N by $k + N$, $k = 0, 1, \ldots$, we obtain the equation

$$
\begin{aligned}
\hat{x}(k + 1|k + N) &= \Phi(k + 1, k)\hat{x}(k|k + N) \\
&\quad + U(k)[\hat{x}(k|k + N) - \hat{x}(k|k)] \quad (6.91)
\end{aligned}
$$

Turning now to the results for optimal fixed-point smoothing, we have from Eqs. (6.60) and (6.61) that

$$\hat{x}(k|j) = \hat{x}(k|j-1) + M(k|j)\tilde{z}(j|j-1)$$

and

$$M(k|j) = \left[\prod_{i=k}^{j-1} A(i)\right] K(j)$$

respectively, for $j = k+1,\ k+2,\ \ldots$. Replacing k by $k+1$ and j by $k+1+N$, we obtain the two results

$$\hat{x}(k+1|k+1+N) = \hat{x}(k+1|k+N)$$
$$+ M(k+1|k+1+N)\tilde{z}(k+1+N|k+N) \quad (6.92)$$

and

$$M(k+1|k+1+N) = \left[\prod_{i=k+1}^{k+N} A(i)\right] K(k+1+N) \quad (6.93)$$

respectively, which hold for $k = 0,\ 1,\ \ldots$.

Substituting for $\hat{x}(k+1|k+N)$ in Eq. (6.92) from Eq. (6.91), we see that

$$\hat{x}(k+1|k+1+N) = \Phi(k+1, k)\hat{x}(k|k+N)$$
$$+ M(k+1|k+1+N)\tilde{z}(k+1+N|k+N)$$
$$+ U(k)[\hat{x}(k|k+N) - \hat{x}(k|k)]$$

Defining

$$C(k+1+N) = \prod_{i=k+1}^{k+N} A(i) \quad (6.94)$$

it follows from Eq. (6.93) that

$$M(k+1|k+1+N) = C(k+1+N)K(k+1+N)$$

and we have

$$\hat{x}(k+1|k+1+N) = \Phi(k+1, k)\hat{x}(k|k+N)$$
$$+ C(k+1+N)K(k+1+N)\tilde{z}(k+1+N|k+N)$$
$$+ U(k)[\hat{x}(k|k+N) - \hat{x}(k|k)]$$

We now have Eqs. (6.84) to (6.86) of the theorem. It is clear that the initial condition is $\hat{x}(0|N)$.

From Theorem 6.1, we know that $\{\hat{x}(k|N), k = N, N-1, \ldots, 0\}$ is a zero mean, Gauss-Markov-2 process for all $N \geq k$. The same is also true if the indexing is reversed, i.e., for $\{\hat{x}(k|N), k = 0, 1, \ldots, N\}$. Since this is valid for all $N \geq k$, it holds for the particular case where N is replaced by $k + N$. Hence $\{\hat{x}(k|k+N), k = 0, 1, \ldots\}$ is a zero mean, Gauss-Markov-2 process for all $N = \text{integer} > 0$.

The matrix equation for recursive computation of the optimal fixed-lag smoothing error covariance matrix,

$$P(k+1|k+1+N) = E[\hat{x}(k+1|k+1+N)\hat{x}'(k+1|k+1+N)]$$

is very easy to obtain. We first rewrite Eq. (6.59) as

$$P(k+1|N) = P(k+1|k) - A^{-1}(k)[P(k|k) - P(k|N)][A'(k)]^{-1}$$

and then replace N by $k + N$ to obtain

$$P(k+1|k+N) = P(k+1|k) - A^{-1}(k)[P(k|k)$$
$$- P(k|k+N)][A'(k)]^{-1} \quad (6.95)$$

Replacing k by $k + 1$ and j by $k + 1 + N$ in Eq. (6.80), we have

$$P(k+1|k+1+N) = P(k+1|k+N) - B(k+1+N)K(k$$
$$+ 1 + N)H(k+1+N)P(k+1+N|k+N)B'(k+1+N)$$

Since

$$B(k+1+N) = \prod_{i=k+1}^{k+N} A(i) = C(k+1+N) \quad (6.96)$$

it follows that

$$P(k+1|k+1+N) = P(k+1|k+N) - C(k+1+N)K(k$$
$$+ 1 + N)H(k+1+N)P(k+1+N|k+N)C'(k+1+N)$$

Substituting into this result from Eq. (6.95) then gives Eq. (6.87) of the theorem. The required initial condition is seen to be $P(0|N)$. This completes the proof.

Discussion   The algorithm for optimal fixed-lag smoothing is obviously more complex than the algorithms for either fixed-interval or fixed-point smoothing.

Let us examine the algorithm more closely to determine the exact nature of this added complexity. First, in order to initiate optimal fixed-lag smoothing, we note that at $k = 0$, $\tilde{z}(N + 1|N)$, $\hat{x}(0|0)$, and $\hat{x}(0|N)$ must be input to Eq. (6.84). The first two of these come from the optimal filter; in particular, the product $K(N + 1)\tilde{z}(N + 1|N)$ is taken directly from the optimal filter while $\hat{x}(0|0) = 0$. However, $\hat{x}(0|N)$ must be obtained from optimal fixed-point smoothing starting with $\hat{x}(0|0) = 0$ and processing the measurements at $k = 1, \ldots, N$ to obtain $\hat{x}(0|1)$, $\hat{x}(0|2)$, $\ldots$, and finally $\hat{x}(0|N)$. Evidently then, the optimal fixed-lag smoothing filter is dormant over the interval $[0,N]$, where N is the value of the lag time, while it "waits" for the initial condition $\hat{x}(0|N)$. After this "waiting period," the optimal smoothing filter is continuously dependent upon the optimal (Kalman) filter to supply it with $K(k + 1 + N)\tilde{z}(k + 1 + N|k + N)$ and $\hat{x}(k|k)$. At this point, we note that $\hat{x}(k + 1|k + 1 + N)$ is a function of $\hat{x}(k|k)$, which means that the latter must be stored for computations at the succeeding time.

In contrast to the Kalman filter and the optimal fixed-interval and fixed-point smoothing filters, the optimal fixed-lag smoothing filter involves two correction terms, each in the form of a gain-times-residual. The first of these involves the familiar Kalman gain matrix $K(k + 1 + N)$ and the measurement residual

$$\tilde{z}(k + 1 + N|k + N) = z(k + 1 + N)$$
$$- H(k + 1 + N)\Phi(k + 1 + N, k + N)\hat{x}(k + N|k + N)$$

The purpose of the gain matrix $C(k + 1 + N)$ can be viewed as that of "reflecting" the new information in the measurement at $k + 1 + N$ "back" to the time $k + 1$ of the estimate. Parenthetically, we remark that according to Eq. (6.96), the two gain matrices $B(k + 1 + N)$ and $C(k + 1 + N)$ are equal. Although this is certainly true, we have chosen to introduce the notation C to distinguish the gain matrices within the two smoothing classifications. The distinction is advantageous because of the fact that k is fixed in fixed-point smoothing, but is variable in fixed-lag smoothing. Comparing the two, we see that

$$B(k) = \prod_{i=k}^{j-1} A(i) \begin{cases} k = \text{constant} \\ j = k + 1, k + 2, \ldots \end{cases}$$

and

$$C(k + 1 + N) = \prod_{i=k+1}^{k+N} A(i) \begin{cases} N = \text{constant} \\ k = 0, 1, \ldots \end{cases}$$

The second correction term in Eq. (6.84) involves the gain matrix $U(k + 1)$ which is given by Eq. (6.86) and the residual $\hat{x}(k|N) - \hat{x}(k|k)$.

Thus, in addition to the first correction which incorporates the new measurement, we have an additional correction term which is proportional to the difference between the optimal fixed-lag smoothed estimate and the optimal filtered estimate both at the time instant k, one unit prior to the time instant $k + 1$ of the estimate being computed, viz., $\hat{x}(k + 1 | k + 1 + N)$. The function of $U(k + 1)$ is then one of "reflecting" this residual into the new estimate. If there is no system disturbance, that is, $Q(k) = 0$, then it is evident from Eq. (6.86) that $U(k + 1) = 0$ and the term plays no role in fixed-lag smoothing. In this connection, the second correction term can be viewed as a "correction for system disturbance."

We note that the first term on the right-hand side of Eq. (6.84), $\Phi(k + 1,\, k)\hat{x}(k|N)$, is analogous to the term $\Phi(k + 1,\, k)\hat{x}(k|k)$ in the Kalman filter, $\hat{x}(k|k)$ in the optimal fixed-interval smoothing filter, and $\hat{x}(k|j - 1)$ in the optimal fixed-point smoothing filter. That is, it corresponds to the uncorrected estimate at the time point of interest.

It is evident that the algorithm for optimal fixed-lag smoothing is forward-time recursive, and could be used on-line in conjunction with the optimal filter, once it has been initialized. Since the estimate lags the measurement by N time units, the smoothing process can be viewed as that of a window of width N moving from left to right on the time scale with the leading edge at the measurement time and the trailing edge at the estimate time as shown in Fig. 6.13.

Noting that the optimal fixed-lag smoothed estimate is given at time $k + 1$ in terms of the measurement at time $k + 1 + N$ and recalling the dependence of fixed-lag smoothing on filtering, we see that the optimal smoothing filter must lag the optimal filter by N units of time. Hence, two time scales are involved. This can be seen from the block diagram for the optimal smoothing filter–optimal filter combination which we give in Fig. 6.14. We emphasize once more that the optimal fixed-lag smoothing filter is inoperative over the interval $[0,N]$.

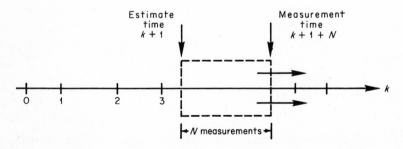

Fig. 6.13 Notion of "moving window" in optimal fixed-lag smoothing.

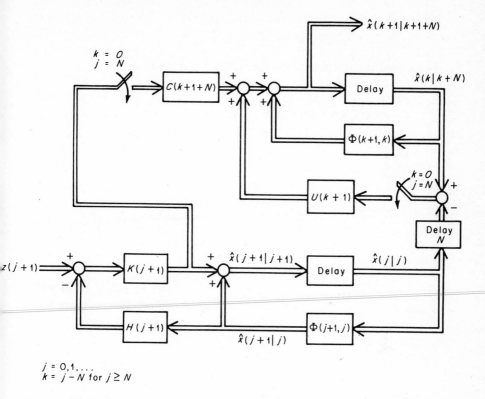

Fig. 6.14 Block diagram for optimal filter and optimal fixed-lag smoothing filter.

Turning now to the question of the gain computations, we consider first Eq. (6.85) which we repeat for convenience,

$$C(k + 1 + N) = \prod_{i=k+1}^{k+N} A(i)$$

Since the index k is variable, we see that determination of $C(k + 1 + N)$ using this relation is computationally inefficient since it requires repetition of previous computations. This inefficiency can be obviated by first noting that

$$C(k + N) = \prod_{i=k}^{k+N-1} A(i)$$

It then follows that

$$C(k + 1 + N) = A^{-1}(k)C(k + N)A(k + N) \tag{6.97}$$

for $k = 0, 1, \ldots,$ where it is recalled that

$$A(i) = P(i|i)\Phi'(i + 1, i)P^{-1}(i + 1|i).$$

The initial condition for Eq. (6.97) is obviously

$$C(N) = \prod_{i=0}^{N-1} A(i) \tag{6.98}$$

which is used to initialize the recursive computation of the smoothing filter gain matrix.

Computation of the other smoothing filter gain matrix $U(k + 1)$, which is given by Eq. (6.86), is straightforward. The computation is not recursive, but is simply carried out separately for each k.

Unfortunately, both gain matrix computations require the inversion of covariance matrices at each time point. In Eq. (6.86), $P^{-1}(k|k)$ is required at each k, while in Eq. (6.97), $P^{-1}(k|k)$ and $P^{-1}(k + 1 + N|k + N)$ are required in determining $A^{-1}(k)$ and $A(k + N)$, respectively.

A similar difficulty arises in the optimal fixed-lag smoothing error covariance equation, Eq. (6.87), where $A^{-1}(k)$ and $[A'(k)]^{-1}$ must be determined.

It is evident that, just as in optimal fixed-interval and fixed-point smoothing, the estimation error covariance matrix computation is not an integral part of the smoothing algorithm, i.e., in the sense that the Kalman filter gain matrix is a function of the prediction error or filter error covariance matrix. This appears to be a general property of optimal smoothing.

Just as the optimal fixed-lag smoothing algorithm depends upon optimal fixed-point smoothing for its initial condition and on optimal filtering for inputs at each time, so does Eq. (6.87) for the optimal fixed-lag smoothing error covariance matrix depend upon error covariance data input from the same two algorithms. In particular, the initial condition $P(0|N)$ for Eq. (6.87) must be obtained by solving Eqs. (6.79) or (6.80) recursively for $P(0|1), P(0|2), \ldots,$ and finally $P(0|N)$. Additionally, Eq. (6.87) requires $P(k|k)$ and $P(k + 1 + N|k + N)$ from the optimal filtering solution for $k = 0, 1, \ldots.$

Example 6.4 We apply the results of Theorem 6.3 to obtain the optimal fixed-lag smoothing filter and its corresponding error variances for the scalar system which we have considered in Examples 6.1 and 6.3. We take a lag of one time unit, and recall again that $\Phi(k + 1, k) = \Gamma(k + 1, k) = H(k + 1) = 1$ for the system. Also, we have $Q(k) = 25$.

From Eq. (6.86), we see that

$$U(k + 1) = Q(k)P^{-1}(k|k) = \frac{25}{P(k|k)}$$

For $N = 1$, Eq. (6.97) becomes

$$C(k + 2) = A^{-1}(k)C(k + 1)A(k + 1)$$

with the initial condition $C(1) = A(0) = P(0|0)P^{-1}(1|0) = {}^{100}\!/_{125} = 0.80$. However, $C(k + 1) = A(k)$ so that

$$C(k + 2) = A(k + 1)$$

for which the values are given in the table in Example 6.1. The two smoothing filter gains which we need are given in the table below for $k = 0, 1, 2$.

k	$C(k + 2)$	$U(k + 1)$
0	0.349	0.250
1	0.302	1.865
2	0.324	2.135

Equation (6.84) for the smoothing filter for this example becomes

$$\hat{x}(k + 1|k + 2) = \hat{x}(k|k + 1) + C(k + 2)K(k + 2)\tilde{z}(k + 2|k + 1)$$
$$+ U(k + 1)[\hat{x}(k|k + 1) - \hat{x}(k|k)]$$

for $k = 0, 1, 2$. The block diagram follows very simply.

The variance of the optimal unit-lag smoothing error is given by

$$P(k + 1|k + 2) = P(k + 1|k) - C^2(k + 2)K(k + 2)P(k + 2|k + 1)$$
$$- \frac{1}{A^2(k)} [P(k|k) - P(k|k + 1)]$$

for $k = 0, 1, 2$, where the initial condition $P(0|1) = 28.6$ is obtained from the table in Example 6.3. Since $C(k + 2) = A(k + 1)$, the variance equation

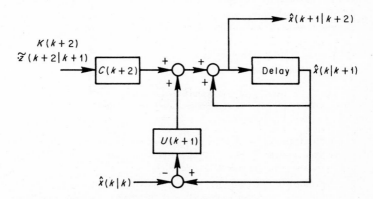

Fig. 6.15 Optimal fixed-lag smoothing filter for Example 6.4.

can also be written

$$P(k + 1|k + 2) = P(k + 1|k) - A^2(k + 1)K(k + 2)P(k + 2|k + 1)$$
$$- \frac{1}{A^2(k)} [P(k|k) - P(k|k + 1)]$$

The following table gives the optimal unit-lag smoothing variance as a function of k. The corresponding values of the error variances for prediction, filtering, and fixed-interval smoothing are included for purposes of comparison. The interesting comparison is between filtering and unit-lag smoothing where the improvement in accuracy which can be obtained by delaying the estimate by one unit is evident.

j	Prediction $P(j\|j - 1)$	Filtering $P(j\|j)$	Fixed-interval smoothing $P(j\|4)$	Unit-lag smoothing $P(j\|j + 1)$
0	. . . . .	100	26.25	28.60
1	125	13.40	9.84	10.04
2	38.4	10.80	8.30	8.51
3	35.8	10.57	8.37	8.37

PROBLEMS

6.1. Develop a block diagram for the double-stage optimal smoothing algorithm of Eq. (6.32). Include in the diagram those operations which are necessary to compute $\hat{x}(k + 1|k + 2)$, which must be input to the algorithm of Eq. (6.32).

6.2. Combine Eqs. (6.12) and (6.32) to obtain an expression for $\hat{x}(k|k + 2)$ for $k = 0$, $1, \ldots$, and develop a block diagram for the resulting smoothing filter. Into which of the three smoothing classifications in Sec. 6.2 does this result fall?

6.3. Determine expressions for the means and covariance matrices which are needed to characterize the gaussian probability density functions $f[\tilde{x}(N - 1|N), \tilde{x}(N|N)]$ and $f[\tilde{x}(k|N)|\tilde{x}(k + 1|N), \tilde{x}(k + 2|N)]$ for $k = N - 2$, $N - 1, \ldots$, 0 which occur in optimal fixed-interval smoothing.

6.4. Show that the stochastic process $\{y(k), k = N, N - 1, \ldots , 0\}$, where

$$y(k) = \begin{bmatrix} \tilde{x}(k|N) \\ \tilde{x}(k|k) \end{bmatrix}$$

is a zero mean Gauss-Markov sequence, and determine an expression for the cross-covariance matrix $P(k) = E[\tilde{x}(k|N)\tilde{x}'(k|k)]$ in terms of $P(k|N)$ and $P(k|k)$, in the optimal fixed-interval smoothing problem.

6.5. Develop a digital computer program flow chart for optimal fixed-interval smoothing including the covariance matrix computation. Assume that $\hat{x}(k|k)$, $\hat{x}(k + 1|k)$, $P(k|k)$, and $P(k + 1|k)$ are available in memory for all $k = 0, 1, \ldots , N$.

6.6. In what way, if any, must the results in Theorem 6.1 be modified if the system model includes a known control input, that is, $x(k+1) = \Phi(k+1, k)x(k) + \Gamma(k+1, k)w(k) + \Psi(k+1, k)u(k)$?

6.7. For the system in Example 5.2, let $P(0|0) = 10$ and $R(k+1) = 10$ for all $k = 0, 1, \ldots$. Assuming that optimal fixed-interval smoothing is carried out for this system over the interval $[0,4]$, determine the optimal fixed-interval smoothing error variance at each time point within the interval and compare these results with those for optimal filtering. Sketch a block diagram for the optimal smoothing filter.

6.8. Assuming that $\hat{x}(k|N)$ is known for some $k = k^*$ in the interval $[0,N]$ along with its estimation error covariance matrix $P(k^*|N) = E[\tilde{x}(k^*|N)\tilde{x}'(k^*|N)]$, develop a forward-time recursive algorithm for optimal fixed-interval smoothing for $k = k^*$, $k^* + 1, \ldots, N$. Give a block diagram for the resulting smoothing filter.

6.9. Prove Corollary 6.1.

6.10. For the class of optimal fixed-point smoothing problems in which there is no system disturbance so that $Q(i) = 0$ for all $i = 0, 1, \ldots$, show that

$$\hat{x}(k|j) = \Phi(k,j)\hat{x}(j|j)$$

and

$$P(k|j) = \Phi(k,j)P(j|j)\Phi'(k,j)$$

for $j = k + 1, k + 2, \ldots$.

6.11. In optimal fixed-point smoothing, does it necessarily always follow that $\operatorname{tr} P(k|j) \leq \operatorname{tr} P(k|j-1)$ or that $\operatorname{tr} P(k|j) \leq \operatorname{tr} P(k|k)$? Explain your answer.

6.12. Consider the problem of optimal fixed-point smoothing for the scalar system $x(k+1) = ax(k)$, $z(k+1) = bx(k+1) + v(k+1)$, $k = 0, 1, \ldots$, where a and b are constants. Under what conditions on a, b, and/or any other parameters in the problem does it follow that $P(0|j) < P(0|j-1)$?

6.13. Consider the scalar system $x(k+1) = 2^{-k}x(k) + w(k)$, $z(k+1) = x(k+1)$, $k = 0, 1, \ldots$, where $x(0)$ has mean zero and variance $\sigma_0^2 = $ positive constant, and $\{w(k), k = 0, 1, \ldots\}$ is a zero mean gaussian white sequence which is independent of $x(0)$ and has a variance of $\sigma_w^2 = $ positive constant. Assuming that optimal fixed-point smoothing is to be employed to determine $\hat{x}(0|j)$, $j = 1, 2, \ldots$, what is the equation for the appropriate smoothing filter? What is the limiting value of $P(0|j)$ as $j \to \infty$? How does this value compare with $P(0|0)$?

6.14. Starting with the relation $\hat{x}(k+1|k+1+N) = E[x(k+1)|z(1), z(2), \ldots, z(k+N), \tilde{z}(k+1+N|k+N)]$, derive Eq. (6.84).

6.15. How must the results in Theorem 6.3 be modified if $x(0)$, $\{w(k), k = 0, 1, \ldots\}$, and $\{v(k+1), k = 0, 1, \ldots\}$ have nonzero mean values?

6.16. What are the equations for optimal fixed-lag smoothing for a unit lag for the system in Prob. 6.7?

6.17. What simplifications if any, in addition to the one noted in the text, occur in the results in Theorem 6.3 if there is no system disturbance, that is, if $Q(k) = 0$ for all $k = 0, 1, \ldots$.

6.18. Show that Theorem 6.3 reduces to Theorem 5.5 when the lag N is made zero.

6.19. For the system model in Prob. 4.13, where the measurement error is time-correlated, develop algorithms for determining the optimal estimates $\hat{x}(k+1|k)$, $\hat{x}(k+1|k+1)$, and $\hat{x}(k|k+1)$ taking, as in part b of Prob. 4.13, the measurement to be

$$\zeta(k) \triangleq z(k + 1) - \theta(k + 1, k)z(k)$$
$$= [H(k + 1)\Phi(k + 1, k) - \theta(k + 1, k)H(k)]x(k)$$
$$+ [H(k + 1)\Gamma(k + 1, k)w(k) + \xi(k)]$$

Note that this procedure avoids the augmented state vector approach in the case of time-correlated measurement errors [6]. Assume that $\bar{x}(0)$, $\bar{w}(k)$, $\bar{v}(0)$, and $\bar{\xi}(k)$ are zero. Also, noting that $\zeta(0) = z(1)$, pay particular attention to the initialization of the algorithms.

REFERENCES

1. Rauch, H. E., Solutions to the Linear Smoothing Problem, *IEEE Trans. Autom. Control*, vol. AC-8, p. 371, 1963. Also Linear Estimation of Sampled Stochastic Processes with Random Parameters, *Tech. Rept.* No. 2108-1, Stanford Electronics Laboratory, Stanford University, Stanford, California, April, 1962.
2. Carlton, A. G., Linear Estimation in Stochastic Processes, *Rept.* No. 311, Applied Physics Laboratory, The Johns Hopkins University, Silver Springs, Maryland, March, 1962.
3. Rauch, H. E., F. Tung, and C. T. Striebel, Maximum Likelihood Estimates of Linear Dynamic Systems, *AIAA J.*, vol. 3, p. 1445, 1965. Also, On the Maximum Likelihood Estimates for Linear Dynamic Systems, *Tech. Rept.* No. 6-90-63-62, Lockheed Missiles and Space Company, Palo Alto, California, June, 1963.
4. Meditch, J. S., Orthogonal Projection and Discrete Optimal Linear Smoothing, *SIAM J. Control*, vol. 5, p. 74, 1967.
5. Fraser, D., Discussion of Optimal Fixed-Point Continuous Linear Smoothing (by J. S. Meditch), *Proc. of the 1967 Joint Automatic Control Conference*, p. 249, University of Pennsylvania, Philadelphia, June, 1967.
6. Bryson, A. E., Jr., and L. J. Henrikson, Estimation Using Sampled-Data Containing Sequentially Correlated Noise, *Tech. Rept.* No. 533, Division of Engineering and Applied Physics, Harvard University, Cambridge, Mass., June, 1967.

7

Optimal Estimation for Continuous Linear Systems, I

7.1 INTRODUCTION

Having treated the problem of optimal estimation for discrete linear systems in the preceding two chapters, we turn now to the same problem for their continuous-time analog.

We shall present two approaches to this latter problem, the first in this chapter, the second in Chap. 8. In the first approach, we shall obtain the algorithms for optimal estimation by considering the limiting behavior of the optimal estimation algorithms for discrete linear systems as the time interval between measurements is made arbitrarily small. This approach is due originally to Kalman [1], who used it to derive the optimal filtering equations. It has subsequently been applied by Rauch et al. [2] to develop the algorithm for optimal fixed-interval smoothing, and by Meditch [3, 4] to obtain the optimal fixed-point and fixed-lag smoothing algorithms.

In the second approach, the problem of optimal estimation for continuous linear systems is solved entirely in the continuous-time domain. The results are the same in both cases as, indeed, they should be.

We shall begin, in this chapter, by giving a formulation of the problem of optimal estimation for continuous linear systems. In order to pursue the first approach which was mentioned above, we shall next formulate a discrete-time estimation problem to which the limiting procedure is applicable. We shall then proceed in a straightforward fashion to develop the algorithms for continuous-time optimal filtering and prediction, and fixed-interval, fixed-point, and fixed-lag smoothing, in that order, by taking the appropriate limits.

7.2 PROBLEM FORMULATION

SYSTEM MODEL

From Sec. 4.4, the system model of interest is characterized by the relations

$$\dot{x} = F(t)x + G(t)w(t) \tag{7.1}$$

and

$$z(t) = H(t)x(t) + v(t) \tag{7.2}$$

for $t \geq t_0$, where x is an n vector, w a p vector, and z and v are m vectors. The matrices $F(t)$, $G(t)$, and $H(t)$ are continuous, and $n \times n$, $n \times p$, and $m \times n$, respectively. Time is denoted by t, the time derivative by the dot, and the fixed initial time by t_0.

The stochastic processes $\{w(t), \ t \geq t_0\}$ and $\{v(t), \ t \geq t_0\}$ are zero mean gaussian white noises with covariance matrices

$$E[w(t)w'(\tau)] = Q(t)\delta(t - \tau)$$

and

$$E[v(t)v'(\tau)] = R(t)\delta(t - \tau)$$

respectively, for all $t, \tau \geq t_0$, where E denotes the expected value, the prime the transpose, and δ the Dirac delta function. The $p \times p$ matrix $Q(t)$ is continuous and positive semidefinite for $t \geq t_0$, while the $m \times m$ matrix $R(t)$ is continuous and positive definite for $t \geq t_0$.

We assume that the above two stochastic processes are independent of each other, so that

$$E[w(t)v'(\tau)] = 0$$

for all $t, \tau \geq t_0$. This assumption constitutes no real restriction, since the procedure which we employ in this chapter is also applicable when the two stochastic processes are correlated. The choice here, however, simplifies many of the manipulations.

The initial state $x(t_0)$ is a zero mean gaussian random n vector; it is independent of $\{w(t), t \geq t_0\}$ and $\{v(t), t \geq t_0\}$; and its $n \times n$ covariance matrix $E[x(t_0)x'(t_0)] = P(t_0)$ is positive semidefinite.

OPTIMAL ESTIMATES

In denoting estimates of the state x for continuous-time systems, we replace the discrete-time instants k and j in our notation by the continuous-time variables t_1 and t, respectively. Then, an estimate of x at some time $t_1 \geq t_0$, based on measurements $z(\tau)$ over the interval $t_0 \leq \tau \leq t$, is denoted by $\hat{x}(t_1|t)$.

If $t_1 > t$, $\hat{x}(t_1|t)$ is a predicted estimate; if $t_1 = t$, it is a filtered estimate; and if $t_1 < t$, it is a smoothed estimate, just as in the discrete-time case.

The estimation error is defined by the expression

$$\tilde{x}(t_1|t) = x(t_1) - \hat{x}(t_1|t)$$

and the quality of an estimate is assessed by the performance measure

$$J[\tilde{x}(t_1|t)] = E\{L[\tilde{x}(t_1|t)]\}$$

where L is an admissible loss function as defined in Sec. 5.2.

An estimate $\hat{x}(t_1|t)$ which minimizes $J[\tilde{x}(t_1|t)]$ is called an *optimal estimate*.

The problem statement follows immediately.

PROBLEM STATEMENT

Given the system of Eqs. (7.1) and (7.2), and the measurements $\{z(\tau), t_0 \leq \tau \leq t\}$, determine an optimal estimate of $x(t_1)$.

7.3 EQUIVALENT DISCRETE–TIME PROBLEM

In order to utilize the limiting procedure of which we spoke in Sec. 7.1 to solve the above problem, we present now the formulation of an equivalent discrete-time problem.

In this section, we shall use t to denote discrete time.† In particular, for any $\Delta t > 0$, we let t denote the discrete-time instants $\{t = t_0 + j\,\Delta t, j = 0, 1, \ldots\}$. Then for any such t, we know from Sec. 4.4 that the discrete-time model of the system of Eq. (7.1) is a zero mean

---

† In this chapter, t will be used to denote both discrete and continuous time. This is a matter of convenience, and should cause no difficulty, since the particular usage will be clear from context.

Gauss-Markov sequence whose defining equation is

$$x(t + \Delta t) = \Phi(t + \Delta t, t)x(t) + \Gamma(t + \Delta t, t)w(t) \tag{7.3}$$

where

$$\Phi(t + \Delta t, t) = I + F(t) \, \Delta t + 0(\Delta t^2) \tag{7.4}$$

$$\Gamma(t + \Delta t, t) = G(t) \, \Delta t + 0(\Delta t^2) \tag{7.5}$$

and $\{w(t), \, t = t_0 + j \, \Delta t, \, j = 0, 1, \ldots\}$ is a zero mean gaussian white sequence whose covariance matrix is

$$E[w(t)w'(\tau)] = \frac{Q(t)}{\Delta t} \, \delta_{jk}$$

where t is the discrete-time index defined above, τ is the discrete-time index $\{\tau = t_0 + k \, \Delta t, \, k = 0, 1, \ldots\}$, $Q(t)$ is positive semidefinite for all $t \geq t_0$, and δ_{jk} is the Kronecker delta. The initial state $x(t_0)$ is a zero mean gaussian random n vector which is independent of $\{w(t), \, t = t_0 + j \, \Delta t, \, j = 0, 1, \ldots\}$ and has covariance matrix $E[x(t_0)x'(t_0)] = P(t_0)$ which is positive semidefinite.

The discrete-time model of the measurement equation, Eq. (7.2), is

$$z(t + \Delta t) = H(t + \Delta t)x(t + \Delta t) + v(t + \Delta t) \tag{7.6}$$

where t is the discrete-time index defined above. The stochastic process $\{v(t + \Delta t), \, t = t_0 + j \, \Delta t, \, j = 0, 1, \ldots\}$ is a zero mean gaussian white sequence which is independent of $x(t_0)$ and has covariance matrix

$$E[v(t + \Delta t)v'(\tau + \Delta t)] = \frac{R(t + \Delta t)}{\Delta t} \, \delta_{jk}$$

where the discrete-time index τ was defined above, and $R(t + \Delta t)$ is positive definite for all $t \geq t_0$.

It is assumed here that the two gaussian white sequences in Eqs. (7.3) and (7.6) are independent.

We let t_1 denote the discrete-time index $t_1 = t_0 + i \, \Delta t, i = 0, 1, \ldots$, and denote estimates of x by $\hat{x}(t_1|t)$. As before, these estimates are classified according to the value of t relative to t_1.

The definitions of estimation error, performance measure, and optimal estimate are the same as in Sec. 7.2, except that t_1 and t are now discrete-time instants.

The equivalent estimation problem is the following.

PROBLEM STATEMENT

Given the system of Eqs. (7.3) and (7.6), and the measurements $\{z(\tau), \, \tau = t_0 + \Delta t, \, t_0 + 2 \, \Delta t, \ldots, t\}$, determine an optimal estimate of $x(t_1)$.

The solution of this problem is obtained by utilizing the results in Chaps. 5 and 6. Since, in the limit as $\Delta t \to 0$, the system in Eqs. (7.3) and (7.6) becomes the system in Eqs. (7.1) and (7.2), respectively, the corresponding limit of the solution of the above problem will yield the desired optimal estimation algorithms.

7.4 OPTIMAL FILTERING AND PREDICTION

We treat first the problem of optimal filtering for which we have the following result.

Theorem 7.1   *Kalman Bucy*

    a. The optimal filtered estimate for the system of Eqs. (7.1) and (7.2) is governed by the relation

$$\dot{\hat{x}} = F(t)\hat{x} + K(t)[z(t) - H(t)\hat{x}] \tag{7.7}$$

    for $t \geq t_0$ where $\hat{x} = \hat{x}(t|t)$, $\hat{x}(t_0|t_0) = 0$, and $K(t)$ is the $n \times m$ filter gain matrix.

    b. $K(t)$ is given by the expression

$$K(t) = P(t|t)H'(t)R^{-1}(t) \tag{7.8}$$

    for $t \geq t_0$ where $P(t|t)$ is the $n \times n$ covariance matrix of the filtering error $\tilde{x}(t|t) = x(t) - \hat{x}(t|t)$.

    c. The stochastic process $\{\tilde{x}(t|t), t \geq t_0\}$ is a zero mean Gauss-Markov process whose covariance matrix is the solution of the matrix differential equation   Matrix Riccatti Equation

$$\dot{P} = F(t)P + PF'(t) - PH'(t)R^{-1}(t)H(t)P + G(t)Q(t)G'(t) \tag{7.9}$$

    for $t \geq t_0$ where $P = P(t|t)$ and $P(t_0|t_0) = P(t_0) = E[x(t_0)x'(t_0)]$.

Proof From Eq. (5.48) in Theorem 5.5, the optimal filter for the system of Eqs. (7.3) and (7.6) is

$$\hat{x}(t + \Delta t|t + \Delta t) = \Phi(t + \Delta t, t)\hat{x}(t|t) + K(t + \Delta t)[z(t + \Delta t)$$
$$- H(t + \Delta t)\Phi(t + \Delta t, t)\hat{x}(t|t)$$

where $\hat{x}(t_0|t_0) = 0$.

Utilizing Eq. (7.4), we see that

$$\Phi(t + \Delta t, t)\hat{x}(t|t) = [I + F(t)\,\Delta t + 0(\Delta t^2)]\hat{x}(t|t)$$

and we can rewrite the above equation in the form

$$\hat{x}(t + \Delta t|t + \Delta t) - \hat{x}(t|t) = F(t)\hat{x}(t|t)\,\Delta t + K(t + \Delta t)[z(t + \Delta t)$$
$$- H(t + \Delta t)\Phi(t + \Delta t, t)\hat{x}(t|t)] + 0(\Delta t^2) \tag{7.10}$$

Dividing through by Δt in Eq. (7.10) and taking the limit as $\Delta t \to 0$, we obtain the result

$$\dot{\hat{x}} = F(t)\hat{x} + \lim_{\Delta t \to 0} \frac{K(t + \Delta t)[z(t + \Delta t) - H(t + \Delta t)\Phi(t + \Delta t, t)\hat{x}(t|t)]}{\Delta t} \tag{7.11}$$

where $\hat{x} = \hat{x}(t|t)$ and $t \geq t_0$.

The limit on the right-hand side of Eq. (7.11) can be evaluated as the product of the two limits

$$\lim_{\Delta t \to 0} \frac{K(t + \Delta t)}{\Delta t} \qquad \text{and} \qquad \lim_{\Delta t \to 0} [z(t + \Delta t) - H(t + \Delta t)\Phi(t + \Delta t, t)\hat{x}]$$

providing these latter two limits exist.

We consider the second limit first. Utilizing Eq. (7.4), we see immediately that

$$\lim_{\Delta t \to 0} [z(t + \Delta t) - H(t + \Delta t)\Phi(t + \Delta t, t)\hat{x}] = z(t) - H(t)\hat{x}(t|t) \tag{7.12}$$

Turning now to the other limit, we have from Eqs. (5.49) and (5.50) of Theorem 5.5 that

$$K(t + \Delta t) = P(t + \Delta t|t)H'(t + \Delta t)\left[H(t + \Delta t)P(t + \Delta t|t)H'(t + \Delta t) \right.$$
$$\left. + \frac{R(t + \Delta t)}{\Delta t} \right]^{-1}$$

$$= P(t + \Delta t|t)H'(t + \Delta t)[H(t + \Delta t)$$
$$P(t + \Delta t|t)H'(t + \Delta t)\,\Delta t + R(t + \Delta t)]^{-1}\,\Delta t \tag{7.13}$$

and

$$P(t + \Delta t|t) = \Phi(t + \Delta t, t)P(t|t)\Phi'(t + \Delta t, t)$$
$$+ \Gamma(t + \Delta t, t)\frac{Q(t)}{\Delta t}\Gamma'(t + \Delta t, t) \tag{7.14}$$

respectively.

Substituting into Eq. (7.14) from Eqs. (7.4) and (7.5), expanding the result, and regrouping terms, we obtain the relation

$$P(t + \Delta t|t) = [I + F(t)\,\Delta t + 0(\Delta t^2)]P(t|t)[I + F(t)\,\Delta t + 0(\Delta t^2)]'$$
$$+ [G(t)\,\Delta t + 0(\Delta t^2)]\frac{Q(t)}{\Delta t}[G(t)\,\Delta t + 0(\Delta t^2)]'$$

$$= P(t|t) + [F(t)P(t|t) + P(t|t)F'(t) + G(t)Q(t)G'(t)]\,\Delta t$$
$$+ 0(\Delta t^2) \tag{7.15}$$

We see from this result that for any $t \geq t_0$ and $\Delta t > 0$,

$$\lim_{\Delta t \to 0} P(t + \Delta t | t) = P(t|t) \tag{7.16}$$

As a consequence of Eq. (7.16) and the assumptions that $H(t)$ and $R(t)$ are continuous, it follows that

$$\lim_{\Delta t \to 0} \frac{K(t + \Delta t)}{\Delta t}$$
$$= P(t|t)H'(t) \lim_{\Delta t \to 0} [H(t + \Delta t)P(t + \Delta t|t)H'(t + \Delta t) \Delta t + R(t + \Delta t)]^{-1}$$
$$= P(t|t)H'(t)R^{-1}(t) \tag{7.17}$$

Hence, in view of Eqs. (7.12) and (7.17), we obtain from Eq. (7.11) the result

$$\dot{\hat{x}} = F(t)\hat{x} + K(t)[z(t) - H(t)\hat{x}]$$

for $t \geq t_0$ with $\hat{x}(t_0|t_0) = 0$, where we have defined

$$K(t) = P(t|t)H'(t)R^{-1}(t)$$

This gives Eqs. (7.7) and (7.8) of the theorem.

We consider next the filtering error. From Eq. (5.64), we have for our problem that

$$\tilde{x}(t + \Delta t | t + \Delta t) = [I - K(t + \Delta t)H(t + \Delta t)]\Phi(t + \Delta t, t)\tilde{x}(t|t)$$
$$+ [I - K(t + \Delta t)H(t + \Delta t)]\Gamma(t + \Delta t, t)w(t)$$
$$- K(t + \Delta t)v(t + \Delta t) \tag{7.18}$$

Substituting into Eq. (7.18) from Eqs. (7.4) and (7.5), we obtain

$$\tilde{x}(t + \Delta t | t + \Delta t) = [I - K(t + \Delta t)H(t + \Delta t)]$$
$$[I + F(t) \Delta t + 0(\Delta t^2)]\tilde{x}(t|t)$$
$$+ [I - K(t + \Delta t)H(t + \Delta t)][G(t) \Delta t + 0(\Delta t^2)]$$
$$w(t) - K(t + \Delta t)v(t + \Delta t)$$
$$= \tilde{x}(t|t) + [F(t) \Delta t - K(t + \Delta t)H(t + \Delta t)]\tilde{x}(t|t)$$
$$+ G(t)w(t) \Delta t - K(t + \Delta t)v(t + \Delta t) + 0(\Delta t^2) \tag{7.19}$$

where we have made use of the fact that $K(t + \Delta t)$ has no zero-order terms in Δt as can be seen from Eqs. (7.13) and (7.15).

Transposing $\tilde{x}(t|t)$ to the left-hand side in Eq. (7.19), dividing through by Δt, and taking $\lim \Delta t \to 0$, we obtain

$$\dot{\tilde{x}} = [F(t) - K(t)H(t)]\tilde{x} + G(t)w(t) - K(t)v(t) \tag{7.20}$$

for $t \geq t_0$, where use has been made of the limit in Eq. (7.17) and the definition of $K(t)$.

Since the stochastic process $\{\tilde{x}(t + \Delta t | t + \Delta t), t = t_0 + j \, \Delta t, j = 0, 1, \ldots\}$ is a zero mean Gauss-Markov sequence, it follows from the results in Sec. 4.4 that the corresponding limiting process which is defined by Eq. (7.20) is a zero mean Gauss-Markov process.

Finally, we obtain the differential equation for $P(t|t)$. From Eq. (5.51) of Theorem 5.5, we have

$$P(t + \Delta t | t + \Delta t) = P(t + \Delta t | t) - K(t + \Delta t) H(t + \Delta t) P(t + \Delta t | t)$$

Substituting into this result from Eq. (7.15) and recalling that $K(t + \Delta t)$ has no zero-order Δt terms, we write

$$\begin{aligned} P(t + \Delta t | t + \Delta t) = P(t|t) + [F(t)P(t|t) + P(t|t)F'(t) \\ + G(t)Q(t)G'(t)] \, \Delta t \\ - K(t + \Delta t)H(t + \Delta t)P(t|t) + 0(\Delta t^2) \end{aligned}$$

We see immediately that

$$\begin{aligned} \dot{P}(t|t) = F(t)P(t|t) + P(t|t)F'(t) + G(t)Q(t)G'(t) \\ - \lim_{\Delta t \to 0} \frac{K(t + \Delta t)H(t + \Delta t)P(t|t)}{\Delta t} \end{aligned}$$

which, in view of Eq. (7.17), means that

$$\dot{P} = F(t)P + PF'(t) - PH'(t)R^{-1}(t)H(t)P + G(t)Q(t)G'(t) \qquad (7.21)$$

where $P = P(t|t)$.

From Theorem 5.5, the appropriate initial condition for Eq. (7.21) is $P(t_0|t_0) = P(t_0)$ and our proof is complete.

Discussion   Theorem 7.1 is due to Kalman [1], who developed it for the more general case where $\{w(t), t \geq t_0\}$ and $\{v(t), t \geq t_0\}$ are correlated. The development for the more general case differs only in certain details from the one above and is left as an exercise.

The block diagram for the optimal filter in Eq. (7.7) is shown in Fig. 7.1.

Instead of the predicted measurement $\hat{z}(k + 1|k)$ as in the discrete-time filter, we now have the filtered measurement $\hat{z}(t|t) = H(t)\hat{x}(t|t)$, so that the measurement residual which is the input to the filter gain matrix is

$$\tilde{z}(t|t) = z(t) - \hat{z}(t|t)$$

The filter is seen to consist of the model of the system dynamics $\dot{x} = F(t)x$, which is forced by the feedback correction signal $K(t)\tilde{z}(t|t)$.

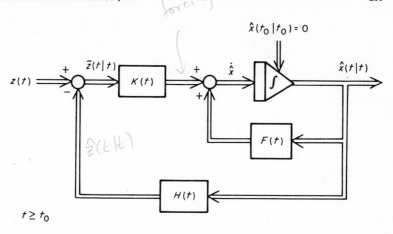

forcing fctn

$\hat{x}(t_0 | t_0) = 0$

$z(t)$ $+$ $\tilde{z}(t|t)$ $K(t)$ $+$ $\dot{\hat{x}}$ $\int$ $\hat{x}(t|t)$

$-$ $+$

$\tilde{z}(t|t)$

$F(t)$

$H(t)$

$t \geq t_0$

Fig. 7.1 Block diagram of optimal filter for continuous linear systems.

Since there is no prediction explicitly present in the filter, there is no prediction error covariance matrix to compute here, only the filtering error covariance matrix which is the solution of Eq. (7.9). This equation is seen to be a system of first-order ordinary differential equations of second degree. It is called a matrix Riccati equation.

Although $P(t|t)$ contains n^2 elements, it is only necessary to consider $n(n+1)/2$ of the equations in the system of Eq. (7.9), since $P(t|t)$ is a covariance matrix, and, therefore, symmetric.

Once $P(t|t)$ is determined, the stochastic process $\{\tilde{x}(t|t),\ t \geq t_0\}$ is completely described by its characteristic function

$$\phi_{\tilde{x}}(s;t) = \exp\left[-\tfrac{1}{2}s'P(t|t)s\right] \quad \textit{Gaussian}$$

where s is an n vector.

As in the discrete-time case, it is not necessary to simulate the filter or even to determine $K(t)$ in order to assess the optimal filter's performance. It is clear that only Eq. (7.9) need be solved for $P(t|t)$ for this purpose.

The reason for choosing $R(t)$ positive definite for all $t \geq t_0$ is evident from Eqs. (7.8) and (7.9). Physically, the assumption means that there is always "some error" present in all the components of the measurement vector $z(t)$.

The optimal filter which is given by Eq. (7.7) is obviously a linear system in which $z(t)$ is the input and $\hat{x}(t|t)$ the output. If we rewrite Eq. (7.7) as

$$\dot{\hat{x}} = [F(t) - K(t)H(t)]\hat{x} + K(t)z(t)$$

and let $\Psi(t,\tau)$ denote the transition matrix of this system, we can express the solution of this differential equation in the form

$$\hat{x}(t|t) = \Psi(t,t_0)\hat{x}(t_0|t_0) + \int_{t_0}^{t} \Psi(t,\tau)K(\tau)z(\tau) \, d\tau$$

Since $\hat{x}(t_0|t_0) = 0$, we have

$$\hat{x}(t|t) = \int_{t_0}^{t} \Psi(t,\tau)K(\tau)z(\tau) \, d\tau$$

Letting $A(t,\tau)$ denote the $n \times m$ matrix $\Psi(t,\tau)K(\tau)$, we can express the optimal filtered estimate as

$$\hat{x}(t|t) = \int_{t_0}^{t} A(t,\tau)z(\tau) \, d\tau$$

and view the optimal filter as a particular linear transformation on the measurement. This point will be pursued further in Chap. 8.

Finally, we remark that $\hat{x}(t|t)$ is unique, a result which follows from the fact that $\hat{x}(k+1|k+1)$ is unique.

Example 7.1 Suppose that messages which are to be transmitted over a communication channel are sample functions from a scalar stochastic process which is defined by the differential equation

$$\dot{x} = -ax + w(t)$$

for $t \geq 0$ where $a = \text{constant} > 0$, $x(0)$ is a zero mean gaussian random variable with variance $\sigma_0^2 = \text{constant} > 0$, and $\{w(t), t \geq 0\}$ is a zero mean gaussian white noise which is independent of $x(0)$ and has a variance $\sigma_w^2 = \text{constant} > 0$. Suppose further that the received signal is of the form

$$z(t) = x(t) + v(t)$$

where $\{v(t), t \geq 0\}$ is a zero mean gaussian white noise which is independent of $x(0)$ and $\{w(t), t \geq 0\}$, and has a variance $\sigma_v^2 = \text{constant} > 0$. Physically, $\{v(t), t \geq 0\}$ is due to transmitter errors and noise, and atmospheric disturbances.

Our problem is to determine the characteristics of a receiver which will give the "best" possible signal-noise separation. Assuming that by "best" here we mean the minimization of the expected value of any admissible loss function of the error $\tilde{x}(t|t) = x(t) - \hat{x}(t|t)$, the solution of the problem is given by Theorem 7.1.

For this problem, we have $F(t) = -a$, $G(t) = H(t) = 1$, $P(0) = \sigma_0^2$, $Q(t) = \sigma_w^2$, and $R(t) = \sigma_v^2$ for all $t \geq t_0 = 0$.

The filtering error variance equation, Eq. (7.9), is

$$\dot{P} = -2aP - \frac{1}{\sigma_v^2}P^2 + \sigma_w^2 \qquad t \geq 0$$

with σ_0^2 the initial condition. This equation can be solved quite readily by first separating the variables to obtain

$$\frac{dP}{P^2 + 2a\sigma_v^2 P - \sigma_v^2\sigma_w^2} = -\frac{1}{\sigma_v^2} \, dt$$

Letting ρ_1 and ρ_2 denote the roots of $P^2 + 2a\sigma_v{}^2 P - \sigma_v{}^2\sigma_w{}^2 = 0$, we have

$$\frac{dP}{(P - \rho_1)(P - \rho_2)} = \frac{1}{(\rho_1 - \rho_2)}\left[\frac{dP}{P - \rho_1} - \frac{dP}{P - \rho_2}\right] = -\frac{1}{\sigma_v{}^2}\,dt$$

where

$$\rho_{1,2} = \left[-a \pm \sqrt{a^2 + \frac{\sigma_w{}^2}{\sigma_v{}^2}}\,\right]\sigma_v{}^2$$

Since

$$\rho_1 - \rho_2 = 2\sqrt{a^2 + \frac{\sigma_w{}^2}{\sigma_v{}^2}}\,\sigma_v{}^2$$

we have

$$\frac{dP}{P - \rho_1} - \frac{dP}{P - \rho_2} = -2\sqrt{a^2 + \frac{\sigma_w{}^2}{\sigma_v{}^2}}\,dt$$

Defining $\mu = \sqrt{a^2 + \sigma_w{}^2/\sigma_v{}^2}$, we obtain

$$\frac{P(t|t) - \rho_1}{P(t|t) - \rho_2} = \alpha e^{-2\mu t}$$

where α is the constant of integration.

Utilizing the fact that $P(0|0) = P(0) = \sigma_0{}^2$, we have

$$\alpha = \frac{\sigma_0{}^2 - \rho_1}{\sigma_0{}^2 - \rho_2}$$

The filtering error variance is then seen to be

$$P(t|t) = \frac{\rho_1 - \rho_2\alpha e^{-2\mu t}}{1 - \alpha e^{-2\mu t}}$$

for $t \geq 0$.

From Eqs. (7.8) and (7.7), we have

$$K(t) = \frac{1}{\sigma_v{}^2}P(t|t)$$

and

$$\dot{\hat{x}} = -a\hat{x} + K(t)[z(t) - \hat{x}]$$

respectively.

For t sufficiently large, we see that

$$P(t|t) \to \rho_1 = \left[-a + \sqrt{a^2 + \frac{\sigma_w{}^2}{\sigma_v{}^w}}\,\right]\sigma_v{}^2$$

and

$$K(t) \to \frac{\rho_1}{\sigma_v{}^2} = -a + \sqrt{a^2 + \frac{\sigma_w{}^2}{\sigma_v{}^2}}$$

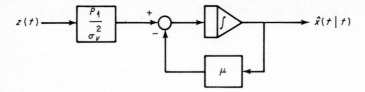

Fig. 7.2   Optimal filter for Example 7.1.

Hence, in the steady state, the optimal receiver is the constant coefficient linear system

$$\dot{\hat{x}} = -a\hat{x} + \frac{\rho_1}{\sigma_v^2} z(t) + \left[a - \sqrt{a^2 + \frac{\sigma_w^2}{\sigma_v^2}} \right] \hat{x}$$

or

$$\dot{\hat{x}} = -\mu\hat{x} + \frac{\rho_1}{\sigma_v^2} z(t)$$

for which we indicate the block diagram in Fig. 7.2.

We observe that there exists an inherent accuracy limitation on the optimal receiver which is set by the value of ρ_1.   Let us examine three particular cases in this connection:

1. If $a^2 \gg \sigma_w^2/\sigma_v^2$, then

$$\rho_1 = \left[-a + a \sqrt{1 + \frac{\sigma_w^2}{a^2\sigma_v^2}} \right] \sigma_v^2 \approx \frac{\sigma_w^2}{2a} \ll \sigma_v^2$$

2. If $a^2 \approx \sigma_w^2/\sigma_v^2$, $\rho_1 \approx 0.414\sigma_v^2$.
3. If $a^2 \ll \sigma_w^2/\sigma_v^2$, we have $\rho_1 \approx \sigma_w\sigma_v$.

OPTIMAL PREDICTION

For some fixed $t_1 \geq t_0$, we consider now the problem of determining the optimal predicted estimate $\hat{x}(t|t_1)$, $t \geq t_1$, of $x(t)$.   The solution of this problem is given below.

Theorem 7.2

a. *Given the optimal filtered estimate $\hat{x}(t_1|t_1)$ for some $t_1 \geq t_0$, the optimal predicted estimate is governed by the expression*

$$\dot{\hat{x}}(t|t_1) = F(t)\hat{x}(t|t_1) \tag{7.22}$$

for all $t \geq t_1$, where the initial condition is $\hat{x}(t_1|t_1)$.

b. *The stochastic process $\{\tilde{x}(t|t_1),\ t \geq t_1\}$, where $\tilde{x}(t|t_1) = x(t) - \hat{x}(t|t_1)$ is the prediction error, is a zero mean Gauss-Markov process whose covariance matrix $P(t|t_1)$ is the solution of the differential equation*

$$\dot{P}(t|t_1) = F(t)P(t|t_1) + P(t|t_1)F'(t) + G(t)Q(t)G'(t) \tag{7.23}$$

for $t \geq t_1$, where $P(t_1|t_1) = E[\tilde{x}(t_1|t_1)\tilde{x}'(t_1|t)]$ is the initial condition.

Proof We apply Theorem 5.4 to the equivalent discrete-time problem to obtain

$$\hat{x}(t + \Delta t | t_1) = \Phi(t + \Delta t, t_1)\hat{x}(t_1 | t_1)$$

and

$$\hat{x}(t | t_1) = \Phi(t, t_1)\hat{x}(t_1 | t_1)$$

where $t + \Delta t \geq t_1$ and $\Delta t > 0$. It follows from these two expressions that

$$\hat{x}(t + \Delta t | t_1) = \Phi(t + \Delta t, t)\Phi(t, t_1)\hat{x}(t_1 | t_1)$$
$$= \Phi(t + \Delta t, t)\hat{x}(t | t_1) \tag{7.24}$$

Substituting into Eq. (7.24) for $\Phi(t + \Delta t, t)$ from Eq. (7.4), we see that

$$\hat{x}(t + \Delta t | t_1) = [I + F(t)\,\Delta t + 0(\Delta t^2)]\hat{x}(t | t_1) \tag{7.25}$$

It is clear that

$$\lim_{\Delta t \to 0} \hat{x}(t + \Delta t | t_1) = \hat{x}(t | t_1)$$

Rewriting Eq. (7.25) as

$$\hat{x}(t + \Delta t | t_1) - \hat{x}(t | t_1) = F(t)\hat{x}(t | t_1)\,\Delta t + 0(\Delta t^2)$$

dividing through by Δt, and taking the limit as $\Delta t \to 0$, we obtain the result

$$\dot{\hat{x}}(t | t_1) = F(t)\hat{x}(t | t_1)$$

for $t \geq t_1 \geq t_0$. The appropriate initial condition is obviously the optimal filtered estimate $\hat{x}(t_1 | t_1)$.

From part *b* of Theorem 5.4, we have that the prediction error stochastic process $\{\tilde{x}(t + \Delta t | t_1), t = t_1 + j\,\Delta t, j = 0, 1, \ldots\}$ is a zero mean Gauss-Markov sequence whose defining relation is

$$\tilde{x}(t + \Delta t | t_1) = \Phi(t + \Delta t, t)\tilde{x}(t | t_1) + \Gamma(t + \Delta t, t)w(t) \tag{7.26}$$

In addition, we have from Eq. (5.40) that the corresponding covariance matrix evolves according to the expression

$$P(t + \Delta t | t_1) = \Phi(t + \Delta t, t)P(t | t_1)\Phi'(t + \Delta t, t)$$
$$+ \Gamma(t + \Delta t, t)\frac{Q(t)}{\Delta t}\Gamma'(t + \Delta t, t) \tag{7.27}$$

Proceeding as before, we put Eq. (7.26) in the form

$$\tilde{x}(t + \Delta t | t_1) - \tilde{x}(t | t_1) = F(t)\tilde{x}(t | t_1)\,\Delta t + G(t)w(t)\,\Delta t + 0(\Delta t^2) \tag{7.28}$$

from which it follows that

$$\dot{\tilde{x}}(t|t_1) = F(t)\tilde{x}(t|t_1) + G(t)w(t)$$

for $t \geq t_1$, which defines the zero mean Gauss-Markov process $\{\tilde{x}(t|t_1), \ t \geq t_1\}$.

Substituting for $\Phi(t + \Delta t, t)$ and $\Gamma(t + \Delta t, t)$ in Eq. (7.27) from Eqs. (7.4) and (7.5), respectively, and expanding the result, we have

$$P(t + \Delta t|t_1) = [I + F(t)\,\Delta t + 0(\Delta t^2)]P(t|t_1)[I + F(t)\,\Delta t + 0(\Delta t^2)]'$$
$$+ [G(t)\,\Delta t + 0(\Delta t^2)]\frac{Q(t)}{\Delta t}[G(t)\,\Delta t + 0(\Delta t^2)]$$
$$= P(t|t_1) + [F(t)P(t|t_1) + P(t|t_1)F'(t)$$
$$+ G(t)Q(t)G'(t)]\,\Delta t + 0(\Delta t^2)$$

from which it follows in the limit as $\Delta t \to 0$ that

$$\dot{P}(t|t_1) = F(t)P(t|t_1) + P(t|t_1)F'(t) + G(t)Q(t)G'(t)$$

giving Eq. (7.23) of the theorem. The appropriate initial condition is obviously $P(t_1|t_1) = E[\tilde{x}(t_1|t_1)\tilde{x}'(t_1|t_1)]$.

Discussion Comparing the filtering and prediction error covariance equations, Eqs. (7.9) and (7.23), respectively, we see that the two differ only by the term $-PH'(t)R^{-1}(t)H(t)P$, which is present in the former. This term can be viewed as a correction term which acts to limit the growth of the filtering error covariance matrix in Eq. (7.9). It is clear that if $R(t) \to \infty$, that is, the measurements are sufficiently noisy as to be considered useless, the two covariance equations are identical.

Even more simply, we note that the optimal predictor is the homogeneous part of the optimal filter, a result that we would intuitively expect in the first place.

Example 7.2 Suppose that the optimal receiver in Example 7.1 has reached steady state and that it is desired to predict the transmitted message after some time $t_1 \gg 0$. Then from Eq. (7.22),

$$\dot{\hat{x}}(t|t_1) = -a\hat{x}(t|t_1)$$

and it follows that

$$\hat{x}(t|t_1) = e^{-a(t-t_1)}\hat{x}(t_1|t_1)$$

for $t \geq t_1 \gg 0$.

We note that the optimal predictor is a pure attenuation, since $0 \leq \exp[-a(t - t_1)] \leq 1$ for all $t \geq t_1$. In the absence of measurements for $t \geq t_1$, the optimal predictor merely reproduces the system's homogeneous response starting with the optimal filtered estimate $\hat{x}(t_1|t_1)$ as its initial condition. Alternately, the optimal predictor simply extrapolates the most recent optimal filtered estimate to obtain the optimal predicted estimate.

The prediction error covariance matrix equation, Eq. (7.23), is

$$\dot{P}(t|t_1) = -2aP(t|t_1) + \sigma_w^2$$

with $P(t_1|t_1) = \rho_1$. Its solution is easily shown to be

$$P(t|t_1) = \left[\rho_1 - \frac{\sigma_w^2}{2a} \right] e^{-2a(t-t_1)} + \frac{\sigma_w^2}{2a}$$

$$= \rho_1 e^{-2a(t-t_1)} + \frac{\sigma_w^2}{2a} [1 - e^{-2a(t-t_1)}]$$

For sufficiently long prediction times, $t \gg t_1$, it is clear that

$$P(t|t_1) \approx \frac{\sigma_w^2}{2a}$$

However, for such prediction times, the prediction error variance has the same value as the variance for the probability distribution of the system's state $x(t)$. This is seen by determining the equilibrium or steady-state solution of

$$\dot{P} = -2aP + \sigma_w^2$$

for $t \geq 0$, where $P = P(t) = E[x^2(t)]$ and $P(0) = \sigma_0^2$.

7.5 OPTIMAL FIXED-INTERVAL SMOOTHING

For the problem of fixed-interval smoothing, we consider the interval $t_0 \leq t \leq t_1$ with t_0 and t_1 fixed. We subdivide the interval into n subintervals each of length $\Delta t = t_1 - t_0/n > 0$. In the limit, we let $n \to \infty$ and $\Delta t \to 0$ such that $n\,\Delta t = t_1 - t_0$.

Since the interval of interest is fixed, the discrete-time index here is $\{t, t = t_0 + j\,\Delta t, j = 0, 1, \ldots, n\}$.

Our concern is with the limiting behavior of $\hat{x}(t + \Delta t|t_1)$ as $\Delta t \to 0$.

With these preliminaries completed, we state and prove the following result.

Theorem 7.3

    a. The optimal fixed-interval smoothed estimate for the system of Eqs. (7.1) and (7.2) is governed by the relation

$$\dot{\hat{x}}(t|t_1) = F(t)\hat{x}(t|t_1) + A(t)[\hat{x}(t|t_1) - \hat{x}(t|t)] \tag{7.29}$$

    for $t_0 \leq t \leq t_1$ where $\hat{x}(t|t)$ is the optimal filtered estimate and $A(t)$ is the $n \times n$ smoothing filter gain matrix. The system in Eq. (7.29) is subject to the boundary condition $\hat{x}(t_1|t_1)$.

b. *The fixed-interval smoothing filter gain matrix is given by the relation*

$$A(t) = G(t)Q(t)G'(t)P^{-1}(t|t) \tag{7.30}$$

where $P(t|t)$ is the optimal filter error covariance matrix.

c. *The stochastic process $\{\tilde{x}(t|t_1),\ t_1 \geq t \geq t_0\}$, where $\tilde{x}(t|t_1) = x(t) - \hat{x}(t|t_1)$ is the fixed-interval smoothing error, is a zero mean Gauss-Markov-2 process whose covariance matrix is the solution of the matrix differential equation*

$$\dot{P}(t|t_1) = [F(t) + A(t)]P(t|t_1) + P(t|t_1)[F(t) + A(t)]'$$
$$- G(t)Q(t)G'(t) \tag{7.31}$$

for $t_0 \leq t \leq t_1$. Equation (7.31) is subject to the boundary condition $P(t_1|t_1) = E[\tilde{x}(t_1|t_1)\tilde{x}'(t_1|t_1)]$, the filter error covariance matrix evaluated at t_1.

Proof Letting t_1 correspond to N, t to k, and $t + \Delta t$ to $k + 1$ in Eqs. (6.57) and (6.58) of Theorem 6.1, we have

$$\hat{x}(t|t_1) = \hat{x}(t|t) + A(t)[\hat{x}(t + \Delta t|t_1) - \hat{x}(t + \Delta t|t)]$$

and

$$A(t) = P(t|t)\Phi'(t + \Delta t,\, t)P^{-1}(t + \Delta t|t) \tag{7.32}$$

respectively.

We rewrite these two equations as

$$\hat{x}(t + \Delta t|t_1) - \hat{x}(t + \Delta t|t) = A^{-1}(t)[\hat{x}(t|t_1) - \hat{x}(t|t)] \tag{7.33}$$

and

$$A^{-1}(t) = P(t + \Delta t|t)[\Phi'(t + \Delta t,\, t)]^{-1}P^{-1}(t|t)$$
$$= P(t + \Delta t|t)\Phi'(t,\, t + \Delta t)P^{-1}(t|t) \tag{7.34}$$

respectively, under the assumption that $P^{-1}(t|t)$ exists. We recall that the inverse of a state transition matrix always exists.

We obtain first an expansion for $A^{-1}(t)$. We see from Eqs. (7.14), (7.4), and (7.5) that

$$P(t + \Delta t|t) = \Phi(t + \Delta t,\, t)P(t|t)\Phi'(t + \Delta t,\, t)$$
$$+ \Gamma(t + \Delta t,\, t)\frac{Q(t)}{\Delta t}\Gamma'(t + \Delta t,\, t)$$
$$= [I + F(t)\,\Delta t + 0(\Delta t^2)]P(t|t)\Phi'(t + \Delta t,\, t)$$
$$+ G(t)Q(t)G'(t)\,\Delta t + 0(\Delta t^2) \tag{7.35}$$

Postmultiplying in Eq. (7.35) by $\Phi'(t, t + \Delta t)$ and noting that

$$\Phi'(t, t + \Delta t) = I - F'(t) \Delta t + 0(\Delta t^2)$$

we obtain

$$P(t + \Delta t|t)\Phi'(t, t + \Delta t) = [I + F(t) \Delta t]P(t|t) + G(t)Q(t)G'(t) \Delta t$$
$$+ 0(\Delta t^2)$$

It now follows that Eq. (7.34) can be written as

$$A^{-1}(t) = I + [F(t) + G(t)Q(t)G'(t)P^{-1}(t|t)] \Delta t + 0(\Delta t^2) \tag{7.36}$$

From Eq. (7.25) with $t_1 = t$, we have

$$\hat{x}(t + \Delta t|t) = [I + F(t) \Delta t + 0(\Delta t^2)]\hat{x}(t|t) \tag{7.37}$$

Substituting Eqs. (7.36) and (7.37) into Eq. (7.33), we have

$$\hat{x}(t + \Delta t|t_1) - [I + F(t) \Delta t + 0(\Delta t^2)]\hat{x}(t|t)$$
$$= [\hat{x}(t|t_1) - \hat{x}(t|t)] + F(t)[\hat{x}(t|t_1) - \hat{x}(t|t)] \Delta t$$
$$+ G(t)Q(t)G'(t)P^{-1}(t|t)[\hat{x}(t|t_1) - \hat{x}(t|t)] \Delta t + 0(\Delta t^2)$$

which reduces to

$$\hat{x}(t + \Delta t|t_1) = \hat{x}(t|t_1) + F(t)\hat{x}(t|t_1) \Delta t + G(t)Q(t)G'(t)P^{-1}(t|t)$$
$$[\hat{x}(t|t_1) - \hat{x}(t|t)] \Delta t + 0(\Delta t^2)$$

Transposing $\hat{x}(t|t_1)$ to the left-hand side, dividing through by Δt, and taking the limit as $\Delta t \rightarrow 0$, we obtain

$$\dot{\hat{x}}(t|t_1) = F(t)\hat{x}(t|t_1) + G(t)Q(t)G'(t)P^{-1}(t|t)[\hat{x}(t|t_1) - \hat{x}(t|t)] \tag{7.38}$$

By redefining $A(t)$ as

$$A(t) = G(t)Q(t)G'(t)P^{-1}(t|t) \tag{7.39}$$

we establish Eqs. (7.29) and (7.30) of the theorem.

It follows from Theorem 6.1 that Eq. (7.38) is subject to the boundary condition $\hat{x}(t_1|t_1)$.

From Theorem 6.1, we know that the fixed-interval smoothing error stochastic process $\{\tilde{x}(t|t_1), t = t_0 + j \Delta t, j = n, n - 1, \ldots, 0\}$ is a zero mean Gauss-Markov-2 sequence whose defining relation is

$$\tilde{x}(t|t_1) = \tilde{x}(t|t) + A(t)[\tilde{x}(t + \Delta t|t_1) - \tilde{x}(t + \Delta t|t)] \tag{7.40}$$

The boundary condition for Eq. (7.40) is $\tilde{x}(t_1|t_1) = x(t_1) - \hat{x}(t_1|t_1)$.

From Eq. (6.59), the covariance matrix for the sequence is given by the expression

$$P(t|t_1) = P(t|t) + A(t)[P(t + \Delta t|t_1) - P(t + \Delta t)]A'(t) \tag{7.41}$$

where the appropriate boundary condition is $P(t_1|t_1) = E[\tilde{x}(t_1|t_1)\tilde{x}'(t_1|t_1)]$.

Let us now consider the limiting behavior of Eq. (7.40). First, we have that

$$\tilde{x}(t + \Delta t|t_1) - \tilde{x}(t + \Delta t|t) = A^{-1}(t)[\tilde{x}(t|t_1) - \tilde{x}(t|t)] \qquad (7.42)$$

For $t_1 = t$, Eq. (7.28) can be written

$$\tilde{x}(t + \Delta t|t) = \tilde{x}(t|t) + F(t)\tilde{x}(t|t) \, \Delta t + G(t)w(t) \, \Delta t + 0(\Delta t^2) \qquad (7.43)$$

Substituting into Eq. (7.42) from Eqs. (7.36) and (7.43), we obtain

$$\tilde{x}(t + \Delta t|t_1) - \tilde{x}(t|t) - F(t)\tilde{x}(t|t) \, \Delta t - G(t)w(t) \, \Delta t + 0(\Delta t^2)$$
$$= [\tilde{x}(t|t_1) - \tilde{x}(t|t)] + F(t)[\tilde{x}(t|t_1) - \tilde{x}(t|t)] \, \Delta t$$
$$+ G(t)Q(t)G'(t)P^{-1}(t|t)[\tilde{x}(t|t_1) - \tilde{x}(t|t)] \, \Delta t + 0(\Delta t^2)$$

which can be simplified to

$$\tilde{x}(t + \Delta t|t_1) - \tilde{x}(t|t_1) = [F(t) + G(t)Q(t)G'(t)P^{-1}(t|t)]\tilde{x}(t|t_1) \, \Delta t$$
$$- G(t)Q(t)G'(t)P^{-1}(t|t)\tilde{x}(t|t) \, \Delta t + G(t)w(t) \, \Delta t + 0(\Delta t^2)$$

Dividing through by Δt and taking the limit as $\Delta t \to 0$, we obtain

$$\dot{\tilde{x}}(t|t_1) = [F(t) + A(t)]\tilde{x}(t|t_1) - A(t)\tilde{x}(t|t) + G(t)w(t)$$

which defines the zero mean Gauss-Markov-2 process $\{\tilde{x}(t|t_1), t_1 \geq t \geq t_0\}$.

Finally, we consider the limiting behavior of Eq. (7.41) as $\Delta t \to 0$ to obtain the matrix differential equation for the fixed-interval smoothing error covariance matrix. We rewrite the equation in the form

$$P(t + \Delta t|t_1) = P(t + \Delta t|t) + A^{-1}(t)[P(t|t_1) - P(t|t)][A'(t)]^{-1} \quad (7.44)$$

From Eqs. (7.32), (7.14), and (7.5),

$$A'(t) = P^{-1}(t + \Delta t|t)\Phi(t + \Delta t, t)P(t|t)$$
$$= [P^{-1}(t|t)\Phi^{-1}(t + \Delta t, t)P(t + \Delta t|t)]^{-1}$$
$$= \{P^{-1}(t|t)\Phi(t, t + \Delta t)[\Phi(t + \Delta t, t)P(t|t)\Phi'(t + \Delta t, t)$$
$$+ G(t)Q(t)G'(t) \, \Delta t + 0(\Delta t^2)]\}^{-1}$$
$$= [\Phi'(t + \Delta t, t)$$
$$+ P^{-1}(t|t)\Phi(t, t + \Delta t)G(t)Q(t)G'(t) \, \Delta t + 0(\Delta t^2)]^{-1}$$

From Eq. (7.4) and the fact that

$$\Phi(t, t + \Delta t) = I - F(t) \, \Delta t + 0(\Delta t^2)$$

we see that

$$A'(t) = [I + F'(t)\,\Delta t + P^{-1}(t|t)G(t)Q(t)G'(t)\,\Delta t + 0(\Delta t^2)]^{-1}$$

from which it follows immediately that

$$[A'(t)]^{-1} = I + [F'(t) + P^{-1}(t|t)G(t)Q(t)G'(t)]\,\Delta t + 0(\Delta t^2) \qquad (7.45)$$

Utilizing Eqs. (7.36) and (7.45), we have

$$
\begin{aligned}
A^{-1}(t)&P(t|t_1)[A'(t)]^{-1}\\
&= \{P(t|t_1) + [F(t) + G(t)Q(t)G'(t)P^{-1}(t|t)]P(t|t_1)\,\Delta t + 0(\Delta t^2)\}\\
&\qquad \{I + [F'(t) + P^{-1}(t|t)G(t)Q(t)G'(t)]\,\Delta t + 0(\Delta t^2)\}\\
&= P(t|t_1) + \{[F(t) + G(t)Q(t)G'(t)P^{-1}(t|t)]P(t|t_1)\\
&\qquad + P(t|t_1)[F'(t) + P^{-1}(t|t)G(t)Q(t)G'(t)]\}\,\Delta t + 0(\Delta t^2) \qquad (7.46)
\end{aligned}
$$

In exactly the same way, we obtain

$$
\begin{aligned}
A^{-1}(t)P(t|t)[A'(t)]^{-1} &= P(t|t) + \{[F(t) + G(t)Q(t)G'(t)P^{-1}(t|t)]P(t|t)\\
&\qquad + P(t|t)[F'(t) + P^{-1}(t|t)G(t)Q(t)G'(t)]\}\,\Delta t + 0(\Delta t^2)
\end{aligned}
$$

which can be simplified to

$$
\begin{aligned}
A^{-1}(t)P(t|t)[A'(t)]^{-1} &= P(t|t)\\
&+ [F(t)P(t|t) + P(t|t)F'(t) + 2G(t)Q(t)G'(t)]\,\Delta t + 0(\Delta t^2) \qquad (7.47)
\end{aligned}
$$

It now follows from Eqs. (7.15) and (7.47) that

$$
\begin{aligned}
P(t + \Delta t|t) &- A^{-1}(t)P(t|t)[A'(t)]^{-1}\\
&= P(t|t) + [F(t)P(t|t) + P(t|t)F'(t) + G(t)Q(t)G'(t)]\,\Delta t - P(t|t)\\
&\qquad - [F(t)P(t|t) + P(t|t)F'(t) + 2G(t)Q(t)G'(t)]\,\Delta t + 0(\Delta t^2)\\
&= -G(t)Q(t)G'(t)\,\Delta t + 0(\Delta t^2) \qquad\qquad\qquad\qquad\qquad (7.48)
\end{aligned}
$$

Substituting Eqs. (7.46) and (7.48) into Eq. (7.44), we obtain the result

$$
\begin{aligned}
P(t + \Delta t|t_1) &= -G(t)Q(t)G'(t)\,\Delta t + P(t|t_1)\\
&+ \{[F(t) + G(t)Q(t)G'(t)P^{-1}(t|t)]P(t|t_1) + P(t|t_1)\\
&\qquad [F(t) + P^{-1}(t|t)G(t)Q(t)G'(t)]P(t_1|t_1)\}\,\Delta t + 0(\Delta t^2)
\end{aligned}
$$

Hence,

$$
\begin{aligned}
\dot{P}(t|t_1) &= [F(t) + G(t)Q(t)G'(t)P^{-1}(t|t)]P(t|t_1) + P(t|t_1)[F(t)\\
&\qquad + P^{-1}(t|t)G(t)Q(t)G'(t)] - G(t)Q(t)G'(t) \qquad (7.49)
\end{aligned}
$$

for $t_0 \leq t \leq t_1$. Utilizing the definition of $A(t)$ which was given in Eq. (7.39), we obtain Eq. (7.31) of the theorem.

From Theorem 6.1, the boundary condition for Eq. (7.49) is obviously $P(t_1|t_1) = E[\tilde{x}(t_1|t_1)\tilde{x}'(t_1|t_1)]$.

Discussion   The algorithm in Theorem 7.3 was developed by Rauch et al. [2] utilizing Kalman's limiting procedure [1].   A slightly different version of this algorithm had previously been developed by Bryson and Frazier [5], who utilized techniques from the calculus of variations and maximum likelihood estimation.   In their paper, Rauch et al. showed that the two algorithms are equivalent.

The block diagram for the optimal fixed-interval smoothing filter is shown in Fig. 7.3.   As in the case of the optimal filter, the smoothing filter is seen to consist of the system dynamics model, $\dot{x} = F(t)x$, forced by a feedback correction signal.   The correction signal here, however, does not involve the measurement data explicitly.   Instead, the input "measurement" is the optimal filtered estimate $\hat{x}(t|t)$, and the "residual" is $\hat{x}(t|t_1) - \hat{x}(t|t)$, the difference between two optimal estimates.

Additionally, the smoothing filter is subject to a terminal condition instead of an initial condition.   This means, of course, that integration is carried out backward in time from t_1 to t_0 to obtain the optimal fixed-interval smoothed estimate.   In order to do this, it is necessary to have $\hat{x}(t|t)$ previously determined for all t, $t_0 \leq t \leq t_1$.   Hence, the filtering problem must be solved before fixed-interval smoothing can be initiated. As a consequence, fixed-interval smoothing is only useful for post-experimental data analysis.   It cannot be used for on-line, real-time data processing as can the optimal filter.   This, of course, is simply a consequence of the discrete-time smoothing algorithm from which the above results are derived.

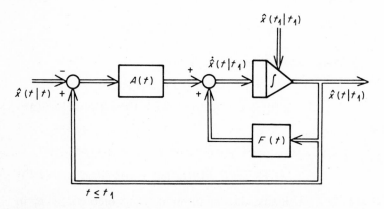

Fig. 7.3 Block diagram of optimal fixed-interval smoothing filter.

Recalling Eq. (7.30) for the smoothing filter gain matrix,

$$A(t) = G(t)Q(t)G'(t)P^{-1}(t|t)$$

we note that the filtering error covariance matrix $P(t|t)$ must be known for $t_0 \leq t \leq t_1$ in order to implement the smoothing filter. This is not an additional computational requirement above and beyond the one that $\hat{x}(t|t)$ be known for $t_0 \leq t \leq t_1$, since $P(t|t)$ must be computed anyhow in the process of determining $\hat{x}(t|t)$. However, one rather obvious problem does exist here. Namely, if $P(t|t)$ is singular at some point or over some subinterval in $[t_0,t_1]$, $A(t)$ does not exist at these time points. A related problem exists even if $P(t|t)$ is nonsingular. If the determinant of $P(t|t)$ is small, the matrix may be "nearly singular" or "ill-conditioned," in which case the attendant numerical errors, such as roundoff errors, may lead to incorrect values for $P^{-1}(t|t)$ in applications.

Rather than computing $P^{-1}(t|t)$ at each t, $t_0 \leq t \leq t_1$, one possible approach is to develop a differential equation which could be solved directly to obtain $P^{-1}(t|t)$. This can be done by taking the time derivative of the identity

$$P(t|t)P^{-1}(t|t) = I$$

to obtain

$$\frac{dP}{dt}P^{-1} + P\frac{dP^{-1}}{dt} = 0$$

Then,

$$\frac{dP^{-1}}{dt} = -P^{-1}\frac{dP}{dt}P^{-1} \tag{7.50}$$

Substituting into Eq. (7.50) from Eq. (7.9), we have

$$\frac{dP^{-1}}{dt} = -P^{-1}[F(t)P + PF'(t) - PH'(t)R^{-1}(t)H(t)P$$
$$+ G(t)Q(t)G'(t)]P^{-1}$$
$$= -P^{-1}F(t) - F'(t)P^{-1} + H'(t)R^{-1}(t)H(t)$$
$$- P^{-1}G(t)Q(t)G'(t)P^{-1}(t) \tag{7.51}$$

Letting $M(t) = P^{-1}(t|t)$, we can express Eq. (7.51) in the form

$$-\dot{M} = F'(t)M + MF(t) + MG(t)Q(t)G'(t)M - H'(t)R^{-1}(t)H(t) \tag{7.52}$$

Assuming that $P(t_1|t_1)$ is nonsingular, the boundary condition for Eq. (7.52) is $M(t_1) = P^{-1}(t_1|t_1)$.

The optimal fixed-interval smoothing filter gain matrix is then

$$A(t) = G(t)Q(t)G'(t)M(t)$$

We note from the theorem that computation of $P(t|t_1)$ is not an integral part of the mechanization of the optimal smoothing filter in the sense that computation of $P(t|t)$ is for the optimal filter. Indeed, $P(t|t_1)$ cannot be determined until $A(t)$ is, while in optimal filtering, $K(t)$ cannot be determined until $P(t|t)$ is.

Finally, we remark on the Markov-2 nature of the error process $\{\bar{x}(t_1|t), t_0 \leq t \times t_1\}$. We recall that the process's defining relation is

$$\dot{\bar{x}}(t|t_1) = [F(t) + A(t)]\bar{x}(t|t_1) - A(t)\bar{x}(t|t) + G(t)w(t)$$

which is subject to the boundary condition $\bar{x}(t_1|t_1) = x(t_1) - \hat{x}(t_1|t_1)$, a zero mean gaussian random n vector with known covariance matrix $P(t_1|t_1)$.

We know that $\{\bar{x}(t|t), t_0 \leq t \leq t_1\}$ is a zero mean Gauss-Markov process with

$$\dot{\bar{x}}(t|t) = [F(t) - K(t)H(t)]\bar{x}(t|t) + G(t)w(t) - K(t)v(t)$$

for which the boundary condition is also $\bar{x}(t_1|t_1) = x(t_1) - \hat{x}(t_1|t_1)$.

Defining the $2n$ vector

$$y(t) = \begin{bmatrix} \bar{x}(t|t_1) \\ \hline \bar{x}(t|t) \end{bmatrix}$$

we combine the above two equations to obtain

$$\dot{y} = \left[\begin{array}{c|c} F(t) + A(t) & -A(t) \\ \hline 0 & F(t) - K(t)H(t) \end{array}\right] y + \begin{bmatrix} G(t) \\ \hline G(t) \end{bmatrix} w(t)$$
$$- \begin{bmatrix} 0 \\ \hline K(t) \end{bmatrix} v(t)$$

It follows from our results in Sec. 4.4 that $\{y(t), t_0 \leq t \leq t_1\}$ is a zero mean Gauss-Markov process.

Strictly speaking, we know from our previous work that the description of $\{\bar{x}(t|t_1), t_0 \leq t \leq t_1\}$ in terms of its mean (zero) and its covariance matrix $P(t|t_1)$ is incomplete, since the process is not Markov-1. The correct description requires that we consider $\{y(t), t_0 \leq t \leq t_1\}$. However, for most practical purposes, it is sufficient to determine $P(t|t_1)$ in performing error analyses. This point has already been noted in connection with fixed-interval smoothing for discrete linear systems.

Example 7.3   As a particular class of problems in optimal fixed-interval smoothing we consider those cases where $Q(t) = 0$ for $t_0 \leq t \leq t_1$, that is, there is no system disturbance (cf. Example 6.2).   We assume that $P^{-1}(t|t)$ exists for $t_0 \leq t \leq t_1$.

It follows from Eq. (7.30) that $A(t) = 0$ for $t_0 \leq t \leq t_1$.   Hence, the optimal smoothing filter is defined by

$$\dot{\hat{x}}(t|t_1) = F(t)\hat{x}(t|t_1) \tag{7.53}$$

whose solution can be written

$$\hat{x}(t|t_1) = \Phi(t,t_1)\hat{x}(t_1|t_1)$$

where $t_0 \leq t \leq t_1$ and $\Phi(t,t_1)$ is the state transition matrix of the system in Eq. (7.53).   In addition,

$$\dot{P}(t|t_1) = F(t)P(t|t_1) + P(t|t_1)F'(t) \tag{7.54}$$

from which it follows that

$$P(t|t_1) = \Phi(t,t_1)P(t_1|t_1)\Phi'(t,t_1)$$

In this class of problems, optimal fixed-interval smoothing obviously consists of backward extrapolation of $\hat{x}(t_1|t_1)$ and $P(t_1|t_1)$.

In practice, one would not compute $\Phi(t,t_1)$ for all $t_0 \leq t \leq t_1$ to determine $\hat{x}(t|t_1)$ and $P(t|t_1)$ as indicated above except in trivial cases.   Instead, Eqs. (7.53) and (7.54) would be numerically integrated backward in time from t_1 starting with the appropriate boundary conditions in order to obtain $\hat{x}(t|t_1)$ and $P(t|t_1)$, respectively.

Example 7.4   Let us suppose that for some $t_1 \gg 0$, the optimal filter in Example 7.1 has reached the steady state and that we wish to obtain a smoothed estimate of the state over a fixed interval $[t_0,t_1]$ where $t_0 \gg 0$, but $t_0 < t_1$.

For simplicity we consider case 3 (page 258) of the example where $a^2 \ll \sigma_w^2/\sigma_v^2$ and the steady-state filtering error variance is

$$P(t|t) = \rho_1 = \sigma_w\sigma_v$$

To review, we have $F(t) = -a$, $G(t) = 1$, and $Q(t) = \sigma_w^2$.

From Eq. (7.30),

$$A(t) = G(t)Q(t)G'(t)P^{-1}(t|t)$$

$$= \sigma_w^2 \frac{1}{\rho_1} = \frac{\sigma_w}{\sigma_v}$$

Since $a^2 \ll \sigma_w^2/\sigma_v^2$, we have $\mu \approx \sigma_w/\sigma_v$ in Example 7.1.   Then, from Eq. (7.29), the optimal fixed-interval smoothing filter is defined by the relation

$$\dot{\hat{x}}(t|t_1) = -a\hat{x}(t|t_1) + \mu[\hat{x}(t|t_1) - \hat{x}(t|t)]$$

$$= (\mu - a)\hat{x}(t|t_1) - \mu\hat{x}(t|t)$$

Again, since $\mu \approx \sigma_w/\sigma_v \gg a$, we can write

$$\dot{\hat{x}}(t|t_1) \approx \mu[\hat{x}(t|t_1) - \hat{x}(t|t)] \tag{7.55}$$

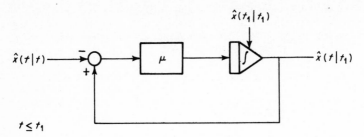

Fig. 7.4 Optimal fixed-interval smoothing filter for Example 7.4.

for $t_0 \leq t \leq t_1$ which we recall is subject to the boundary condition $\hat{x}(t_1|t_1)$. The block diagram for the smoothing filter of Eq. (7.55) is shown in Fig. 7.4 where it is emphasized that the filter operates in reverse time from t_1.

For this example, the smoothing error variance equation, Eq. (7.31), is

$$\dot{P}(t|t_1) = 2(\mu - a)P(t|t_1) - \sigma_w^2$$

for $t_0 \leq t \leq t_1$, where $P(t_1|t_1) = \rho_1 = \sigma_w \sigma_v$. Since $\mu \gg a$, we use the approximation

$$\dot{P}(t|t_1) \approx 2\mu P(t|t_1) - \sigma_w^2 \tag{7.56}$$

The solution of Eq. (7.56) is

$$P(t|t_1) = \beta e^{2\mu t} + \frac{\sigma_w^2}{2\mu}$$

$$= \beta e^{2\mu t} + \frac{\sigma_w \sigma_v}{2}$$

where β is the constant of integration. It then follows from the boundary condition that

$$\beta = \frac{\sigma_w \sigma_v}{2} e^{-2\mu t_1}$$

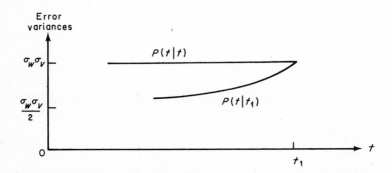

Fig. 7.5 Comparison of optimal filtering and fixed-interval smoothing error variance for system in Example 7.4.

Hence,

$$P(t|t_1) = \frac{\sigma_w \sigma_v}{2} [1 + e^{2\mu(t-t_1)}]$$

Since $t \leq t_1$, it follows that $P(t|t_1) \leq \sigma_w \sigma_v = P(t|t)$ with equality holding only for $t = t_1$.

The comparison of the filtering and fixed-interval smoothing error variances is given in the curves in Fig. 7.5. In this connection we note that for $t \ll t_1$,

$$P(t|t_1) \approx \frac{\sigma_w \sigma_v}{2}$$

assuming that, for the particular value of t which is chosen, the optimal filter is in the steady state; that is, $P(t|t) = \sigma_w \sigma_v$.

7.6 OPTIMAL FIXED-POINT SMOOTHING

If one is interested in a smoothed estimate of the system state $x(t)$ at only one time point in some interval $[t_0, t_1]$, $t_0 < t_1$, it is obvious that the optimal fixed-interval smoothing algorithm of Theorem 7.3 could be used for this purpose. Operation of the smoothing filter would terminate at the desired time point instead of continuing, backward in time, to t_0. However, just as for discrete-time problems, this procedure has two disadvantages. First, it is post-experimental in nature, and, therefore, cannot be used for real-time data processing. Secondly, it is inefficient, since it must compute the entire time history of the smoothed estimate $\hat{x}(t|t_1)$ from t_1 back to the time point of interest. Hence, it is desirable to have available an algorithm for optimal fixed-point smoothing for continuous linear systems which can operate on-line.

In this section, we develop such an algorithm, starting with the results in Theorem 6.2. We assume that the fixed time point of interest is some $t_1 \geq t_0$ and that the optimal filtered estimate $\hat{x}(t_1|t_1)$ has been obtained. Our concern then is with the optimal estimate $\hat{x}(t_1|t)$, $t \geq t_1$.

The discrete time index here will be $\{t, t = t_1 + j\,\Delta t, j = 0, 1, \ldots\}$ where $\Delta t > 0$.

Theorem 7.4 [3]

    a. *The optimal fixed-point smoothed estimate for the system of Eqs. (7.1) and (7.2) is given by the expression*

$$\dot{\hat{x}}(t_1|t) = B(t)K(t)[z(t) - H(t)\hat{x}(t|t)] \tag{7.57}$$

for $t \geq t_1$ where $K(t)$ is the optimal filter gain matrix, $\hat{x}(t_1|t_1)$ is the initial condition, and $B(t)$ is the $n \times n$ fixed-point smoothing filter gain matrix.

b. $B(t)$ *is the solution of the matrix linear differential equation*

$$\dot{B} = -B[F(t) + A(t)] \tag{7.58}$$

where $t \geq t_1$, $B(t_1) = I$, *and* $A(t) = G(t)Q(t)G'(t)P^{-1}(t|t)$.

c. *The stochastic process* $\{\tilde{x}(t_1|t_2),\ t \geq t_1\}$, *where* $\tilde{x}(t|t_1) = x(t_1) - \hat{x}(t_1|t)$ *is the fixed-point smoothing error, is a zero mean Gauss-Markov-2 process whose covariance matrix is the solution of the matrix linear differential equation*

$$\dot{P}(t_1|t) = -B(t)K(t)R(t)K'(t)B'(t) \tag{7.59}$$

$t \geq t_1$, *for which the initial condition is* $P(t_1|t_1) = E[\tilde{x}(t_1|t_1)\tilde{x}'(t_1|t_1)]$.

Proof Letting t_1 correspond to k, $t + \Delta t$ to j, and t to $j - 1$ in Eqs. (6.76) and (6.77) of Theorem 6.2, we have

$$\hat{x}(t_1|t + \Delta t) = \hat{x}(t_1|t) + B(t + \Delta t)[\hat{x}(t + \Delta t|t + \Delta t) - \hat{x}(t + \Delta t|t)] \tag{7.60}$$

and

$$B(t + \Delta t) = \prod_{\tau = t_1}^{t} A(\tau) \tag{7.61}$$

where

$$A(\tau) = P(\tau|\tau)\Phi'(\tau + \Delta t, \tau)P^{-1}(\tau + \Delta t|\tau) \tag{7.62}$$

$t = t_1 + j\,\Delta t$, $j = 0, 1, \ldots$, and $\tau = t_1 + i\,\Delta t$, $i = 0, 1, \ldots$. From the optimal filtering solution, we recall that

$$\hat{x}(t + \Delta t|t + \Delta t) - \hat{x}(t + \Delta t|t)$$
$$= K(t + \Delta t)[z(t + \Delta t) - H(t + \Delta t)\Phi(t + \Delta t, t)\hat{x}(t|t)]$$

Hence, Eq. (7.60) can be written as

$$\hat{x}(t_1|t + \Delta t) - \hat{x}(t_1|t)$$
$$= B(t + \Delta t)K(t + \Delta t)[z(t + \Delta t) - H(t + \Delta t)\Phi(t + \Delta t, t)\hat{x}(t|t)]$$

It follows then that

$$\dot{\hat{x}}(t_1|t) = B(t)K(t)[z(t) - H(t)\hat{x}(t|t)] \tag{7.63}$$

for $t \geq t_1$ if

$$\lim_{\Delta t \to 0} B(t + \Delta t) = B(t)$$

exists. From Theorem 6.2, the appropriate initial condition for Eq. (7.63) is $\hat{x}(t_1|t_1)$.

We now show that this limit exists and obtain the differential equation for $B(t)$.

From Eqs. (7.61) and (7.62)

$$B(t + \Delta t) = \left[\prod_{\tau = t_1}^{t - \Delta t} A(\tau) \right] A(t)$$

$$= B(t)A(t) \qquad (7.64)$$

where

$$A(t) = P(t|t)\Phi'(t + \Delta t, t)P^{-1}(t + \Delta t|t)$$

Then, since

$$\lim_{\Delta t \to 0} A(t) = I$$

it follows from Eq. (7.64) that

$$\lim_{\Delta t \to 0} B(t + \Delta t) = B(t)$$

Rewriting Eq. (7.64) in the form

$$B(t + \Delta t)A^{-1}(t) = B(t)$$

and substituting into this result from Eq. (7.36), we obtain

$$B(t + \Delta t)\{I + [F(t) + G(t)Q(t)G'(t)P^{-1}(t|t)]\,\Delta t + 0(\Delta t^2)\} = B(t)$$

Expanding and rearranging terms,

$$B(t + \Delta t) - B(t) = -B(t + \Delta t)\{[F(t) + G(t)Q(t)G'(t)P^{-1}(t|t)]\,\Delta t$$
$$+ 0(\Delta t^2)\}$$

Finally, dividing through by Δt and taking the limit as $\Delta t \to 0$, we have

$$\dot{B} = -B[F(t) + G(t)Q(t)G'(t)P^{-1}(t|t)]$$

for $t \geq t_1$. Noting the definition of $A(t)$ then gives us Eq. (7.58) of the theorem.

Setting $t = t_1$ in Eq. (7.61), we get

$$B(t_1 + \Delta t) = A(t_1)$$

Since

$$\lim_{\Delta t \to 0} B(t_1 + \Delta t) = B(t_1)$$

it is clear that

$$B(t_1) = \lim_{\Delta t \to 0} A(t_1)$$

But from Eq. (7.62),

$$\lim_{\Delta t \to 0} A(t_1) = \lim_{\Delta t \to 0} P(t_1|t_1)\Phi'(t_1 + \Delta t, t_1)P^{-1}(t_1 + \Delta t|t_1)$$
$$= I$$

Hence, the required initial condition is $B(t_1) = I$.

The defining equation for the Gauss-Markov-2 process $\{\tilde{x}(t_1|t), t \geq t_1\}$, where $\tilde{x}(t_1|t) = x(t_1) - \hat{x}(t_1|t)$ is the fixed-point smoothing error, can be developed utilizing the same procedure as in the proof of Theorem 7.3. We leave the details as an exercise and develop the appropriate differential equation for $P(t_1|t) = E[\tilde{x}(t_1|t)\tilde{x}'(t_1|t)]$.

From Eq. (6.80) of Theorem 6.2, we have, with k replaced by t_1, j by $t + \Delta t$, and $j - 1$ by t, the result

$$P(t_1|t + \Delta t) - P(t_1|t) = -B(t + \Delta t)K(t + \Delta t)H(t + \Delta t)$$
$$P(t + \Delta t|t)B'(t + \Delta t)$$

Dividing through by Δt and letting $\Delta t \to 0$, we see that

$$\dot{P}(t_1|t) = -B(t)P(t|t)H'(t)R^{-1}(t)H(t)P(t|t)B'(t) \qquad (7.65)$$

for $t \geq t_1$, where we have made use of the previously established facts that

$$\lim_{\Delta t \to 0} B(t + \Delta t) = B(t)$$

$$\lim_{\Delta t \to 0} \frac{K(t + \Delta t)}{\Delta t} = P(t|t)H'(t)R^{-1}(t)$$

and

$$\lim_{\Delta t \to 0} P(t + \Delta t|t) = P(t|t)$$

We recall from Eq. (7.8) that

$$K(t) = P(t|t)H'(t)R^{-1}(t)$$

from which it is clear that

$$H(t)P(t|t) = R(t)K'(t)$$

Substituting these two results into Eq. (7.65), we obtain the desired result

$$\dot{P}(t_1|t) = -B(t)K(t)R(t)K'(t)B'(t)$$

The initial condition is obviously $P(t_1|t_1) = E[\tilde{x}(t_1|t_1)\tilde{x}'(t_1|t_1)]$.

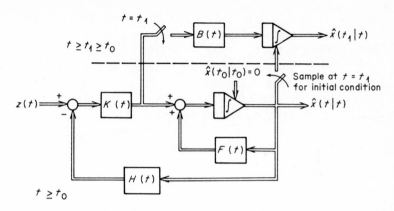

Fig. 7.6 Block diagram of optimal fixed-point smoothing filter.

Discussion The block diagram for the optimal fixed-point smoothing filter is presented in Fig. 7.6, where the optimal filter representation is included to emphasize the dependence of the former on the latter.

The smoothing filter is, of course, inoperative until $t \geq t_1$, while the optimal filter starts at $t = t_0$. It is clear that both filters operate on the same time scale after t_1.

The utility of the smoothing filter for determining the optimal estimate of x at one particular time in an on-line application is quite evident. The operation is independent of whether or not the time at which measurements are no longer available is known a priori.

However, it is not necessary to continue smoothing until measurements are no longer available. Instead, $P(t_1|t)$ could be computed along with $\hat{x}(t_1|t)$ and smoothing discontinued whenever, say, the trace of $P(t_1|t)$, which is the mean square fixed-point smoothing error, dropped below a specified threshold. The smoothing filter would then be available for fixed-point smoothing, beginning at some new t_1. In this connection, it is seen that computation of $P(t_1|t)$ is quite simple numerically, since, from Eq. (7.59),

$$P(t_1|t) = P(t_1|t_1) - \int_{t_1}^{t} B(\tau)K(\tau)R^{-1}(\tau)K'(\tau)B'(\tau) \, d\tau$$

where $t \geq t_1$.

In contrast to the expression for the optimal fixed-interval smoothing filter gain matrix $A(t)$, Eq. (7.30), which is a particular matrix product, we note that computation of the fixed-point smoothing filter gain matrix requires the solution of a matrix linear differential equation, Eq. (7.58). In particular, since $B(t)$ is not necessarily symmetric, Eq. (7.58) is a system of n^2 equations. Moreover, the coefficient in Eq. (7.58) includes $P^{-1}(t|t)$ with its attendant computational problems. The difficulty can be avoided, in part, by utilizing Eq. (7.52) to solve directly for $M(t) = P^{-1}(t|t)$, $t \geq t_1$. In this case, the initial condition for Eq. (7.52) is $M(t_1) = P^{-1}(t_1|t_1)$. Clearly, $P(t_1|t_1)$ must be nonsingular in this case.

Alternately, it is possible to obtain a fixed-point smoothing filter gain matrix formulation which avoids $P^{-1}(t|t)$ entirely. The result is given in Prob. 7.10; it is also developed in Sec. 8.4.

Example 7.5 Let us consider the problem of fixed-point smoothing for the same class of systems as we treated in Example 7.3, viz., those where there is no system noise, so that $Q(t) = 0$ for all $t \geq t_0$.

Assuming that $P(t|t)$ is nonsingular for all $t \geq t_1$, we have that

$$G(t)Q(t)G'(t)P^{-1}(t|t) = 0$$

so that Eq. (7.58) becomes

$$\dot{B} = -BF(t) \tag{7.66}$$

with $B(t_1) = I$. It is easily verified that the solution of this equation is

$$B(t) = \Phi(t_1,t) \tag{7.67}$$

for $t \geq t_1$, where $\Phi(t_1,t)$ is the state transition matrix of the system $\dot{x} = F(t)x$.

Substituting into Eq. (7.57) from Eq. (7.67), we have

$$\dot{\hat{x}}(t_1|t) = \Phi(t_1,t)K(t)[z(t) - H(t)\hat{x}(t|t)]$$

However, from Eq. (7.7),

$$K(t)z(t) = \dot{\hat{x}}(t|t) - F(t)\hat{x}(t|t)$$

$$\dot{\hat{x}}(t_1|t) = \Phi(t_1,t)[\dot{\hat{x}}(t|t) - F(t)\hat{x}(t|t)]$$

$$= \Phi(t_1,t)\dot{\hat{x}}(t|t) - \Phi(t_1,t)F(t)\hat{x}(t|t)$$

Since $\Phi(t_1,t)$ is the solution of Eq. (7.66), we see that

$$\dot{\hat{x}}(t_1|t) = \Phi(t_1,t)\dot{\hat{x}}(t|t) + \dot{\Phi}(t_1,t)\hat{x}(t|t)$$

$$= \frac{d}{dt}[\Phi(t_1,t)\hat{x}(t|t)]$$

and it follows immediately that

$$\hat{x}(t_1|t) = \Phi(t_1,t)\hat{x}(t|t) \tag{7.68}$$

for $t \geq t_1$. Hence, when $Q(t) = 0$, the optimal fixed-point smoothed estimate is the most recent optimal filtered estimate propagated back to the time point of interest through the state transition matrix.

Turning now to the fixed-point smoothing error covariance equation, Eq. (7.59), and substituting into it from Eqs. (7.67) and (7.8), we have

$$\dot{P}(t_1|t) = -\Phi(t_1,t)P(t|t)H'(t)R^{-1}(t)H(t)P(t|t)\Phi'(t_1,t)$$

But from Eq. (7.9),

$$-P(t|t)H'(t)R^{-1}(t)H(t)P(t|t) = \dot{P}(t|t) - F(t)P(t|t) - P(t|t)F'(t)$$

when $Q(t) = 0$. Then, it is clear that

$$\dot{P}(t_1|t) = \Phi(t_1,t)\dot{P}(t|t)\Phi'(t_1,t) - \Phi(t_1,t)F(t)P(t|t)\Phi'(t_1,t) - \Phi(t_1,t)P(t|t)F'(t)\Phi'(t_1,t)$$
$$= \Phi(t_1,t)\dot{P}(t|t)\Phi'(t_1,t) + \dot{\Phi}(t_1,t)P(t|t)\Phi'(t_1,t) + \Phi(t_1,t)P(t|t)\dot{\Phi}'(t_1,t)$$
$$= \frac{d}{dt}[\Phi(t_1,t)P(t|t)\Phi'(t_1,t)]$$

where we have made use of the results in Eqs. (7.66) and (7.67). Consequently,

$$P(t_1|t) = \Phi(t_1,t)P(t|t)\Phi'(t_1,t) \tag{7.69}$$

for $t \geq t_1$.

As a specific problem illustrating the use of Eqs. (7.68) and (7.69), we consider the question of determining the initial concentration of a reacting substance in a first-order chemical reaction. In such a reaction, the rate at which the substance is being used up is proportional to the amount of the substance instantaneously present. Letting x denote the concentration, we have

$$\dot{x} = -ax \tag{7.70}$$

where $a = $ constant > 0. We take $t_0 = 0$ and assume that the initial concentration can be approximated as a gaussian random variable with mean $\bar{x}(0)$ and variance $\sigma_0{}^2$.

For purposes of quantitative analysis, we would like to reduce the uncertainty $\sigma_0{}^2$ associated with our knowledge of the initial concentration. We do this by monitoring the concentration during the reaction and applying fixed-point smoothing. We assume that our measurement process can be modeled by the relation

$$z(t) = x(t) + v(t)$$

where $\{v(t), t \geq 0\}$ is a scalar zero mean gaussian white noise which is independent of $x(0)$ and has a variance $\sigma_v{}^2$.

Since we are interested in $x(0)$, we have $t_1 = t_0 = 0$.

The filtering problem which we must first solve here is the same as in Example 7.1, except that now $Q(t) = \sigma_w{}^2 = 0$ and $x(0)$ has nonzero mean.

From Example 7.1, we have

$$\rho_1 = (-a + \sqrt{a^2})\sigma_v{}^2 = 0$$
$$\rho_2 = (-a - \sqrt{a^2})\sigma_v{}^2 = -2a\sigma_v{}^2$$
$$\mu = \sqrt{a^2} = a$$

and

$$\alpha = \frac{\sigma_0{}^2}{\sigma_0{}^2 + 2a\sigma_v{}^2}$$

so that

$$P(t|t) = \frac{(2a\sigma_0{}^2\sigma_v{}^2/\sigma_0{}^2 + 2a\sigma_v{}^2)e^{-2\alpha t}}{1 - (\sigma_0{}^2/\sigma_0{}^2 + 2a\sigma_v{}^2)e^{-2\alpha t}} = \frac{2a\sigma_0{}^2\sigma_v{}^2 e^{-2\alpha t}}{(\sigma_0{}^2 + 2a\sigma_v{}^2) - \sigma_0{}^2 e^{-2\alpha t}}$$

and

$$K(t) = \frac{1}{\sigma_v{}^2} P(t|t)$$

The optimal filter equation is

$$\dot{\hat{x}}(t|t) = -a\hat{x}(t|t) + K(t)[z(t) - \hat{x}(t|t)]$$

for $t \geq 0$ with $\hat{x}(0|0) = \bar{x}(0)$ to account for the fact that $x(0)$ has nonzero mean.

For the system of Eq. (7.70), we have

$$\Phi(t_1,t) = \Phi(0,t) = e^{at}$$

where $t \geq 0$.

From Eqs. (7.68) and (7.69), it then follows that

$$\hat{x}(0|t) = e^{at}\hat{x}(t|t)$$

and

$$P(0|t) = \frac{2a\sigma_0{}^2\sigma_v{}^2}{(\sigma_0{}^2 + 2a\sigma_v{}^2) - \sigma_0{}^2 e^{-2at}}$$

$$= \frac{2a\sigma_v{}^2}{[1 + 2a(\sigma_v{}^2/\sigma_0{}^2)] - e^{-2at}}$$

respectively.

For a sufficiently large $\sigma_0{}^2$, we see that

$$P(0|t) \approx \frac{2a\sigma_v{}^2}{1 - e^{-2at}}$$

Noting that the time constant for the chemical reaction is $T = 1/a$ and assuming that the reaction is essentially completed in four time constants, $t = 4T$, we have

$$P(0|T) \approx \frac{2a\sigma_v{}^2}{1 - e^{-8}} \cong 2a\sigma_v{}^2$$

as a performance limit on estimation of the initial concentration when $\sigma_0{}^2$ is arbitrarily large.

7.7 OPTIMAL FIXED–LAG SMOOTHING

We conclude this chapter by examining the problem of optimal fixed-lag smoothing for systems of the type in Eqs. (7.1) and (7.2). We are concerned with the estimate $\hat{x}(t|t + T)$, $t \geq t_0$, where $T = \text{constant} > 0$ and is called the lag. We note that T is the fixed increment of time by which the estimate lags behind the time of the most recent measurement.

We recall from the discussion in Sec. 6.2 that fixed-lag smoothing is of interest in telemetry and communication problems where one is willing to allow a lag in the estimate.

Theorem 7.5 [4]

a. *The optimal fixed-lag smoothed estimate for the system of Eqs.* (7.1) *and* (7.2) *is given by the expression*

$$\dot{\hat{x}}(t|t + T) = F(t)\hat{x}(t|t + T) + C(t + T)K(t + T)$$
$$[z(t + T) - H(t + T)\hat{x}(t + T|t + T)]$$
$$+ A(t)[\hat{x}(t|t + T) - \hat{x}(t|t)] \quad (7.71)$$

for $t \geq t_0$ *where* $K(t) = P(t|t)H'(t)R^{-1}(t)$,

$$A(t) = G(t)Q(t)G'(t)P^{-1}(t|t)$$

$C(t + T)$ *is the* $n \times n$ *fixed-lag smoothing filter gain matrix, and the initial condition is the optimal fixed-point smoothed estimate* $\hat{x}(t_0|t_0 + T)$.

b. $C(t + T)$ *is the solution of the matrix linear differential equation*

$$\dot{C}(t + T) = [F(t) + A(t)]C(t + T)$$
$$- C(t + T)[F(t + T) + A(t + T)] \quad (7.72)$$

where $t \geq t_0$ *and the initial condition is* $C(t_0 + T) = B(t_0 + T)$, *the optimal fixed-point smoothing filter gain matrix evaluated at* $t_0 + T$.

c. *The stochastic process* $\{\tilde{x}(t|t + T), t \geq t_0\}$, *where*

$$\tilde{x}(t|t + T) = x(t) - \hat{x}(t|t + T)$$

is the fixed-lag smoothing error, is a zero mean Gauss-Markov-2 process whose covariance matrix is the solution of the matrix linear differential equation

$$\dot{P}(t|t + T) = [F(t) + A(t)]P(t|t + T) + P(t|t + T)[F(t) + A(t)]'$$
$$- C(t + T)K(t + T)R(t + T)K'(t + T)C'(t + T)$$
$$- G(t)Q(t)G'(t) \quad (7.73)$$

$t \geq t_0$, *for which the initial condition is*

$$P(t_0|t_0 + T) = E[\tilde{x}(t_0|t_0 + T)\tilde{x}'(t_0|t_0 + T)]$$

the optimal fixed-point smoothing error covariance matrix evaluated at $t_0 + T$.

Proof The proof is left as an exercise. All the techniques and limits which are required have been presented in the proofs of the preceding theorems in this chapter.

Discussion As in the discrete-time case, optimal continuous-time fixed-lag smoothing involves two correction terms in the smoothing filter, and

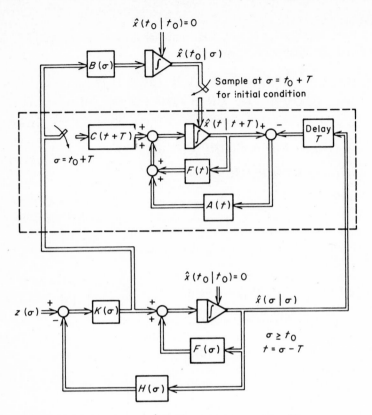

Fig. 7.7 Block diagram of optimal fixed-lag smoothing filter.

depends upon both optimal filtering and optimal fixed-point smoothing for input data.   The extent of this dependence is given in Table 7.1 and emphasized still further in the block diagram in Figure 7.7.

It is noted in the figure that two time scales are involved here.   The σ time scale is for the optimal filter and optimal fixed-point filter.   The

Table 7.1   Fixed-lag smoothing filter input data requirements

Input source	Input data required
Optimal filter	(a) $K(t + T)[z(t + T) - H(t + T)\hat{x}(t + T\|t + T)]$ for all $t \le t_0$ (b) $\hat{x}(t\|t)$ for all $t \ge t_0$
Optimal fixed-point smoothing filter	(a) $\hat{x}(t_0\|t_0 + T)$ as the initial condition; requires fixed-point smoothing over the interval $t_0 \le t \le t_0 + T$

t time scale is for the fixed-lag smoothing filter. The fixed-lag smoothing filter is inoperative until $\sigma \geq t_0 + T$. The reason for this is that it must "wait" T units of time until the measurements have reached the time point $t_0 + T$ before it can start processing data. In the meantime, the fixed-point smoothing filter is processing data to obtain the needed initial condition $\hat{x}(t_0|t_0 + T)$. It is clear that the fixed-point smoothing filter need not be operative for $\sigma > t_0 + T$, while the optimal filter must be.

The term $A(t) = G(t)Q(t)G'(t)P^{-1}(t|t)$, which is recognized as the optimal fixed-interval smoothing filter gain matrix, is present in Eqs. (7.71) through (7.73). Here, however, it must be computed forward in time. The inversion of $P(t|t)$ can once more be avoided by utilizing Eq. (7.52) with the initial condition $M(t_0) = P^{-1}(t_0|t_0)$. Clearly, $P(t_0|t_0) = P(t_0)$ must be nonsingular for this purpose.

The fixed-lag smoothing filter gain matrix equation, Eq. (7.72), is a system of n^2 linear ordinary differential equations. The initial conditions for these equations are the n^2 elements of $B(t_0 + T)$ which can only be obtained by solving Eq. (7.58) over the interval $t_0 \leq t \leq t_0 + T$ first.

A similar situation with respect to the initial conditions also exists for the covariance equation, Eq. (7.73).

Example 7.6 We consider fixed-lag smoothing for a simple communication system in which the message process can be modeled as the output of an integrator with a zero mean gaussian white noise input. The assumed message process equation is

$$\dot{x} = w(t)$$

for $t \geq 0$, where $x(0)$ is a zero mean gaussian random variable with variance σ_0^2, and $\{w(t), t \geq 0\}$ is a zero mean gaussian white noise which is independent of $x(0)$ and has a variance σ_w^2.

We assume that the model for the received signal is

$$z(t) = x(t) + v(t)$$

for $t \geq 0$, where $\{v(t), t \geq 0\}$ is a zero mean gaussian white noise which is independent of $x(0)$ and $\{w(t), t \geq 0\}$, and has a variance σ_v^2. Also, we assume that $\sigma_0^2 \gg \sigma_w \sigma_v$.

We note that the model here is the special case of the one in Example 7.1 where $a = 0$. Referring to that example, we have

$$\rho_1 = \sigma_w \sigma_v \qquad \rho_2 = -\sigma_w \sigma_v$$

$$\mu = \frac{\sigma_w}{\sigma_v} \qquad \alpha = \frac{\sigma_0^2 - \sigma_w \sigma_v}{\sigma_0^2 + \sigma_w \sigma_v} \approx 1$$

The optimal filtering error variance is seen to be

$$P(t|t) = \sigma_w \sigma_v \frac{1 + e^{-2\mu t}}{1 - e^{-2\mu t}} = \sigma_w \sigma_v \frac{e^{\mu t} + e^{-\mu t}}{e^{\mu t} - e^{-\mu t}}$$

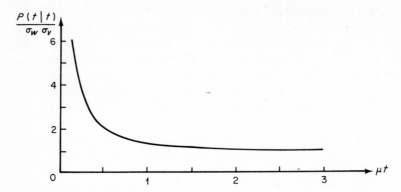

Fig. 7.8 Optimal filtering error variance as a function of time for Example 7.6.

or

$$P(t|t) = \sigma_w \sigma_v \operatorname{ctnh} \mu t$$

for $t \geq 0$. Also, we have

$$K(t) = \frac{1}{\sigma_v^2} P(t|t) = \mu \operatorname{ctnh} \mu t \qquad (7.74)$$

A sketch of $P(t|t)/\sigma_w \sigma_v$ versus μt is given in Fig. 7.8, where we note that for $\mu t \geq 2$, $P(t|t)/\sigma_w \sigma_v \approx 1.0$.

We obtain next the gain and error covariance expressions for optimal fixed-point smoothing which we shall need. From Eq. (7.58), we have

$$\dot{B} = -\sigma_w^2 \frac{1}{\sigma_w \sigma_v} \tanh \mu t \, B = -(\mu \tanh \mu t) B$$

for $t \geq 0$, with $B(0) = 1$. The solution is

$$B(t) = \frac{1}{\cosh \mu t} = \operatorname{sech} \mu t$$

which is obtained by separation of variables.

For $t_1 = 0$, Eq. (7.59) assumes here the form

$$\dot{P}(0|t) = -R(t)B^2(t)K^2(t)$$
$$= -\sigma_v^2 \operatorname{sech}^2 \mu t (\mu^2 \operatorname{ctnh}^2 \mu t)$$
$$= -\sigma_w^2 \operatorname{csch}^2 \mu t$$

for $t \geq 0$. It follows that

$$P(0|t) = -\sigma_w^2 \frac{1}{\mu} \int \operatorname{csch}^2 \mu t \, d(\mu t)$$
$$= \sigma_w \sigma_v \operatorname{ctnh} \mu t \qquad (7.75)$$

In passing, we note that $P(0|t) = P(t|t)$ for all $t \geq 0$. This means that, by fixed-point smoothing, it is possible to estimate the initial state to within the same accuracy as the estimate of the current state, having started at $t = 0$ with an arbitrarily large initial error variance $\sigma_0{}^2 \gg \sigma_w \sigma_v$.

We now have the necessary information to compute $C(t + T)$ and $P(t|t + T)$. We note that

$$A(t) = G(t)Q(t)G'(t)P^{-1}(t|t)$$

$$= \sigma_w{}^2 \frac{1}{\sigma_w \sigma_v} \tanh \mu t$$

$$= \mu \tanh \mu t \tag{7.76}$$

Substituting into Eq. (7.72), we obtain

$$\dot{C}(t + T) = \mu[\tanh \mu t - \tanh \mu(t + T)]C(t + T)$$

for $t \geq 0$, where $C(T) = B(T) = \mathrm{sech}\, \mu T$. Then,

$$\frac{dC}{C} = [\tanh \mu t - \tanh \mu(t + T)]\, d(\mu t)$$

which, when integrated, gives

$$C(t + T) = \gamma \frac{\cosh \mu t}{\cosh \mu(t + T)}$$

$$= \gamma \cosh \mu t \, \mathrm{sech}\, \mu(t + T)$$

where γ is the constant of integration. For $t = 0$, $C(T) = B(T) = \mathrm{sech}\, \mu T$. Hence

$$\gamma \cosh 0 \, \mathrm{sech}\, \mu T = \mathrm{sech}\, \mu T$$

from which $\gamma = 1$, so that

$$C(t + T) = \cosh \mu T \, \mathrm{sech}\, \mu(t + T) \tag{7.77}$$

for $t \geq 0$.

For our problem, Eq. (7.73) reduces to

$$\dot{P}(t|t + T) = 2A(t)P(t|t + T) - R(t + T)C^2(t + T)K^2(t + T) - G^2(t)Q(t)$$

Substituting into this relation from Eqs. (7.74), (7.76), and (7.77), we have

$$\dot{P}(t|t + T) = (2\mu \tanh \mu t)P(t|t + T)$$

$$- \sigma_v{}^2 \cosh^2 \mu t \, \mathrm{sech}^2 \mu(t + T) \cdot \mu^2 \mathrm{ctnh}^2 \mu(t + T) - \sigma_w{}^2$$

which we can express as

$$\dot{P}(t|t + T) - 2\mu \tanh \mu t P(t|t + T) = -\sigma_w{}^2[1 + \cosh^2 \mu t \, \mathrm{csch}^2 \mu(t + T)] \tag{7.78}$$

Equation (7.78) is in the standard form of a first-order linear ordinary differential equation whose solution can be written in the form

$$P(t|t + T) = \frac{1}{c(t)} \int c(t)h(t)\, dt + \frac{\eta}{c(t)}$$

where

$$h(t) = -\sigma_w^2[1 + \cosh^2 \mu t \, \mathrm{csch}^2 \mu(t + T)]$$

$$\eta = \text{constant of integration}$$

and $c(t)$ is the integrating factor

$$c(t) = \exp \int (-2\mu \tanh \mu t)\, dt$$

$$= \exp [-2\int \tanh \mu t \, d(\mu t)]$$

$$= \exp (-2 \ln \cosh \mu t)$$

$$= \frac{1}{\cosh^2 \mu t} = \mathrm{sech}^2 \mu t$$

Hence,

$$P(t|t + T) = -\sigma_w^2 \cosh^2 \mu t \int \mathrm{sech}^2 \mu t[1 + \cosh^2 \mu t \, \mathrm{csch}^2 \mu(t + T)]\, dt + \eta \cosh^2 \mu t$$

$$= -\sigma_w \sigma_v \cosh^2 \mu t \int [\mathrm{sech}^2 \mu t + \mathrm{csch}^2 \mu(t + T)]\, d(\mu t) + \eta \cosh^2 \mu t$$

$$= -\sigma_w \sigma_v \cosh^2 \mu t[\tanh \mu t - \mathrm{ctnh} \, \mu(t + T)] + \eta \cosh^2 \mu t$$

For $t = 0$, $P(0|T) = \sigma_w \sigma_v \, \mathrm{ctnh} \, \mu T$, as seen from Eq. (7.75). Hence, $\eta = 0$, and we have

$$P(t|t + T) = \sigma_w \sigma_v \cosh^2 \mu t[\mathrm{ctnh} \, \mu(t + T) - \tanh \mu t]$$

for $t \geq 0$.

An interesting comparison of $P(t|t)$ and $P(t|t + T)$ can be obtained by considering values of t for which $\mu t \geq 2$. Then, we have

$$P(t|t) \approx \sigma_w \sigma_v.$$

Also,

$$\cosh^2 \mu t \approx \frac{e^{2\mu t}}{4}$$

$$\mathrm{ctnh} \, \mu(t + T) \approx 1 + 2e^{-2\mu T}e^{-2\mu t}$$

and

$$\tanh \mu t \approx 1 - 2e^{-2\mu t}$$

Then,

$$P(t|t + T) \approx \tfrac{1}{4}\sigma_w \sigma_v e^{2\mu t}[1 + 2e^{-2\mu T}e^{-2\mu t} - 1 + 2e^{-2\mu t}] = \tfrac{1}{2}\sigma_w \sigma_v(1 + e^{-2\mu T})$$

Hence, for $\mu t \geq 2$, we have

$$P(t|t + T) = \tfrac{1}{2}(1 + e^{-2\mu T})P(t|t)$$

It is clear from this expression that fixed-lag smoothing can give a reduction

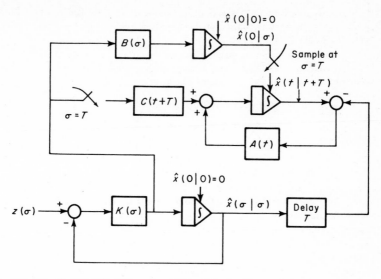

Fig. 7.9 Optimal fixed-lag smoothing filter for Example 7.6.

in estimation error variance of up to 50 percent over that possible with optimal filtering.

For this example, the appropriate filtering and smoothing equations are

$$\dot{\hat{x}}(\sigma|\sigma) = K(\sigma)[z(\sigma) - \hat{x}(\sigma|\sigma)] \qquad \sigma \geq 0$$

$$\dot{\hat{x}}(0|\sigma) = B(\sigma)K(\sigma)[z(\sigma) - \hat{x}(\sigma|\sigma)] \qquad 0 \leq \sigma \leq T$$

and

$$\dot{\hat{x}}(t|t + T) = C(t + T)K(t + T)[z(t + T) - \hat{x}(t + T|t + T)]$$
$$+ A(t)[\hat{x}(t|t + T) - \hat{x}(t|t)] \qquad t \geq 0$$

where $\hat{x}(0|0) = 0$ and $t = \sigma - T$.

The block diagram for the operations which are represented by these equations is given in Fig. 7.9.

It should be noted that some approximation must be used for $K(\sigma)$ when σ is small, since $K(0) = \infty$. This situation arises as a consequence of our assumption that $\sigma_0{}^2 \gg \sigma_w \sigma_v$.

PROBLEMS

7.1. Apply the limiting procedure used in this chapter to the results in Prob. 5.11 to show that the equations for optimal filtering are

$$\dot{\hat{x}} = F(t)\hat{x} + K(t)[z(t) - H(t)\hat{x}]$$

$$K(t) = [P(t|t)H'(t) + G(t)S(t)]R^{-1}(t)$$

$$\dot{P} = [F(t) - G(t)S(t)R^{-1}(t)]P + P[F(t) - G(t)S(t)R^{-1}(t)]'$$
$$- PH'(t)R^{-1}(t)H(t)P + G(t)[Q(t) - S(t)R^{-1}(t)S'(t)]G'(t)$$

for the system of Eqs. (7.1) and (7.2) when $E[w(t)v'(\tau)] = S(t)\delta(t - \tau)$. In these equations, $t \geq t_0$, $\hat{x} = \hat{x}(t|t)$, $\hat{x}(t_0|t_0) = 0$, $P = P(t|t)$, and $P(t_0|t_0) = P(t_0) = E[x(t_0)x'(t_0)]$. (See Reference [1].)

7.2. What changes are necessary in the results of Theorems 7.1 and 7.2 if $x(t_0)$, $\{w(t), t \geq t_0\}$, and $\{v(t), t \geq t_0\}$ have nonzero means and a *known* control input $u(t)$ is present? Note that Eq. (7.1) is of the form

$$\dot{x} = F(t)x + G(t)w(t) + C(t)u(t)$$

for this problem.

7.3. If an arbitrary gain matrix $K^*(t)$ is used in Eq. (7.7) instead of the optimal one which is given by Eq. (7.8), what is the differential equation for the corresponding error covariance matrix? Is the filtering error process Gauss-Markov in this case?

7.4. Assuming that a step size $h > 0$ is used in the solution of Eq. (7.9) using a digital computer, does one use (a) $Q(t)$ and $R(t)$, (b) $Q(t)/h$ and $R(t)/h$, or (c) $Q(t)h$ and $R(t)h$ in the computations? Explain your answer.

7.5. In a communication system, the transmitted message is of the form $x(t) = V \sin \omega_0 t$, where $t \geq 0$, $\omega_0 = $ constant, and V is a zero mean gaussian random variable with variance σ_0^2. The received signal is $z(t) = x(t) + v(t)$, where $\{v(t), t \geq 0\}$ is zero mean gaussian white noise which is independent of V, and has a constant variance σ_w^2. Set up, but do not solve, the equations which are necessary to realize the optimal filter for $\hat{x}(t|t)$, and sketch the filter block diagram.

7.6. What are the equations for optimal fixed-interval smoothing if $x(t_0)$, $\{w(t), t \geq t_0\}$, and $\{v(t), t \geq t_0\}$ have nonzero means, and a *known* control input $u(t)$ is present? (See Prob. 7.2.)

7.7. Determine the differential equation for the cross-covariance matrix $P_{12}(t|t_1) = E[\tilde{x}(t|t_1)\tilde{x}'(t|t)]$ for the case of optimal fixed-interval smoothing.

7.8. For some arbitrary $t_1 > 0$, determine the relations for optimal fixed-interval smoothing for the system in Example 7.6, and sketch the smoothing filter's block diagram. Show that $P(t|t_1) \leq P(t|t)$ for $0 \leq t \leq t_1$.

7.9. What is the differential equation which defines the fixed-point smoothing error stochastic process $\{\tilde{x}(t_1|t), t \geq t_1\}$? Show that this process is Gauss-Markov-2.

7.10. Show that the results in Theorem 7.4 can also be expressed in the form

$$\dot{\hat{x}}(t_1|t) = B^*(t)H'(t)R^{-1}(t)[z(t) - H(t)\hat{x}(t|t)]$$
$$\dot{B}^* = B^*[F(t) - K(t)H(t)]'$$
$$B^*(t_1) = P(t_1|t_1)$$
$$\dot{P}(t_1|t) = -B^*(t)H'(t)R^{-1}(t)H(t)B^{*'}(t)$$

where $t \geq t_1$. Note that this formulation avoids computation of $P^{-1}(t|t)$.

7.11. Consider the system of Example 7.1 with $a = 0$ and $\sigma_0^2 = \sigma_w \sigma_v$.

(a) For optimal filtering for this problem, show that

$$P(t|t) = \sigma_0^2 \qquad \text{and} \qquad K(t) = \mu = \sigma_w \sigma_v \qquad \text{for all } t \geq 0$$

(b) Noting that the optimal filter is a constant coefficient linear system, determine its transfer function and show how the filter can be realized as a simple RC network.

(c) For optimal fixed-point smoothing with $t_1 = 0$, show that $B(t) = e^{-\mu t}$ and $P(0|t) = (\sigma_0^2/2)(1 + e^{-2\mu t})$ for $t \geq 0$. Note that a reduction of up to 50 percent in the initial error variance is possible.

(d) For optimal fixed-lag smoothing show that $C(t + T) = e^{-\mu T}$ and $P(t|t + T) = (\sigma_0^2/2)(1 + e^{-2\mu T})$ for $t \geq 0$. In this case, note that a reduction of up to 50 percent in the estimation error variance is possible for all $t \geq 0$.

(e) Give a block diagram for the combined optimal filter, optimal fixed-point smoothing filter, and optimal fixed-lag smoothing filter.

7.12. Develop the proof for Theorem 7.5.

7.13. For zero lag, $T = 0$, show that the results in Theorem 7.5 reduce to those in Theorem 7.1.

REFERENCES

1. Kalman, R. E., New Methods in Wiener Filtering Theory, in J. L. Bogdaroff and F. Kozin, eds., "Proceedings of the First Symposium on Engineering Applications of Random Function Theory and Probability," John Wiley & Sons, Inc., New York, 1963, pp. 270–388. Also, New Methods and Results in Linear Prediction and Filtering Theory, *Tech. Rept.* 61-1, Research Institute for Advanced Studies (RIAS), Martin Company, Baltimore, 1961.

2. Rauch, H. E., F. Tung, C. T. Striebel, Maximum Likelihood Estimates of Linear Dynamic Systems, *AIAA J.*, vol. 3, p. 1445, 1965.

3. Meditch, J. S., Optimal Fixed-Point Continuous Linear Smoothing, *Proc. 1967 Joint Automatic Control Conf.*, p. 249, University of Pennsylvania, Philadelphia, June, 1967. Also, *Tech. Rept.* 66-108, IPAC Systems Laboratory, Northwestern University, Evanston, Illinois, December, 1966.

4. ———, On Optimal Linear Smoothing Theory, *Inform. Control*, vol. 10, p. 598, 1967. Also, *Tech. Rept.* 67-105, IPAC Systems Laboratory, Northwestern University, Evanston, Illinois, March, 1967.

5. Bryson, A. E., Jr. and M. Frazier, Smoothing for Linear and Nonlinear Dynamic Systems, TDR 63-119, Aero. Systems Div., Wright-Patterson Air Force Base, Ohio, February, 1963, pp. 353–364.

8
Optimal Estimation for Continuous Linear Systems, II

8.1 INTRODUCTION

In this chapter, we present a second approach to the estimation problem for continuous linear systems in which the problem is solved entirely in the continuous-time domain without any resort to limiting procedures. The presentation here supplements the one in Chap. 7.

After we have formulated the problem, we shall develop a matrix integral equation which is a necessary and sufficient condition for optimal estimation. The integral equation, termed the Wiener-Hopf equation, will be derived utilizing classical variational methods. We shall then solve the Wiener-Hopf equation to obtain the algorithms for optimal filtering and optimal fixed-point smoothing. We leave the other cases as exercises, since the results are known from Chap. 7.

8.2 WIENER-HOPF INTEGRAL EQUATION

PRELIMINARIES

We consider first a slightly different continuous-time optimal estimation problem from the one in Sec. 7.2. Specifically, we take the state to be a continuous-time zero mean gaussian process $\{x(t),\ t \geq t_0\}$, where x is an n vector and t_0 is the given initial time.   Evidently, the stochastic process which is defined by Eq. (7.1) is a special case in the present formulation.

The measurement process is more restricted.   In particular, we take $\{z(\tau),\ \tau \geq t_0\}$ to be a continuous-time zero mean gaussian process which is defined by the relation

$$z(\tau) = H(\tau)x(\tau) + v(\tau) \tag{8.1}$$

In Eq. (8.1), z is an m vector, $H(\tau)$ is a continuous $m \times n$ matrix, and $\{v(\tau),\ \tau \geq t_0\}$ is a zero mean gaussian white noise with

$$E[v(\tau)v'(\sigma)] = R(\tau)\delta(\tau - \delta)$$

for all $\tau, \sigma \geq t_0$, where $R(\tau)$ is continuous and positive definite for all $\tau \geq t_0$. We assume that $\{x(t),\ t \geq t_0\}$ and $\{v(t),\ t \geq t_0\}$ may be correlated.

As before, we denote an estimate of $x(t_1)$ for some $t_1 \geq t_0$, based on a given set of measurements $\{z(\tau),\ t_0 \leq \tau \leq t\}$ for some $t \geq t_0$, by $\hat{x}(t_1|t)$. The estimation error is again

$$\tilde{x}(t_1|t) = x(t_1) - \hat{x}(t_1|t)$$

We restrict the performance measure for estimation to be the mean square error so that

$$J[\tilde{x}(t_1|t)] = E[\tilde{x}'(t_1|t)\tilde{x}(t_1|t)] \tag{8.2}$$

Further, we require that the optimal estimate, i.e., the one which minimizes $J[\tilde{x}(t_1|t)]$, be of the form

$$\hat{x}(t_1|t) = \int_{t_0}^{t} A(t_1,\tau)z(\tau)\ d\tau \tag{8.3}$$

where $A(t_1,\tau)$ is an $n \times m$ matrix which is continuously differentiable in both of its arguments.

Equation (8.3) simply means that we consider only linear estimates as candidates for the optimal estimate.   The intuitive basis for this restriction is found in Theorem 5.2, the properties of optimal estimates on page 164, Eq. (5.12) as developed for optimal estimation for discrete-time gaussian processes, and the discussion in Sec. 7.4 (page 256). Indeed, Eq. (8.3) is simply the continuous-time analog of Eq. (5.12) and has the interpretation that the optimal estimate is the output of a time-varying linear system with the measurement vector as the input.   This is indicated schematically in Fig. 8.1.

From an alternate point of view, we can argue that we choose to restrict our attention to estimates of the form in Eq. (8.3) because of

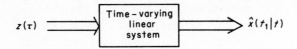

Fig. 8.1 Representation of optimal estimator as a time-varying linear system.

their simplicity.   Hopefully, this will facilitate the optimization and lead to simple and efficient algorithms for estimation.   This viewpoint is the one adopted in classical control theory wherein the structure of the controller is fixed a priori and its parameters are then adjusted to optimize a relevant set of performance characteristics.

We point out that by utilizing more advanced methods than are available to us here, it can be shown [1, 2] that if $\{x(t), t \geq t_0\}$ and $\{z(\tau), \tau \geq t_0\}$ are zero mean and gaussian, the estimate which is optimal not only for the mean square error performance measure in Eq. (8.2), but for all performance measures

$$J[\hat{x}(t_1|t)] = E\{L[\hat{x}(t_1|t)]\}$$

where L is an admissible loss function, must be of the form in Eq. (8.3).

In the sequel, we shall refer to $A(t_1,\tau)$ as a *system matrix*, recognizing that it defines an input-output relation between $z(\tau)$, $t_0 \leq \tau \leq t$, and $\hat{x}(t_1|t)$, $t_1 \geq t_0$, respectively.

Our problem statement now follows very simply.

PROBLEM STATEMENT

Given the measurements $\{z(\tau), t_0 \leq \tau \leq t\}$, *find an estimate of* $x(t_1)$, $t_1 \geq t_0$, *of the form in Eq. (8.3), which minimizes the mean square error as given by Eq. (8.2).*

Clearly, the problem is that of determining the system matrix $A(t_1,\tau)$, $t_0 \leq \tau \leq t$.

FUNDAMENTAL THEORY

The solution of the estimation problem posed above is given below in Theorem 8.1.   This theorem was obtained by Kalman and Bucy [1] utilizing the concept of orthogonal projection in Hilbert space.   We shall prove it here by applying classical variational methods.

Very simply, the theorem provides a necessary and sufficient condition for optimal estimation in the form of a matrix integral equation termed the *Wiener-Hopf equation*.   As a consequence, the estimation problem is reduced to one of solving a particular integral equation.

Theorem 8.1   *A necessary and sufficient condition for $\hat{x}(t_1|t)$ to be an optimal estimate is that the system matrix $A(t_1,\tau)$ satisfy the Wiener-Hopf equation*

$$E[x(t_1)z'(\sigma)] - \int_{t_0}^{t} A(t_1,\tau)E[z(\tau)z'(\sigma)]\,d\tau = 0 \tag{8.4}$$

for all $t_0 \leq \sigma \leq t$.

Proof   From the definition of estimation error and Eq. (8.3),

$$\tilde{x}(t_1|t) = x(t_1) - \hat{x}(t_1|t) = x(t_1) - \int_{t_0}^{t} A^*(t_1,\tau)z(\tau)\,d\tau$$

where $A^*(t,\tau)$ is any system matrix. It then follows that the covariance matrix of $\tilde{x}(t_1|t)$ is

$$\begin{aligned}
E[\tilde{x}(t_1|t)\tilde{x}'(t_1|t)] = E\,\Big\{&\Big[x(t_1) - \int_{t_0}^{t} A^*(t_1,\tau)z(\tau),\,d\tau\Big] \\
&\Big[x(t_1) - \int_{t_0}^{t} A^*(t_1,\sigma)z(\sigma)\,d\sigma\Big]'\Big\} \\
= E[x(t_1)x'(t_1)] &- \int_{t_0}^{t} E[x(t_1)z'(\sigma)]A^{*\prime}(t_1,\sigma)\,d\sigma \\
&- \int_{t_0}^{t} A^*(t_1,\tau)E[z(\tau)x'(t_1)]\,d\tau \\
&+ \int_{t_0}^{t} A^*(t_1,\sigma)\,d\tau \int_{t_0}^{t} E[z(\tau)z'(\sigma)]A^{*\prime}(t_1,\sigma)\,d\sigma \tag{8.5}
\end{aligned}$$

The mean square error is the trace of this matrix. Since the third term on the right-hand side of Eq. (8.5) is the transpose of the second term, the trace of the sum of these two terms is simply twice the trace of the second one. This means that we can express the mean square error as

$$\begin{aligned}
J[\tilde{x}(t_1|t)] = \mathrm{tr}\,\Big\{&E[x(t_1)x'(t_1)] - 2\int_{t_0}^{t} E[x(t_1)z'(\sigma)]A^{*\prime}(t_1,\sigma)\,d\sigma \\
&+ \int_{t_0}^{t} A^*(t_1,\tau)\,d\tau \int_{t_0}^{t} E[z(\tau)z'(\sigma)]A^{*\prime}(t_1,\sigma)\,d\sigma\Big\} \tag{8.6}
\end{aligned}$$

where tr denotes the trace. Equation (8.6) gives the mean square error as a functional in $A(t,\tau)$.

We now assume that there exists a system matrix $A(t_1,\tau)$ for which Eq. (8.6) is minimized. We construct another system matrix of the form

$$A^*(t_1,\tau) = A(t_1,\tau) + \epsilon A_\epsilon(t_1,\tau) \tag{8.7}$$

where $A_\epsilon(t_1,\tau)$ is any arbitrary system matrix and ϵ is a scalar parameter. The term $\epsilon A_\epsilon(t_1,\tau)$ is called the variation of $A(t_1,\tau)$.

If Eq. (8.7) is substituted into Eq. (8.6), the mean square error becomes a function of ϵ for a fixed $A_\epsilon(t,\tau)$. Then, if $A(t,\tau)$ is a system matrix which minimizes the mean square error, this error must be stationary for $\epsilon = 0$. That is, the partial derivative of the mean square error with respect to ϵ must vanish for $\epsilon = 0$. This approach will lead to a necessary condition on $A(t_1,\tau)$. After we have developed this condition, we shall show that it is also sufficient.

NECESSITY Substituting Eq. (8.7) into Eq. (8.6), we obtain

$$
\begin{aligned}
J[\tilde{x}(t_1|t)] = \text{tr} \Big\{ &E[x(t_1)x'(t_1)] \\
&- 2 \int_{t_0}^{t} E[x(t_1)z'(\sigma)][A(t_1,\sigma) + \epsilon A_\epsilon(t_1,\sigma)]' \, d\sigma \\
&+ \int_{t_0}^{t} [A(t_1,\tau) + \epsilon A_\epsilon(t_1,\tau)] \, d\tau \\
&\qquad \int_{t_0}^{t} E[z(\tau)z'(\sigma)][A(t_1,\sigma) + \epsilon A_\epsilon(t_1,\sigma)]' \, d\sigma \Big\} \quad (8.8)
\end{aligned}
$$

It then follows that

$$
\begin{aligned}
\frac{\partial J}{\partial \epsilon} = \text{tr} \Big\{ &-2 \int_{t_0}^{t} E[x(t_1)z'(\sigma)]A_\epsilon'(t_1,\sigma) \, d\sigma \\
&+ \int_{t_0}^{t} [A(t_1,\tau) + \epsilon A_\epsilon(t_1,\tau)] \, d\tau \int_{t_0}^{t} E[z'(\tau)z'(\sigma)]A_\epsilon'(t_1,\sigma) \, d\sigma \\
&+ \int_{t_0}^{t} A_\epsilon(t_1,\tau) \, d\tau \int_{t_0}^{t} E[z(\tau)z'(\sigma)][A(t_1,\sigma) + \epsilon A_\epsilon(t_1,\sigma)]' \, d\sigma \Big\}
\end{aligned}
$$

Since the last term on the right-hand side in this expression is the transpose of the second one, the trace of their sum is twice the trace of the first one.   Hence

$$
\begin{aligned}
\frac{\partial J}{\partial \epsilon} = 2 \, \text{tr} \Big\{ &\int_{t_0}^{t} [A(t_1,\tau) + \epsilon A_\epsilon(t_1,\tau)] \, d\tau \int_{t_0}^{t} E[z(\tau)z'(\sigma)]A_\epsilon'(t_1,\sigma) \, d\sigma \\
&\qquad\qquad - \int_{t_0}^{t} E[x(t_1)z'(\sigma)]A_\epsilon'(t_1,\sigma) \, d\sigma \Big\}
\end{aligned}
$$

Setting

$$
\frac{\partial J}{\partial \epsilon} \Big|_{\epsilon=0} = 0
$$

we obtain

$$
\begin{aligned}
\text{tr} \Big\{ &\int_{t_0}^{t} A(t_1,\tau) \, d\tau \int_{t_0}^{t} E[z(\tau)z'(\sigma)]A_\epsilon'(t_1,\sigma) \, d\sigma \\
&\qquad - \int_{t_0}^{t} E[x(t_1)z'(\sigma)]A_\epsilon'(t_1,\sigma) \, d\sigma \Big\} = 0
\end{aligned}
$$

Rearranging the integrations in the first term in the above expression and multiplying through by -1, we have the result

$$
\text{tr} \Big[\int_{t_0}^{t} \Big\{ E[x(t_1)z'(\sigma)] - \int_{t_0}^{t} A(t_1,\tau)E[z(\tau)z'(\sigma)] \, d\tau \Big\} A_\epsilon'(t_1,\sigma) \, d\sigma \Big] = 0
$$

$$(8.9)$$

Equation (8.9) must hold for *all* system matrices $A_\epsilon(t_1,\tau)$. Therefore, the equation holds if and only if $A(t_1,\tau)$ satisfies the relation

$$E[x(t_1)z'(\sigma)] - \int_{t_0}^{t} A(t_1,\tau)E[z(\tau)z'(\sigma)]\,d\tau = 0$$

for all $t_0 \leq \sigma \leq t$.

SUFFICIENCY   Expanding Eq. (8.8), we have

$$
\begin{aligned}
J[\tilde{x}(t_1|t)] = \text{tr} \Big\{ &E[x(t_1)x'(t_1)] - 2\int_{t_0}^{t} E[x(t_1)z'(\sigma)]A'(t_1,\sigma)\,d\sigma \\
&- 2\epsilon \int_{t_0}^{t} E[x(t_1)z'(\sigma)]A_\epsilon'(t_1,\sigma)\,d\sigma \\
&+ \int_{t_0}^{t} A(t_1,\tau)\,d\tau \int_{t_0}^{t} E[z(\tau)z'(\sigma)]A'(t_1,\sigma)\,d\sigma \\
&+ \epsilon \int_{t_0}^{t} A_\epsilon(t_1,\tau)\,d\tau \int_{t_0}^{t} E[z(\tau)z'(\sigma)]A'(t_1,\sigma)\,d\sigma \\
&+ \epsilon \int_{t_0}^{t} A(t_1,\tau)\,d\tau \int_{t_0}^{t} E[z(\tau)z'(\sigma)]A_\epsilon'(t_1,\sigma)\,d\sigma \\
&+ \epsilon^2 \int_{t_0}^{t} A_\epsilon(t_1,\tau)\,d\tau \int_{t_0}^{t} E[z(\tau)z'(\sigma)]A_\epsilon'(t_1,\sigma)\,d\sigma \Big\}
\end{aligned}
$$
(8.10)

Comparing the first, second, and fourth terms in this expression with Eq. (8.6), we see that these three terms are equal to the mean square error for the system function $A(t,\tau)$, that is, the minimum mean square error. We denote these terms by $J^0[\tilde{x}(t_1|t)]$.

Since the matrix in the fifth term is the transpose of that in the sixth, the trace of their sum is twice that of the latter. Hence, Eq. (8.10) can be simplified to

$$
\begin{aligned}
J[\tilde{x}(t_1|t)] = J^0[\tilde{x}(t_1|t)] &- 2\epsilon \text{ tr} \Big[\int_{t_0}^{t} \Big\{ E[x(t_1)z'(\sigma)] \\
&- \int_{t_0}^{t} A(t_1,\tau)E[z(\tau)z'(\sigma)]\,d\tau \Big\} A_\epsilon'(t_1,\sigma)\,d\sigma \Big] \\
&+ \epsilon^2 \text{ tr} \Big\{ \iint_{t_0}^{t} A_\epsilon(t_1,\tau)E[z(\tau)z'(\sigma)]A_\epsilon'(t_1,\sigma)\,d\sigma\,d\tau \Big\}
\end{aligned}
$$
(8.11)

From Eq. (8.1),

$$
\begin{aligned}
E[z(\tau)z'(\sigma)] = H(\tau)E[x(\tau)x'(\sigma)]H'(\sigma) &+ H(\tau)E[x(\tau)v'(\sigma)] \\
&+ E[v(\tau)x'(\sigma)]H'(\sigma) + R(\tau)\delta(\tau - \sigma)
\end{aligned}
$$

Substitution of this expression into the last term in Eq. (8.11) will lead to a nonnegative contribution from its first three terms and a positive contribution from its last term, $R(\tau)\delta(\tau - \sigma)$, since $R(\tau)$ is positive definite for all $\tau \geq t_0$. Consequently, the last term in Eq. (8.11) is positive.

It is now clear that if

$$E[x(t_1)z'(\sigma)] - \int_{t_0}^{t} A(t_1,\tau)E[z(\tau)z'(\sigma)] \, d\tau = 0$$

for all $t_0 \le \sigma \le t$, the second term in Eq. (8.11) vanishes, and we have

$$J[\tilde{x}(t_1|t)] > J^0[\tilde{x}(t_1|t)]$$

This means that the mean square error is not decreased by perturbing $A(t,\tau)$ if the Wiener-Hopf equation is satisfied. This completes the proof.

For convenience, the Wiener-Hopf equation is often written in the form

$$E[\tilde{x}(t_1|t)z'(\sigma)] = 0 \tag{8.12}$$

for all $t_0 \le \sigma \le t$.

This can be shown as follows. First, Eq. (8.4) can be expressed as

$$E\left\{[x(t_1) - \int_{t_0}^{t} A(t_1,\tau)z(\tau) \, d\tau]z'(\sigma)\right\} = 0$$

Then, since

$$\hat{x}(t_1|t) = \int_{t_0}^{t} A(t_1,\tau)z(\tau) \, d\tau$$

and $\tilde{x}(t_1|t) = x(t_1) - \hat{x}(t_1|t)$, Eq. (8.12) follows immediately.

We have also the following corollary to Theorem 8.1.

Corollary 8.1 $E[\tilde{x}(t_1|t)\hat{x}'(t_1|t)] = 0.$ $\tag{8.13}$

Proof   The proof is very easy. We note first that

$$E[\tilde{x}(t_1|t)\hat{x}'(t_1|t)] = E\left\{\tilde{x}(t_1|t)\left[\int_{t_0}^{t} A(t_1,\tau)z(\tau) \, d\tau\right]'\right\}$$

$$= E\left[\int_{t_0}^{t} \tilde{x}(t_1|t)z'(\tau)A'(t_1,\tau) \, d\tau\right]$$

$$= \int_{t_0}^{t} E[\tilde{x}(t_1|t)z'(\tau)]A'(t_1,\tau) \, d\tau$$

But from Eq. (8.12), $E[\tilde{x}(t_1|t)z'(\tau)] = 0$ for all $t_0 \le \tau \le t$, and the corollary is established.

The corollary states that the optimal estimate is uncorrelated with the corresponding estimation error. Since both vectors are gaussian, this means further that they are independent.

Discussion   As with integral equations in general, solution of the Wiener-Hopf equation is a nontrivial matter.   If $\{x(t_1),\ t_1 \geq t_0\}$ and $\{z(\tau),\ t_0 \leq \tau \leq t\}$ are stationary processes, the equation can be solved in general terms utilizing Wiener's technique of "spectral factorization" as given, for example, in Lee [3] and Newton et al. [4].   Usually, however, it is necessary to consider very specific cases, as we do in Secs. 8.3 and 8.4.

An intuitive argument for the Wiener-Hopf equation is very easy to give.   First, it is claimed that the optimal estimate must be linear and of the form in Eq. (8.3) as the continuous-time analog of Eq. (5.9). Then, in view of the continuous-time analog of the third property of optimal estimates on page 164, the estimation error

$$\tilde{x}(t_1|t) = x(t_1) - \int_{t_0}^{t} A(t_1,\tau)z(\tau)\, d\tau$$

is independent of the set of available measurements, viz., $\{z(\sigma),\ t_0 \leq \sigma \leq t\}$.   This means that $\{\tilde{x}(t_1|t),\ t_1 \geq t_0\}$ and $\{z(\sigma),\ t_0 \leq \sigma \leq t\}$ are uncorrelated and therefore that

$$E\left\{\left[x(t_1) - \int_{t_0}^{t} A(t_1,\tau)z(\tau)\, d\tau \right] z'(\sigma)\right\} = 0$$

for $t_0 \leq \sigma \leq t$.

Corollary 8.1 is then simply the continuous-time analog of Eq. (5.10).

We conclude this section with the following lemma, which will be needed in the derivation of the results in Secs. 8.3 and 8.4.   The lemma states that the cross-covariance matrix of the difference of two optimal estimates is zero.

Lemma 8.1   *Let $\hat{x}(t_1|t)$ and $\hat{x}^*(t_1|t)$ be optimal estimates of $x(t_1)$.   Then,*

$$E\{[\hat{x}(t_1|t) - \hat{x}^*(t_1|t)][\hat{x}(t_1|t) - \hat{x}^*(t_1|t)]'\} = 0 \tag{8.14}$$

Proof   Let $\tilde{x}(t_1|t) = x(t_1) - \hat{x}(t_1|t)$ and $\tilde{x}^*(t_1|t) = x(t_1) - \hat{x}^*(t_1|t)$. From Corollary 8.1,

$$E[\tilde{x}(t_1|t)\hat{x}'(t_1|t)] = 0 \tag{8.15}$$

$$E[\tilde{x}(t_1|t)\hat{x}^{*\prime}(t_1|t)] = 0 \tag{8.16}$$

Subtracting Eq. (8.16) from Eq. (8.15), we obtain

$$E\{\tilde{x}(t_1|t)[\hat{x}(t_1|t) - \hat{x}^*(t_1|t)]'\} = 0 \tag{8.17}$$

In exactly the same manner, $E[\tilde{x}^*(t_1|t)\hat{x}'(t_1|t)] = 0$ and $E[\tilde{x}^*(t_1|t)\hat{x}^{*\prime}(t_1|t)] = 0$, so that

$$E\{\tilde{x}^*(t_1|t)[\hat{x}(t_1|t) - \hat{x}^*(t_1|t)]'\} = 0 \tag{8.18}$$

Subtracting Eq. (8.17) from Eq. (8.18),

$$E\{[\tilde{x}^*(t_1|t) - \tilde{x}(t_1|t)][\hat{x}(t_1|t) - \hat{x}^*(t_1|t)]'\} = 0$$

But $\tilde{x}^*(t_1|t) - \tilde{x}(t_1|t) = \hat{x}(t_1|t) - \hat{x}^*(t_1|t)$, and the proof is complete.

8.3 OPTIMAL FILTERING

We shall now solve the Wiener-Hopf equation where $t_1 = t$, that is, for the optimal filtering problem, and where $\{x(t),\ t \geq t_0\}$ is the Gauss-Markov process which is defined by Eq. (7.1).   Here, in contrast to the formulation in Sec. 7.2 and in the solution of the optimal filtering problem in Sec. 7.4, we assume that $\{w(t),\ t \geq t_0\}$ and $\{v(t),\ t \geq t_0\}$ are correlated with

$$E[w(t)v'(\tau)] = S(t)\delta(t - \tau)$$

for all $t, \tau \geq t_0$, where $S(t)$ is a continuous $p \times m$ matrix.

The Wiener-Hopf equation for optimal filtering for the case where $S(t) = 0$ was solved by Kalman and Bucy [1].   The development here differs from theirs only in minor details.

The optimal filtering algorithm which is developed below is commonly termed the Kalman-Bucy filter.

FILTER

We consider the Wiener-Hopf equation over the interval $t_0 \leq \sigma < t$, where $t_1 = t$,

$$E[x(t)z'(\sigma)] - \int_{t_0}^{t} A(t,\tau)E[z(\tau)z'(\sigma)]\,d\tau = 0 \tag{8.19}$$

It is important for the reader to note that for the present, we consider σ only in the interval $[t_0,t)$.   The situation where $\sigma = t$ will be considered separately.   The reason for this will become apparent as we proceed.

Taking the partial derivative in Eq. (8.19) with respect to t, we have the two terms

$$\frac{\partial}{\partial t} E[x(t)z'(\sigma)] = E[\dot{x}(t)z'(\sigma)] = E\{[F(t)x(t) + G(t)w(t)]z'(\sigma)\}$$
$$= F(t)E[x(t)z'(\sigma)] + G(t)E[w(t)z'(\sigma)] \tag{8.20}$$

and

$$\frac{\partial}{\partial t} \int_{t_0}^{t} A(t,\tau)E[z(\tau)z'(\sigma)]\,d\tau = \int_{t_0}^{t} \frac{\partial A(t,\tau)}{\partial t} E[z(\tau)z'(\sigma)]\,d\tau$$
$$+ A(t,t)E[z(t)z'(\sigma)] \tag{8.21}$$

both for $t_0 \leq \sigma < t$.

We consider the last term in Eq. (8.20). From Eq. (7.2),

$$z'(\sigma) = x'(\sigma)H'(\sigma) + v'(\sigma) \tag{8.22}$$

so that

$$E[w(t)z'(\sigma)] = E[w(t)x'(\sigma)]H'(\sigma) + E[w(t)v'(\sigma)]$$

Since $\sigma < t$, $E[w(t)v'(\sigma)] = 0$, and we have

$$E[w(t)z'(\sigma)] = E[w(t)x'(\sigma)]H'(\sigma) \tag{8.23}$$

The solution of Eq. (7.1) is

$$x(\sigma) = \Phi(\sigma,t_0)x(t_0) + \int_{t_0}^{\sigma} \Phi(\sigma,\tau)G(\tau)w(\tau)\,d\tau \tag{8.24}$$

where $\Phi(\sigma,\tau)$ is the state transition matrix. Utilizing this result, we obtain

$$E[w(t)x'(\sigma)] = E[w(t)x'(t_0)]\Phi'(\sigma,t_0) + \int_{t_0}^{\sigma} E[w(t)w'(\tau)]G'(\tau)\Phi'(\sigma,\tau)\,d\tau$$

The first term on the right-hand side vanishes, since $x(t_0)$ is independent of $\{w(t),\ t \geq t_0\}$. In the second term, $E[w(t)w'(\tau)] = Q(t)\delta(t - \tau)$. However, since t is outside the range of integration, this term also vanishes. Hence,

$$E[w(t)x'(\sigma)] = 0 \tag{8.25}$$

for $t_0 \leq \sigma < t$.

Substituting this result into Eq. (8.23), we obtain

$$E[w(t)z'(\sigma)] = 0$$

for $t_0 \leq \sigma < t$. Consequently, the last term in Eq. (8.20) is zero, and we have

$$\frac{\partial}{\partial t}E[x(t)z'(\sigma)] = F(t)E[x(t)z'(\sigma)] \tag{8.26}$$

for $t_0 \leq \sigma < t$.

Let us now consider the second term on the right-hand side of Eq. (8.21). Utilizing Eq. (7.2), we see that

$$\begin{aligned} E[z(t)z'(\sigma)] &= E\{[H(t)x(t) + v(t)]z'(\sigma)\} \\ &= H(t)E[x(t)z'(\sigma)] + E[v(t)z'(\sigma)] \end{aligned} \tag{8.27}$$

From Eq. (8.22),

$$E[v(t)z'(\sigma)] = E[v(t)x'(\sigma)]H'(\sigma) + E[v(t)v'(\sigma)] \tag{8.28}$$

Since $\sigma < t$, $E[v(t)v'(\sigma)] = 0$.

Utilizing Eq. (8.24), we see that

$$E[v(t)x'(\sigma)] = E[v(t)x'(t_0)]\Phi'(\sigma,t_0) + \int_{t_0}^{\sigma} E[v(t)w'(\tau)]G'(\tau)\Phi'(\sigma,\tau)\,d\tau$$

The first term vanishes since $x(t_0)$ is independent of $\{v(t),\ t \geq t_0\}$. Although $E[v(t)w'(\tau)] = S(t)\delta(t - \tau)$, the second term also vanishes since t is outside the range of integration. Hence,

$$E[v(t)x'(\sigma)] = 0$$

for $t_0 \leq \sigma < t$. As a result, Eq. (8.28) reduces to

$$E[v(t)z'(\sigma)] = 0$$

for $t_0 \leq \sigma < t$, which means that Eq. (8.27) becomes

$$E[z(t)z'(\sigma)] = H(t)E[x(t)z'(\sigma)] \tag{8.29}$$

for $t_0 \leq \sigma < t$.

Consequently, Eq. (8.21) can be written

$$\frac{\partial}{\partial t}\int_{t_0}^{t} A(t,\tau)E[z(\tau)z'(\sigma)]\,d\tau = \int_{t_0}^{t} \frac{\partial A(t,\tau)}{\partial t} E[z(\tau)z'(\sigma)]\,d\tau$$

$$+ A(t,t)H(t)E[x(t)z'(\sigma)] \tag{8.30}$$

where $t_0 \leq \sigma < t$.

Combining Eqs. (8.26) and (8.30), we have that the partial derivative of Eq. (8.19) with respect to t is

$$F(t)E[x(t)z'(\sigma)] - \int_{t_0}^{t} \frac{\partial A(t,\tau)}{\partial t} E[z(\tau)z'(\sigma)]\,d\tau$$

$$- A(t,t)H(t)E[x(t)z'(\sigma)] = 0$$

for $t_0 \leq \sigma < t$. Substituting into this result from Eq. (8.19) for $E[x(t)z'(\sigma)]$, we obtain

$$[F(t) - A(t,t)H(t)]\int_{t_0}^{t} A(t,\tau)E[z(\tau)z'(\sigma)]\,d\tau$$

$$- \int_{t_0}^{t} \frac{\partial A(t,\tau)}{\partial t} E[z(\tau)z'(\sigma)]\,d\tau = 0$$

or, equivalently,

$$\int_{t_0}^{t} [F(t)A(t,\tau) - \frac{\partial A(t,\tau)}{\partial t}$$

$$- A(t,t)H(t)A(t,\tau)]E[z(\tau)z'(\sigma)]\,d\tau = 0 \tag{8.31}$$

for $t_0 \leq \sigma < t$.

In order that Eq. (8.31) be satisfied, it is sufficient that $A(t,\tau)$ satisfy the partial differential equation

$$F(t)A(t,\tau) - \frac{\partial A(t,\tau)}{\partial t} - A(t,t)H(t)A(t,\tau) = 0 \qquad (8.32)$$

over the interval $t_0 \leq \tau \leq t$, that is, that the bracketed term in Eq. (8.31) vanish over the range of integration.

Equation (8.32) can also be shown to be a necessary condition for Eq. (8.31) to hold as follows. Let $B(t,\tau)$ denote the bracketed term in Eq. (8.31). Then, if $A(t,\tau)$ satisfies the Wiener-Hopf equation, Eq. (8.19), so does $A(t,\tau) + B(t,\tau)$. This means that

$$\hat{x}(t|t) = \int_{t_0}^{t} A(t,\tau)z(\tau)\,d\tau$$

and

$$\hat{x}^*(t|t) = \int_{t_0}^{t} [A(t,\tau) + B(t,\tau)]z(\tau)\,d\tau$$

$$= \hat{x}(t|t) + \int_{t_0}^{t} B(t,\tau)z(\tau)\,d\tau$$

are optimal estimates of $x(t)$. However, by Lemma 8.1, the covariance matrix of the difference of these two estimates must vanish; i.e.,

$$E\{[\hat{x}^*(t|t) - \hat{x}(t|t)][\hat{x}^*(t|t) - \hat{x}(t|t)]'\} = 0$$

Hence,

$$\int_{t_0}^{t}\int_{t_0}^{t} B(t,\tau_1)E[z(\tau_1)z'(\tau_2)]B'(t,\tau_2)\,d\tau_1\,d\tau_2 = 0 \qquad (8.33)$$

However,

$$E[z(\tau_1)z'(\tau_2)] = H(\tau_1)E[x(\tau_1)x'(\tau_2)]H'(\tau_2) + H(\tau_1)E[x(\tau_1)v'(\tau_2)]$$
$$+ E[v(\tau_1)x'(\tau_2)]H'(\tau_2) + R(\tau_1)\delta(\tau_1 - \tau_2) \qquad (8.34)$$

The first three terms on the right-hand side in Eq. (8.34), when substituted into Eq. (8.33), will lead to a nonnegative contribution to the integration, while the fourth term will give a positive contribution, since $R(\tau_1)$ is positive definite. Hence, in order for Eq. (8.33) to be satisfied, it is necessary that $B(t,\tau) = 0$ for $t_0 \leq \tau \leq t$.

Returning now to the filter equation, Eq. (8.3) with $t_1 = t$, and taking the partial derivative with respect to t, we have

$$\dot{\hat{x}}(t|t) = \int_{t_0}^{t} \frac{\partial A(t,\tau)}{\partial t} z(\tau)\,d\tau + A(t,t)z(t)$$

However, it is clear from Eq. (8.32) that

$$\frac{\partial A(t,\tau)}{\partial t} = F(t)A(t,\tau) - A(t,t)H(t)A(t,\tau)$$

Consequently,

$$\hat{x}(t|t) = \int_{t_0}^{t} [F(t)A(t,\tau) - A(t,t)H(t)A(t,\tau)]z(\tau)\, d\tau + A(t,t)z(t)$$

$$= F(t)\int_{t_0}^{t} A(t,\tau)z(\tau)\, d\tau + A(t,t)\left[z(t) - H(t)\int_{t_0}^{t} A(t,\tau)z(\tau)\, d\tau\right]$$

$$= F(t)\hat{x}(t|t) + A(t,t)[z(t) - H(t)\hat{x}(t|t)]$$

Defining $K(t) = A(t,t)$, we write finally

$$\dot{\hat{x}} = F(t)\hat{x} + K(t)[z(t) - H(t)\hat{x}] \tag{8.35}$$

for $t \geq t_0$, where $\hat{x} = \hat{x}(t|t)$. Equation (8.35) defines the filter. It remains for us to establish the initial condition and the expression for the $n \times m$ gain matrix $K(t)$. Clearly,

$$\hat{x}(t_0|t_0) = \int_{t_0}^{t_0} A(t_0,\tau)z(\tau)\, d\tau$$

or

$$\hat{x}(t_0|t_0) = 0$$

GAIN MATRIX

We now consider the Wiener-Hopf equation for $\sigma = t$,

$$E[x(t)z'(t)] - \int_{t_0}^{t} A(t,\tau)E[z(\tau)z'(t)]\, d\tau = 0 \tag{8.36}$$

First, we see that

$$E[x(t)z'(t)] = E\{x(t)[H(t)x(t) + v(t)]'\}$$

$$= E[x(t)x'(t)]H'(t) + E[x(t)v'(t)] \tag{8.37}$$

From Eq. (8.24),

$$x(t) = \Phi(t,t_0)x(t_0) + \int_{t_0}^{t} \Phi(t,\tau)G(\tau)w(\tau)\, d\tau$$

and it follows that

$$E[x(t)v'(t)] = \int_{t_0}^{t} \Phi(t,\tau)G(\tau)E[w(\tau)v'(t)]\, d\tau \tag{8.38}$$

since $x(t_0)$ is independent of $\{v(t),\ t \geq t_0\}$. However,

$$E[w(\tau)v'(t)] = S(t)\delta(t - \tau)$$

so that

$$E[x(t)v'(t)] = G(t)S(t) \tag{8.39}$$

Equation (8.37) now becomes

$$E[x(t)z'(t)] = E[x(t)x'(t)]H'(t) + G(t)S(t) \tag{8.40}$$

We have also that

$$E[z(\tau)z'(t)] = E\{z(\tau)[H(t)x(t) + v(t)]'\}$$
$$= E[z(\tau)x'(t)]H'(t) + E[z(\tau)v'(t)]$$
$$= E[z(\tau)x'(t)]H'(t) + E\{[H(\tau)x(\tau) + v(\tau)]v'(t)\}$$
$$= E[z(\tau)x'(t)]H'(t) + H(\tau)E[x(\tau)v'(t)] + R(t)\delta(t - \tau)$$

$$(8.41)$$

Substituting Eqs. (8.40) and (8.41) into Eq. (8.36), we obtain

$$E[x(t)x'(t)]H'(t) + G(t)S(t) - \int_{t_0}^{t} A(t,\tau)E[z(\tau)x'(t)]H'(t)\, d\tau$$
$$- \int_{t_0}^{t} A(t,\tau)H(\tau)E[x(\tau)v'(t)]\, d\tau - \int_{t_0}^{t} A(t,\tau)R(t)\delta(t - \tau)\, d\tau = 0$$

$$(8.42)$$

It is clear that the fourth term vanishes since the integrand vanishes almost everywhere on $[t_0, t]$. This is a consequence of the fact that

$$E[x(\tau)v'(t)] = \begin{cases} 0 & t \neq \tau \\ G(t)S(t) & t = \tau \end{cases}$$

as seen from Eqs. (8.38) and (8.39).

Also,

$$\int_{t_0}^{t} A(t,\tau)R(t)\delta(t - \tau)\, d\tau = A(t,t)R(t) = K(t)R(t)$$

and we can express Eq. (8.42) in the form

$$K(t)R(t) = E[x(t)x'(t)]H'(t)$$
$$\qquad\qquad - \int_{t_0}^{t} A(t,\tau)E[z(\tau)x'(t)]\, d\tau\, H'(t) + G(t)S(t)$$
$$= E\{[x(t) - \int_{t_0}^{t} A(t,\tau)z(\tau)\, d\tau]x'(t)\}H'(t) + G(t)S(t)$$
$$= E[\tilde{x}(t|t)x'(t)]H'(t) + G(t)S(t)$$

Since $x(t) = \tilde{x}(t|t) + \hat{x}(t|t)$,

$$E[\tilde{x}(t|t)x'(t)] = E\{\tilde{x}(t|t)[\tilde{x}(t|t) + \hat{x}(t|t)]'\}$$
$$= E[\tilde{x}(t|t)\tilde{x}'(t|t)]$$
$$= P(t|t)$$

where $P(t|t)$ is defined to be the covariance matrix of the filtering error, and the term $E[\tilde{x}(t|t)\hat{x}'(t|t)]$ vanishes by virtue of Corollary 8.1.

We now have

$$K(t)R(t) = P(t|t)H'(t) + G(t)S(t)$$

which we solve for $K(t)$ to obtain

$$K(t) = [P(t|t)H'(t) + G(t)S(t)]R^{-1}(t) \tag{8.43}$$

for $t \geq t_0$. The need for $R(t)$ to be positive definite is obvious so far as computation of the optimal filter gain matrix is concerned. However, the reader should note that without the positive definiteness of $R(t)$, it would not have been possible to prove the necessity of Eq. (8.30) in developing the filter equation in the first place.

At this point, we recall that the Wiener-Hopf equation was considered over the interval $t_0 \leq \sigma < t$ to derive the filter equation. The case in which $\sigma = t$ was then employed to compute the filter gain matrix.

FILTERING ERROR COVARIANCE MATRIX

The filtering error and its corresponding time rate of change are

$$\tilde{x}(t|t) = x(t) - \hat{x}(t|t) \tag{8.44}$$

and

$$\dot{\tilde{x}}(t|t) = \dot{x}(t) - \dot{\hat{x}}(t|t) \tag{8.45}$$

respectively.

Substituting into Eq. (8.45) from Eqs. (7.1), (7.2), and (8.35), we get

$$\begin{aligned}
\dot{\tilde{x}} &= F(t)x + G(t)w(t) - F(t)\hat{x} - K(t)[z(t) - H(t)\hat{x}] \\
&= F(t)\tilde{x} + G(t)w(t) - K(t)[H(t)x + v(t) - H(t)\hat{x}] \\
&= [F(t) - K(t)H(t)]\tilde{x} + G(t)w(t) - K(t)v(t) \tag{8.46}
\end{aligned}$$

for $t \geq t_0$. From Eq. (8.44), $\tilde{x}(t_0|t_0) = x(t_0) - \hat{x}(t_0|t_0) = x(t_0)$. Hence, $\tilde{x}(t_0|t_0)$ is a zero mean gaussian random n vector with covariance matrix

$$\begin{aligned}
P(t_0|t_0) &= E[\tilde{x}(t_0|t_0)\tilde{x}'(t_0|t_0)] \\
&= E[x(t_0)x'(t_0)] \\
&= P(t_0)
\end{aligned}$$

Since $\{w(t), t \geq t_0\}$ and $\{v(t), t \geq t_0\}$ are zero mean gaussian white noises, it follows from Sec. 4.4 that the stochastic process defined by Eq. (8.46) is a zero mean Gauss-Markov process. Hence, the process is characterized by its covariance matrix which we now determine.

Letting $\Psi(t,\tau)$ denote the state transition matrix of the system in Eq. (8.46) and defining $C(t) = F(t) - K(t)H(t)$, we can write the solution of the equation as

$$\tilde{x}(t|t) = \Psi(t,t_0)\tilde{x}(t_0|t_0) + \int_{t_0}^{t} \Psi(t,\tau)[G(\tau)w(\tau) - K(\tau)v(\tau)]\, d\tau$$

Then,

$$P(t|t) = E[\tilde{x}(t|t)\tilde{x}'(t|t)]$$

$$= \Psi(t,t_0)E[\tilde{x}(t_0|t_0)\tilde{x}'(t_0|t_0)]\Psi'(t,t_0)$$

$$+ E\left\{\int_{t_0}^t \Psi(t,\tau)[G(\tau)w(\tau) - K(\tau)v(\tau)]\, d\tau\right.$$

$$\left.\int_{t_0}^t [w'(\sigma)G'(\sigma) - v'(\sigma)K'(\sigma)]\Psi'(t,\sigma)\, d\sigma\right\} \quad (8.47)$$

where the cross terms involving $\tilde{x}(t_0|t_0)$ and $\{w(t), t \geq t_0\}$ and $\{v(t), t \geq t_0\}$ vanish, since $\tilde{x}(t_0|t_0) = x(t_0)$ is independent of these two processes.

Recalling that $E[\tilde{x}(t_0|t_0)\tilde{x}'(t_0|t_0)] = P(t_0)$, $E[w(t)w'(\tau)] = Q(t)\delta(t - \tau)$, $E[v(t)v'(\tau)] = R(t)\delta(t - \tau)$ and $E[w(t)v'(\tau)] = S(t)\delta(t - \tau)$, we can put Eq. (8.47) in the form

$$P(t|t) = \Psi(t,t_0)P(t_0)\Psi'(t,t_0) + \int_{t_0}^t \Psi(t,\tau)[G(\tau)Q(\tau)G'(\tau)$$

$$+ K(\tau)R(\tau)K'(\tau) - G(\tau)S(\tau)K'(\tau)$$

$$- K(\tau)S'(\tau)G'(\tau)]\Psi'(t,\tau)\, d\tau \quad (8.48)$$

for $t \geq t_0$.

Equation (8.48) yields the filtering error covariance matrix for any filter gain matrix. However, for the optimal filter gain matrix, $K(\tau) = [P(\tau|\tau)H'(\tau) + G(\tau)S(\tau)]R^{-1}(\tau)$, it becomes an integral equation in $P(t|t)$ and is, therefore, not practical for computing $P(t|t)$.

The most logical thing to do, then, is to attempt to reduce Eq. (8.48) to a differential equation. Differentiating with respect to t,

$$\dot{P} = \dot{\Psi}(t,t_0)P(t_0)\Psi(t,t_0) + \Psi(t,t_0)P(t_0)\dot{\Psi}(t,t_0)$$

$$+ \Psi(t,t)[G(t)Q(t)G'(t) + K(t)R(t)K'(t)$$

$$- G(t)S(t)K'(t) - K(t)S'(t)G'(t)]\Psi'(t,t)$$

$$+ \int_{t_0}^t \dot{\Psi}(t,\tau)[G(\tau)Q(\tau)G'(\tau) + K(\tau)R(\tau)K'(\tau)$$

$$- G(\tau)S(\tau)K'(\tau) - K(\tau)S'(\tau)G'(\tau)]\Psi'(t,\tau)\, d\tau$$

$$+ \int_{t_0}^t \Psi(t,\tau)[G(\tau)G(\tau)G'(\tau) + K(\tau)R(\tau)K'(\tau)$$

$$- G(\tau)S(\tau)K'(\tau) - K(\tau)S'(\tau)G'(\tau)]\dot{\Psi}'(t,\tau)\, d\tau \quad (8.49)$$

However, $\dot{\Psi}(t,\tau) = C(t)\Psi(t,\tau)$ for all t, $\tau \geq t_0$, with $\Psi(\tau,\tau) = I$ for all $\tau \geq t_0$. Also, from Eq. (8.48),

$$\int_{t_0}^t \Psi(t,\tau)[G(\tau)Q(\tau)G'(\tau) + K(\tau)R(\tau)K'(\tau)$$

$$- G(\tau)S(\tau)K'(\tau) - K(\tau)S'(\tau)G'(\tau)]\Psi'(t,\tau)\, d\tau$$

$$= P(t|t) - \Psi(t,t_0)P(t_0)\Psi'(t,t_0)$$

Substituting these results into Eq. (8.49), we obtain

$$
\begin{aligned}
\dot{P} &= C(t)\Psi(t,t_0)P(t_0)\Psi'(t,t_0) + \Psi(t,t_0)P(t_0)\Psi'(t,t_0)C'(t) \\
&\quad + G(t)Q(t)G'(t) + K(t)R(t)K'(t) \\
&\quad - G(t)S(t)K'(t) - K(t)S'(t)G'(t) \\
&\quad + C(t)[P(t|t) - \Psi(t,t_0)P(t_0)\Psi'(t,t_0)] \\
&\qquad\qquad + [P(t|t) - \Psi(t,t_0)P(t_0)\Psi'(t,t_0)]C'(t) \\
&= C(t)P(t|t) + P(t|t)C'(t) + K(t)R(t)K'(t) \\
&\quad - G(t)S(t)K'(t) - K(t)S'(t)G'(t) + G(t)Q(t)G'(t)
\end{aligned}
$$

Recalling that $C(t) = F(t) - K(t)H(t)$, we now have

$$
\begin{aligned}
\dot{P} &= [F(t) - K(t)H(t)]P + P[F(t) - K(t)H(t)]' + K(t)R(t)K'(t) \\
&\quad - G(t)S(t)K'(t) - K(t)S'(t)G'(t) + G(t)Q(t)G'(t) \quad (8.50)
\end{aligned}
$$

for $t \geq t_0$, where $P = P(t|t)$. The solution of Eq. (8.50) subject to the initial condition $P(t_0|t_0) = P(t_0)$ is the filtering error covariance matrix for the filter in Eq. (8.35). It is stressed that Eq. (8.50) is valid for any filter gain matrix $K(t)$. If the optimal $K(t)$ is used, Eq. (8.50) can be simplified as follows. By rearranging terms, the equation can be written as

$$
\begin{aligned}
\dot{P} &= F(t)P - K(t)H(t)P + PF'(t) - PH'(t)K'(t) + K(t)R(t)K'(t) \\
&\quad - G(t)S(t)K'(t) - K(t)S'(t)G'(t) + G(t)Q(t)G'(t) \\
&= F(t)P + PF'(t) - K(t)[PH'(t) + G(t)S(t)]' \\
&\quad - [PH'(t) + G(t)S(t)]K'(t) + K(t)R(t)K'(t) + G(t)Q(t)G'(t)
\end{aligned}
$$

Substituting $K(t) = [P(t|t)H'(t) + G(t)S(t)]R^{-1}(t)$ into this result, we obtain

$$
\begin{aligned}
\dot{P} &= F(t)P + PF'(t) \\
&\quad - [PH'(t) + G(t)S(t)]R^{-1}(t)[PH'(t) + G(t)S(t)]' + G(t)Q(t)G'(t)
\end{aligned}
$$

Since P is the unknown, a more convenient form for this equation is

$$
\begin{aligned}
\dot{P} &= [F(t) - G(t)S(t)R^{-1}(t)H(t)]P + P[F(t) - G(t)S(t)R^{-1}(t)H(t)]' \\
&\quad - PH'(t)R^{-1}(t)H(t)P + G(t)[Q(t) - S(t)R^{-1}(t)S'(t)]G'(t) \quad (8.51)
\end{aligned}
$$

where $t \geq t_0$ and the initial condition is $P(t_0|t_0) = P(t_0)$.

SUMMARY

We summarize the results of this section in the following theorem.

Theorem 8.2

a. *The continuous optimal filter for the case where the gaussian white noises are correlated with $E[w(t)v'(\tau)] = S(t)\delta(t - \tau)$ is governed by the equation*

$$\dot{\hat{x}} = F(t)\hat{x} + K(t)[z(t) - H(t)\hat{x}] \tag{8.52}$$

for $t \geq t_0$ where $\hat{x} = \hat{x}(t|t)$, the initial condition is $\hat{x}(t_0|t_0) = 0$, and the $n \times m$ filter gain matrix is given by

$$K(t) = [P(t|t)H'(t) + G(t)S(t)]R^{-1}(t) \tag{8.53}$$

where $P(t|t)$ is the $n \times n$ covariance matrix of the filtering error $\tilde{x}(t|t) = x(t) - \hat{x}(t|t)$, $t \geq t_0$.

b. *The stochastic process $\{\tilde{x}(t|t), t \geq t_0\}$ is a zero mean Gauss-Markov process whose covariance matrix is the solution of the matrix differential equation*

$$\dot{P} = [F(t) - G(t)S(t)R^{-1}(t)H(t)]P$$
$$+ P[F(t) - G(t)S(t)R^{-1}(t)H(t)]' - PH'(t)R^{-1}(t)H(t)P$$
$$+ G(t)[Q(t) - S(t)R^{-1}(t)S'(t)]G'(t) \tag{8.54}$$

for $t \geq t_0$, where $P = P(t|t)$ with $P(t_0|t_0) = P(t_0) = E[x(t_0)x'(t_0)]$

Comparing these results with those obtained in Theorem 7.1, we observe that the filter equation is of the same form regardless of whether or not $\{w(t), t \geq t_0\}$ and $\{v(t), t \geq t_0\}$ are independent, but that the gain matrices are different. However, if the two processes are independent, then $S(t) = 0$ for $t \geq t_0$, and the above results reduce to those in Theorem 7.1 as indeed they should.

EXAMPLE

We illustrate the use of the results in Theorem 8.2 with the following example.

Example 8.1 Let us suppose that the system disturbance vector $w(t)$ and the measurement error vector $v(t)$ both have the same number of elements, i.e., $p = m$, and that the gaussian white noise which disturbs the system also corrupts the measurements. This means that the three covariance matrices involved are equal; i.e., $Q(t) = R(t) = S(t)$ for all $t \geq t_0$ and are all $m \times m$. We assume that $Q(t)$ is positive definite for all $t \geq t_0$.

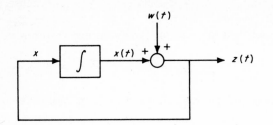

Fig. 8.2 Block diagram for scalar first-order system for Example 8.1.

It follows immediately that $S(t)R^{-1}(t) = Q(t)Q^{-1}(t) = I$, the $m \times m$ identity matrix, and $Q(t) - S(t)R^{-1}(t)S(t) = Q(t) - Q(t) = 0$, the $m \times m$ null matrix. In this case, the gain matrix is

$$K(t) = P(t|t)H'(t)Q^{-1}(t) + G(t)$$

and the filtering error covariance equation, Eq. (8.54), becomes

$$\dot{P} = [F(t) - G(t)H(t)]P + P[F(t) - G(t)H(t)]' - PH'(t)Q^{-1}(t)H(t)P$$

As a specific case, let us consider the scalar constant coefficient system

$$\dot{x} = x + w(t)$$

and

$$z(t) = x(t) + w(t)$$

for $t \geq 0$ with $E[w(t)] = 0$, $E[w^2(t)] = \sigma^2 = $ constant, $E[x(0)] = 0$, $E[x^2(0)] = \sigma_0^2 = $ constant, and $E[x(0)w(t)] = 0$. The system block diagram is indicated in Fig. 8.2.

We have $F(t) = G(t) = H(t) = 1$, so that $F(t) - G(t)H(t) = 0$. Hence, the variance equation is

$$\dot{P} = -\frac{1}{\sigma^2}P^2$$

where P is the scalar variance of the filtering error. Its solution follows from separation of variables, and we obtain

$$P(t|t) = \frac{\sigma_0^2\sigma^2}{\sigma^2 + \sigma_0^2 t}$$

In normalized form, this can be written

$$\frac{P(t|t}{\sigma_0^2} = \frac{1}{1 + \left(\dfrac{\sigma_0}{\sigma}\right)^2 t}$$

for which a sketch is shown in Fig. 8.3.

The filter gain in this case is simply

$$K(t) = \frac{\sigma_0^2}{\sigma^2 + \sigma_0^2 t} + 1$$

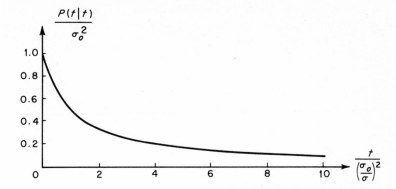

Fig. 8.3 Normalized optimal filtering error variance as a function of normalized time.

and the filter equation is obviously

$$\dot{\hat{x}} = \hat{x} + K(t)[z(t) - \hat{x}]$$

As $t \to \infty$, we note that $P(t) \to 0$, while $K(t) \to 1$. In the steady state, we achieve a zero variance filtering error, but the gain does not go to zero as we might expect intuitively. In particular, we see that the filter equation becomes $\dot{\hat{x}} = z(t)$ in the steady state; i.e., the filtering operation becomes one of integration. This result is plausible, for if we examine the system equations, we observe that $\dot{x} = z(t)$.

We note that even though the system $\dot{x} = x + w(t)$ is unstable, the filter is well-behaved and we obtain a zero variance error for sufficiently long filtering times.

Suppose now that in order to simplify the filter, we choose $K(t) = 1$ for all $t \geq 0$, i.e., we decide to use the steady state optimal gain during the entire operating time of the filter. The filtered estimate, no longer optimal, is then

$$\hat{x}(t|t) = \int_0^t z(\tau)\, d\tau$$

To determine the corresponding error variance, we set $F(t) = G(t) = H(t) = K(t) = 1$ and $Q(t) = R(t) = S(t) = \sigma^2$ in Eq. (8.47) and obtain the result $\dot{P} = 0$. This means that $P(t) = \sigma_0^2$ for all $t \geq 0$. If σ_0^2 is large, the filter is of little value, although it is relatively easy to implement since it involves no time-varying gain. On the other hand, if σ_0^2 is an acceptable error variance for all $t \geq 0$, then this suboptimal filter can be used.

8.4 OPTIMAL FIXED-POINT SMOOTHING

PROBLEM FORMULATION

We shall now solve the Wiener-Hopf equation to develop the algorithm for optimal fixed-point smoothing. The results which we develop here, in addition to substantiating those in Chap. 7, will provide an alternate differential equation for the smoothing filter gain matrix and two alternate differential equations for the smoothing error covariance matrix.

The procedure which we follow is similar to that in the preceding section. The present results were derived in 1967 by Meditch [5].

We seek the optimal estimate $\hat{x}(t_1|t)$ for some fixed $t_1 \geq t_0$, where t is variable and $t \geq t_1$. For $t_0 \leq \sigma \leq t_1$, we can consider the problem as one of filtering, and assume that Eq. (8.52) has been solved over this interval to obtain $\hat{x}(t_1|t_1)$. For simplicity, we consider only the case where $S(t) = 0$ for all $t \geq t_0$.

For purposes of fixed-point smoothing, we take t_1 as the initial time and $\hat{x}(t_1|t_1)$ as the initial estimate. The Wiener-Hopf equation for this case can be written as

$$E[x(t_1)z'(\sigma)] - \int_{t_1}^{t} A(t,\tau)E[z(\tau)z'(\sigma)] \, d\tau = 0$$

for $t_1 \leq \sigma \leq t$, where the optimal fixed-point smoothed estimate is

$$\hat{x}(t_1|t) = \int_{t_1}^{t} A(t,\tau)z(\tau) \, d\tau \tag{8.55}$$

In order to distinguish between the system matrices for the optimal fixed-point smoothing filter and the optimal filter of the preceding section, we shall denote the latter by $A^0(t,\tau)$, so that

$$\hat{x}(t|t) = \int_{t_0}^{t} A^0(t,\tau)z(\tau) \, d\tau \tag{8.56}$$

for all $t \geq t_0$ and

$$E[x(t)z'(\sigma)] - \int_{t_0}^{t} A^0(t,\tau)E[z(\tau)z'(\sigma)] \, d\tau = 0 \tag{8.57}$$

for all $t_0 \leq \sigma \leq t$. Equations (8.56) and (8.57) are, of course, also valid for $t_1 \leq \sigma \leq t$ if we replace t_0 by t_1 in them.

FILTER

We begin, as before, by considering the Wiener-Hopf equation over the interval $t_1 \leq \sigma < t$, and take the partial derivative with respect to t to obtain

$$\int_{t_1}^{t} \frac{\partial A(t,\tau)}{\partial t} E[z(\tau)z'(\sigma)] \, d\tau + A(t,t)E[z(t)z'(\sigma)] = 0 \tag{8.58}$$

From Eq. (8.29), we have

$$E[z(t)z'(\sigma)] = H(t)E[x(t)z'(\sigma)]$$

for $t_1 \leq \sigma < t$. Consequently, Eq. (8.58) can be written as

$$\int_{t_1}^{t} \frac{\partial A(t,\tau)}{\partial t} E[z(\tau)z'(\sigma)] \, d\tau + A(t,t)H(t)E[x(t)z'(\sigma)] = 0$$

for $t_1 \leq \sigma < t$. Substituting into this expression for $E[x(t)z'(\sigma)]$ from Eq. (8.57), we obtain

$$\int_{t_1}^{t} \frac{\partial A(t,\tau)}{\partial t} E[z(\tau)z'(\sigma)] \, d\tau + A(t,t)H(t) \int_{t_1}^{t} A^0(t,\tau)E[z(\tau)z'(\sigma)] \, d\tau = 0$$

or equivalently,

$$\int_{t_1}^{t} \left[\frac{\partial A(t,\tau)}{\partial t} + A(t,t)H(t)A^0(t,\tau) \right] E[z(\tau)z'(\sigma)] \, d\tau = 0 \qquad (8.59)$$

for $t_1 \leq \sigma < t$.

In exactly the same way that we showed that Eq. (8.32) is a necessary and sufficient condition for Eq. (8.31) to be satisfied, we can show that

$$\frac{\partial A(t,\tau)}{\partial t} + A(t,t)H(t)A^0(t,\tau) = 0 \qquad (8.60)$$

for $t_1 \leq \tau \leq t$ is a necessary and sufficient condition for Eq. (8.59) to hold.

Taking the partial derivative with respect to t in Eq. (8.55), we see that

$$\dot{\hat{x}}(t_1|t) = \int_{t_1}^{t} \frac{\partial A(t,\tau)}{\partial t} z(\tau) \, d\tau + A(t,t)z(t)$$

Substituting for $\partial A(t,\tau)/\partial t$ in this expression from Eq. (8.60), we obtain

$$\dot{\hat{x}}(t_1|t) = A(t,t)z(t) - \int_{t_1}^{t} A(t,t)H(t)A^0(t,\tau)z(\tau) \, d\tau$$

$$= A(t,t)z(t) - A(t,t)H(t) \int_{t_1}^{t} A^0(t,\tau)z(\tau) \, d\tau$$

which, in view of Eq. (8.56) considered over the interval $t_1 \leq \tau \leq t$, reduces to

$$\dot{\hat{x}}(t_1|t) = A(t,t)[z(t) - H(t)\hat{x}(t|t)] \qquad (8.61)$$

for all $t \geq t_1$.

Equation (8.61) specifies the structure of the optimal fixed-point smoothing filter. The initial condition for this equation is obviously the optimal filtered estimate $\hat{x}(t_1|t_1)$.

The solution to Eq. (8.61) can be written formally as

$$\hat{x}(t_1|t) = \hat{x}(t_1|t_1) + \int_{t_1}^{t} A(\tau,\tau)[z(\tau) - H(\tau)\hat{x}(\tau|\tau)] \, d\tau$$

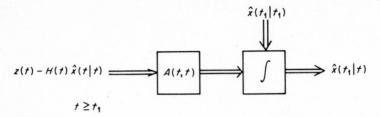

Fig. 8.4 Block diagram of optimal fixed-point smoothing filter.

for $t \geq t_1$. The interpretation here, as we know from Chap. 7, is that the filter consists of the integration of the product of a time-varying gain with the measurement residual as indicated in Fig. 8.4.

We emphasize again that the optimal fixed-point smoothing filter requires that the optimal filter, described in the preceding section, be operated concurrently.

GAIN MATRIX

We now develop an expression for the optimal fixed-point smoothing filter gain matrix $A(t,t)$ which we recall is an $n \times m$ matrix. Just as in Section 8.3, we begin by considering the appropriate form of the Wiener-Hopf equation for $\sigma = t$,

$$E[x(t_1)z'(t)] - \int_{t_1}^{t} A(t,\tau)E[z(\tau)z'(t)] \, d\tau = 0 \qquad (8.62)$$

Since we have assumed in this section that $\{w(t), \ t \geq t_0\}$ and $\{v(t), \ t \geq t_0\}$ are independent, we have $S(t) = 0$ for all $t \geq t_0$. Hence, Eq. (8.40) can be written as

$$E[x(t_1)z'(t)] = E[x(t_1)x'(t)]H'(t) \qquad (8.63)$$

for our purposes here, where $t \geq t_1$.

Also, Eq. (8.41) reduces to

$$E[z(\tau)z'(t)] = E[z(\tau)x'(t)]H'(t) + R(t)\delta(t - \tau) \qquad (8.64)$$

since $E[x(\tau)v'(t)] = 0$ for all $t, \tau \geq t_0$, whenever $S(t) = 0$ for $t \geq t_0$.

Substituting Eqs. (8.63) and (8.64) into Eq. (8.62), we obtain

$$E[x(t_1)x'(t)]H'(t) - \int_{t_1}^{t} A(t,\tau)E[z(\tau)x'(t)]H'(t) \, d\tau - A(t,t)R(t) = 0$$

It now follows that

$$A(t,t)R(t) = E\{[x(t_1) - \int_{t_1}^{t} A(t,\tau)z(\tau) \, d\tau]x'(t)\}H'(t)$$

$$= E[\tilde{x}(t_1|t)x'(t)]H'(t)$$

where we have made use of Eq. (8.55) and the definition of estimation error.

Then, trivially,

$$A(t,\tau) = E[\tilde{x}(t_1|t)x'(t)]H'(t)R^{-1}(t) \tag{8.65}$$

for $t \geq t_1$, since $R(t)$ is positive definite for all $t \geq t_0$.

Explicit evaluation of the $n \times n$ matrix $E[\tilde{x}(t_1|t)x'(t)]$ is cumbersome and tedious. However, it is relatively easy to establish a linear matrix differential equation whose solution yields this matrix. First, we define

$$B^*(t) = E[\tilde{x}(t_1|t)x'(t)] \tag{8.66}$$

Then, clearly,

$$\dot{B}^* = E[\dot{\tilde{x}}(t_1|t)x'(t)] + E[\tilde{x}(t_1|t)\dot{x}'(t)] \tag{8.67}$$

where the dot denotes the derivative with respect to t.

Since $\tilde{x}(t_1|t) = x(t_1) - \hat{x}(t_1|t)$, it follows from Eqs. (8.61) and (7.2), and the definition of filtering error that

$$\begin{aligned}
\dot{\tilde{x}}(t_1|t) &= -\dot{\hat{x}}(t_1|t) \\
&= -A(t,t)[z(t) - H(t)\hat{x}(t|t)] \\
&= -A(t,t)[H(t)x(t) + v(t) - H(t)\hat{x}(t|t)] \\
&= -A(t,t)[H(t)\tilde{x}(t|t) + v(t)]
\end{aligned} \tag{8.68}$$

Consequently,

$$E[\dot{\tilde{x}}(t_1|t)x'(t)] = -A(t,t)\{H(t)E[\tilde{x}(t|t)x'(t)] + E[v(t)x'(t)]\} \tag{8.69}$$

For $S(t) = 0$, it follows from Eq. (8.39) that the second term in braces in Eq. (8.69) vanishes for all $t \geq t_1$.

From the fact that $x(t) = \tilde{x}(t|t) + \hat{x}(t|t)$, Corollary 8.1, and the definition of $P(t|t)$,

$$\begin{aligned}
E[\tilde{x}(t|t)x'(t)] &= E\{\tilde{x}(t|t)[\tilde{x}(t|t) + \hat{x}(t|t)]'\} \\
&= E[\tilde{x}(t|t)\tilde{x}'(t|t)] \\
&= P(t|t)
\end{aligned}$$

Therefore, Eq. (8.69) simplifies to

$$E[\dot{\tilde{x}}(t_1|t)x'(t)] = -A(t,t)H(t)P(t|t) \tag{8.70}$$

This takes care of the first term on the right-hand side of Eq. (8.67).

Turning now to the second term, we utilize Eqs. (7.1) and (8.66) to obtain

$$E[\tilde{x}(t_1|t)\dot{x}'(t)] = E[\tilde{x}(t_1|t)x'(t)]F'(t) + E[\hat{x}(t_1|t)w'(t)]G'(t)$$
$$= B^*(t)F'(t) + E[\tilde{x}(t_1|t)w'(t)]G'(t) \tag{8.71}$$

However, $E[\tilde{x}(t_1|t)w'(t)] = 0$ for $t > t_1$. This is shown in the following way. Utilizing the definition of $\tilde{x}(t_1|t)$, we have

$$E[\tilde{x}(t_1|t)w'(t)] = E[x(t_1)w'(t)] - E[\hat{x}(t_1|t)w'(t)] \tag{8.72}$$

The first term on the right is zero by virtue of Eq. (8.25). Substituting into the second term from Eqs. (8.55) and (7.2), we see that

$$E[\hat{x}(t_1|t)w'(t)] = \int_{t_1}^{t} A(t,\tau)E\{[H(\tau)x(\tau) + v(\tau)]w'(t)\}\,d\tau$$
$$= \int_{t_1}^{t} A(t,\tau)H(\tau)E[x(\tau)w'(t)]\,d\tau \tag{8.73}$$

since $E[v(\tau)w'(t)] = 0$ for all t, $\tau \geq t_1$, because $\{w(t),\ t \geq t_0\}$ and $\{v(t),\ t \geq t_0\}$ are independent.

From Eq. (8.24), we have

$$x(\tau) = \Phi(\tau,t_0)x(t_0) + \int_{t_0}^{\tau} \Phi(\tau,\sigma)G(\sigma)w(\sigma)\,d\sigma$$

for all $\tau \geq t_0$. Since $\{w(t),\ t \geq t_0\}$ is an independent stochastic process with $E[w(\sigma)w'(t)] = Q(\sigma)\delta(\sigma - t)$, it follows immediately that

$$E[x(\tau)w'(t)] = \begin{cases} 0 & t_1 \leq \tau < t \\ G(t)Q(t) & \tau = t \end{cases}$$

Substitution of this result into Eq. (8.73) gives us

$$E[\hat{x}(t_1|t)w'(t)] = 0$$

which means that Eq. (8.72) reduces to

$$E[\tilde{x}(t_1|t)w'(t)] = 0$$

for all $t > t_1$, as claimed. Consequently then, Eq. (8.71) becomes

$$E[\tilde{x}(t_1|t)\dot{x}'(t)] = B^*(t)F'(t) \tag{8.74}$$

Substituting Eqs. (8.70) and (8.74) into Eq. (8.67), we obtain the $n \times n$ matrix differential equation

$$\dot{B}^* = -A(t,t)H(t)P(t|t) + B^*F'(t)$$

However, from Eqs. (8.65), (8.66), and (8.53), the latter for $S(t) = 0$, we observe that

$$A(t,t)H(t)P(t|t) = B^*(t)H'(t)R^{-1}(t)H(t)P(t|t)$$
$$= B^*(t)H'(t)K'(t)$$

Hence,

$$\dot{B}^* = -B^*H'(t)K'(t) + B^*F'(t)$$

or equivalently

$$\dot{B}^* = B^*[F(t) - K(t)H(t)]' \tag{8.75}$$

where $t \geq t_1$.

From Eq. (8.66) and Corollary 8.1, the initial condition for Eq. (8.75) is

$$B^*(t_1) = E[\tilde{x}(t_1|t_1)x'(t_1)]$$
$$= E\{\tilde{x}(t_1|t_1)[\tilde{x}(t_1|t_1) + \hat{x}(t_1|t_1)]'\}$$
$$= E[\tilde{x}(t_1|t_1)\tilde{x}'(t_1|t_1)]$$

or

$$B^*(t_1) = P(t_1|t_1) \tag{8.76}$$

Once Eq. (8.75) is solved for $B^*(t)$, $t \geq t_1$, the smoothing filter gain matrix is determined by the expression

$$A(t,t) = B^*(t)H'(t)R^{-1}(t) \tag{8.77}$$

which follows from Eqs. (8.65) and (8.66).

The above formulation for the fixed-point smoothing filter gain matrix is an alternate to the one we obtained in Chap. 7. (See also Prob. 7.10). It is relatively easy to show that it is equivalent to the one given there. We let $B(t)$ be the $n \times n$ matrix

$$B(t) = B^*(t)P^{-1}(t|t) \tag{8.78}$$

under the assumption that $P^{-1}(t|t)$ exists for all $t \geq t_1$. It then follows that Eq. (8.77) can be written

$$A(t,t) = B^*(t)P^{-1}(t|t)P(t|t)H'(t)R^{-1}(t)$$
$$= B(t)K(t) \tag{8.79}$$

where we have made use of Eqs. (8.78) and (8.53), the latter for $S(t) = 0$, $t \geq t_1$.

Differentiating in Eq. (8.78) with respect to t, we obtain

$$\dot{B} = \dot{B}^*P^{-1} + B^*\frac{dP^{-1}}{dt}$$

where we have momentarily dropped the time arguments for convenience. Substituting

$$\frac{dP^{-1}}{dt} = -P^{-1}\dot{P}P^{-1}$$

and Eq. (8.75) into this relation, we have

$$\dot{B} = B^*(F - KH)'P^{-1} - B^*P^{-1}\dot{P}P^{-1} = B^*F'P^{-1} - B^*H'K'P^{-1}$$
$$- B^*P^{-1}\dot{P}P^{-1} \quad (8.80)$$

For $S(t) = 0$, $t \geq t_1$, Eq. (8.54) is

$$\dot{P} = FP + PF' - PH'R^{-1}HP + GQG'$$

where we have again momentarily omitted the time arguments. Substituting this result and $K = PH'R^{-1}$ into Eq. (8.80), expanding, and regrouping terms, we see that

$$\dot{B} = B^*F'P^{-1} - B^*H'K'P^{-1}$$
$$- B^*P^{-1}(FP + PF' - PH'R^{-1}HP + GQG')P^{-1}$$
$$= B^*F'P^{-1} - B^*H'R^{-1}H - B^*P^{-1}F - B^*F'P^{-1} + B^*H'R^{-1}H$$
$$- B^*P^{-1}GQG'P^{-1}$$
$$= -B^*P^{-1}(F - GQG'P^{-1})$$

But, $B = B^*P^{-1}$, and we have immediately the matrix differential equation

$$\dot{B} = -B[F(t) + G(t)Q(t)G'(t)P^{-1}(t|t)] \quad (8.81)$$

for $t \geq t_1$. In view of Eqs. (8.78) and (8.76), we note that the initial condition for Eq. (8.81) is

$$B(t_1) = B^*(t_1)P^{-1}(t_1|t_1) = P(t_1|t_1)P^{-1}(t_1|t_1) = I \quad (8.82)$$

the $n \times n$ identity matrix.

Equations (8.79), (8.81), and (8.82) are the same as those given for the fixed-point smoothing filter gain matrix in Theorem 7.4.

SMOOTHING ERROR COVARIANCE MATRIX

We develop now three matrix linear ordinary differential equations for the fixed-point smoothing error covariance matrix. In the order presented, they involve the matrices $B^*(t)$, $B(t)$, and $A(t,t)$.

From its definition,

$$P(t_1|t) = E[\tilde{x}(t_1|t)\tilde{x}'(t_1|t)]$$

where $t \geq t_1$. Substituting for $\tilde{x}(t_1|t)$ and recalling Corollary 8.1, we have

$$P(t_1|t) = E\{\tilde{x}(t_1|t)[x(t_1) - \hat{x}(t_1|t)]'\} = E[\tilde{x}(t_1|t)x'(t_1)]$$

Differentiating with respect to t and substituting into the result from Eq. (8.68),

$$\dot{P}(t_1|t) = E[\dot{\tilde{x}}(t_1|t)x'(t_1)]$$

$$= -A(t,t)\{H(t)E[\tilde{x}(t|t)x'(t_1)] + E[v(t)x'(t_1)]\}$$

Since $S(t) = 0$, it follows from Eq. (8.39) that $E[v(t)x'(t_1)] = 0$ for all $t \geq t_1$, and the above expression becomes

$$\dot{P}(t_1|t) = -A(t,t)H(t)E[\tilde{x}(t|t)x'(t_1)] \tag{8.83}$$

For $t \geq t_1$, we note that the filtering error $\tilde{x}(t|t)$ can be expressed as

$$\tilde{x}(t|t) = \Psi(t,t_1)\tilde{x}(t_1|t_1) + \int_{t_1}^{t} \Psi(t,\tau)[G(\tau)w(\tau) - K(\tau)v(\tau)]\, d\tau \tag{8.84}$$

where $\Psi(t,\tau)$ is the state transition matrix of the system

$$\dot{\tilde{x}} = [F(t) - K(t)H(t)]\tilde{x} \tag{8.85}$$

Utilizing Eq. (8.84)

$$E[\tilde{x}(t|t)x'(t_1)] = \Psi(t,t_1)E[\tilde{x}(t_1|t_1)x'(t_1)]$$

$$+ \int_{t_1}^{t} \Psi(t,\tau)\{G(\tau)E[w(\tau)x'(t_1)] - K(\tau)E[v(\tau)x'(t_1)]\}\, d\tau$$

However, the two expected values in the integrand on the right-hand side are zero as a result of Eqs. (8.25) and (8.39), the latter since $S(t) = 0$. Then, from the definition of filtering error, $x(t_1) = \tilde{x}(t_1|t_1) + \hat{x}(t_1|t_1)$, and Corollary 8.1,

$$E[\tilde{x}(t|t)x'(t_1)] = \Psi(t,t_1)E\{\tilde{x}(t_1|t_1)[\tilde{x}(t_1|t_1) + \hat{x}(t_1|t_1)]'\}$$

$$= \Psi(t,t_1)P(t_1|t_1) \tag{8.86}$$

Comparing Eq. (8.85) with the transpose of Eq. (8.75), we observe that both have the same state transition or fundamental matrix, namely, $\Psi(t,\tau)$. From Eq. (8.76), $B^{*\prime}(t_1) = P(t_1|t_1)$, and we have

$$B^{*\prime}(t) = \Psi(t,t_1)P(t_1|t_1) \tag{8.87}$$

It now follows immediately from Eqs. (8.86) and (8.87) that

$$E[\tilde{x}(t|t)x'(t_1)] = B^{*\prime}(t)$$

Substituting this result into (8.83) gives

$$\dot{P}(t_1|t) = -A(t,t)H(t)B^{*\prime}(t) \tag{8.88}$$

for $t \geq t_1$. The initial condition for Eq. (8.88) is obviously

$$P(t_1|t_1) = E[\tilde{x}(t_1|t_1)\tilde{x}'(t_1|t_1)] \tag{8.89}$$

which is obtained by evaluating the solution of Eq. (8.54) at $t = t_1$ with $S(t) = 0$.

The three alternate forms of Eq. (8.88) follow immediately from the relations between $A(t,t)$, $B^*(t)$, and $B(t)$.

First, substitution of Eq. (8.77) into Eq. (8.88) yields

$$\dot{P}(t_1|t) = -B^*(t)H'(t)R^{-1}(t)H(t)B^{*\prime}(t) \tag{8.90}$$

Second, from Eq. (8.78), we note that

$$B^*(t) = B(t)P(t|t)$$

and Eq. (8.90) can be written as

$$\begin{aligned}
\dot{P}(t_1|t) &= -B(t)P(t|t)H'(t)R^{-1}(t)H(t)P(t|t)B'(t) \\
&= -B(t)K(t)R(t)K'(t)B'(t)
\end{aligned} \tag{8.91}$$

which coincides with Eq. (7.59) of Theorem 7.4.

Finally, noting from Eq. (8.77) that

$$H(t)B^{*\prime}(t) = R(t)A'(t,t)$$

we express Eq. (8.90) in the form

$$\dot{P}(t_1|t) = -A(t,t)R(t)A'(t,t) \tag{8.92}$$

Discussion and Summary The stochastic process $\{\tilde{x}(t_1|t), \ t \geq t_1\}$ which is defined by Eq. (8.68) is obviously a zero mean Gauss-Markov-2 process, since it can be represented as

$$\tilde{x}(t_1|t) = \tilde{x}(t_1|t_1) - \int_{t_1}^{t} A(\tau,\tau)[H(\tau)\tilde{x}(\tau|\tau) + v(\tau)] \, d\tau$$

and we know that $\{\tilde{x}(\tau|\tau), \ \tau \geq t_1\}$ is a zero mean Gauss-Markov process while $\{v(\tau), \ \tau \geq t_1\}$ is a zero mean gaussian white noise.

Turning now to the question of computing $A(t,t)$ and $P(t_1|t)$, we note that the use of Eqs. (8.77), (8.75), and (8.90) is preferable to that of Eqs. (8.79), (8.81), and (8.91). The reason for this preference is because of the particular matrix inversion which is involved. Specifically, Eq. (8.81) requires that $P(t|t)$ be inverted to compute $C(t)$. In problems involving long filtering and smoothing times, $P(t|t)$ may approach zero, and, therefore, become nearly singular. In the simple examples which we consider, this is no real problem, since we have "infinite" accuracy in our analytical computations. However, with finite word length in a digital computer, round off errors can propagate and lead to meaningless results.

The use of Eq. (8.92) in computing $P(t_1|t)$ is perhaps preferable to either Eqs. (8.90) or (8.91), since $A(t,t)$ will be determined anyhow in analyzing or mechanizing the smoothing filter.

In view of these comments, we choose to summarize the results of this section as a supplement to Theorem 7.4 by presenting the following theorem.

Theorem 8.3

a. *The optimal fixed-point smoothing filter is governed by the equation*

$$\dot{\hat{x}}(t_1|t) = A(t,t)[z(t) - H(t)\hat{x}(t|t)] \tag{8.93}$$

for $t \geq t_1$ subject to the initial condition $\hat{x}(t_1|t_1)$, the optimal filtered estimate at time t_1.

b. *The $n \times n$ smoothing filter gain matrix is given by the relation*

$$A(t,t) = B^*(t)H'(t)R^{-1}(t) \tag{8.94}$$

where the $n \times n$ matrix $B^*(t)$ is the solution of the matrix differential equation

$$\dot{B}^* = B^*[F(t) - K(t)H(t)]' \tag{8.95}$$

for $t \geq t_1$, where $B^*(t_1) = P(t_1|t_1)$ and $K(t) = P(t|t)H'(t)R^{-1}(t)$, the optimal filter gain matrix.

c. *The stochastic process $\{\tilde{x}(t_1|t), \ t \geq t_1\}$ is a zero mean Gauss-Markov-2 process whose covariance matrix is the solution of the matrix differential equation*

$$\dot{P}(t_1|t) = -A(t,t)R(t)A'(t,t) \tag{8.96}$$

for $t \geq t_1$ subject to the initial condition $P(t_1|t_1) = E[\tilde{x}(t_1|t_1)\tilde{x}'(t_1|t_1)]$, the optimal filtering error covariance matrix at time t_1.

We illustrate the use of the above results in the following simple example where we take $t_0 = -\infty$ and $t_1 = 0$. In other words, we initiate fixed-point smoothing after the optimal filter for $\hat{x}(t|t)$ has reached steady state.

Example 8.2   We consider the system

$$\dot{x} = ax + w(t) \tag{8.97}$$

$$z(t) = x(t) + v(t)$$

where a is a real scalar, $\{w(t), t \geq -\infty\}$ is a zero mean scalar gaussian white noise with unit variance, and $\{v(t), t \geq -\infty\}$ is also a zero mean scalar gaussian white noise with unit variance. We assume that the two noise processes are independent of each other. Utilizing the results of Theorem 7.1, the variance equation for the filtering error is

$$\dot{P} = 2aP - P^2 + 1$$

where $P = P(t|t)$. The steady-state solution is obtained by setting $\dot{P} = 0$ to obtain

$$\bar{P}(t|t) = a + \sqrt{a^2 + 1}$$

Since $H(t) = 1$ and $R^{-1}(t) = 1$, it follows that $\bar{K}(t) = \bar{P}(t|t)$.

Let us now consider fixed-point smoothing for four separate cases as a function of the system parameter a with $t \geq t_1 = 0$.

1. $a > 0$ with $|a| \ll 1$. In this case, the system of Eq. (8.97) is unstable. We have

$$F(t) - \bar{K}(t)H(t) = a - a - \sqrt{a^2 + 1} \approx -1$$

and

$$\bar{P}(0|0) = a + \sqrt{a^2 + 1} \approx 1$$

Equation (8.95) is

$$\dot{B}^* = -B^*$$

and its solution is obviously

$$B^*(t) = e^{-t} \qquad t \geq 0$$

recalling that $B^*(0) = \bar{P}(0|0) = 1$. Since $H(t)$ and $R^{-1}(t)$ are both unity, it follows from Eq. (8.94) that the optimal fixed-point smoothing filter gain is $A(t,t) = B^*(t)$, and from Eq. (8.93) that

$$\dot{\hat{x}}(0|t) = e^{-t}[z(t) - \hat{x}(t|t)]$$

The fixed-point smoothing error variance equation, Eq. (8.96), becomes

$$\dot{P}(0|t) = -e^{-2t}$$

Since $P(0|0) = 1$, we obtain

$$P(0|t) = \tfrac{1}{2}(1 + e^{-2t}) \qquad t \geq 0$$

and, for sufficiently large t, it is clear that $P(0|t) \to \tfrac{1}{2}$. We note in this case that the corresponding filtering variance is 1 and that we can effect a 50 percent improvement in accuracy by means of fixed point smoothing.

2. $a > 0$ with $|a| \gg 1$. Here, we have

$$\bar{P}(t|t) \approx 2a \qquad \text{and} \qquad \bar{K}(t) \approx 2a$$

so that $F(t) - \bar{K}(t)H(t) = -a$. Hence, $\dot{B} = -aB$, giving

$$B^*(t) = 2ae^{-at} \qquad t \geq 0$$

since $B^*(0) = \bar{P}(0|0) = 2a$. The variance equation is $\dot{P}(0|t) = -4a^2e^{-2at}$. Its solution is easily seen to be

$$P(0|t) = 2ae^{-2at}$$

For sufficiently large t, $P(0|t) \to 0$. Hence, in theory, we can determine $x(0)$ exactly with fixed-point smoothing, whereas the corresponding filtering error variance is $2a$ with $|a| \gg 1$.

3. $a < 0$ with $|a| \ll 1$. The system of Eq. (8.97) is obviously stable. We note that $\bar{P}(t|t) = a + \sqrt{a^2 + 1} \approx 1$, and the result is the same as that for case 1 above, viz.,

$$P(0|t) = \tfrac{1}{2}(1 + e^{-2t})$$

4. $a < 0$ with $|a| \gg 1$. In this case,

$$\bar{P}(t|t) = a + \sqrt{a^2 + 1} \approx a + |a| = 0$$

since a is negative, and it is obvious that fixed-point smoothing is not needed in order to improve estimation accuracy.

PROBLEMS

8.1. Establish the sufficiency of the Wiener-Hopf equation by examining $\partial^2 J/\partial\epsilon^2$ for $\epsilon = 0$.

8.2. Express the Wiener-Hopf equation in terms of the appropriate correlation functions.

8.3. What is the Wiener-Hopf equation if $\{x(t_1), t_1 \geq t_0\}$ and $\{z(\tau), t_0 \leq \tau \leq t\}$ both have nonzero mean.

8.4. What are the appropriate forms of the Wiener-Hopf equation for the problems of (a) optimal prediction, (b) optimal fixed-interval smoothing, and (c) optimal fixed-lag smoothing?

8.5. Formulate and solve the appropriate Wiener-Hopf equation for the system model of Eqs. (7.1) and (7.2) to obtain the optimal fixed-lead predicted estimate $\hat{x}(t + T|t)$, where $t \geq t_0$ and $T =$ positive constant, the "lead time." Determine also a differential equation for the covariance matrix of the fixed-lead prediction error $\tilde{x}(t + T|t) = x(t + T) - \hat{x}(t + T|t)$.

8.6. For the problem of optimal filtering, show that the optimal filter system matrix is

$$A(t,\tau) = \Psi(t,\tau)K(\tau)$$

for $t, \tau \geq t_0$, where $K(\tau)$ is given by Eq. (8.53) and $\Psi(t,\tau)$ is the state transition matrix of the system

$$\dot{y} = [F(t) - K(t)H(t)]y$$

where y is any real n vector.

8.7. Suppose that the filter gain matrix is determined under the assumption that $S(t) = 0$, that is, using the relations

$$K(t) = P(t|t)H'(t)R^{-1}(t)$$

and

$$\dot{P} = F(t)P + PF'(t) - PH'(t)R^{-1}(t)H(t)P + G(t)Q(t)G'(t)$$

for $t \geq t_0$ where $P = P(t|t)$ and $P(t_0|t_0) = P(t_0)$, but that $S(t)$ is not actually zero. What is the differential equation that must be solved to determine the *actual* filtering error covariance matrix?

8.8. If $R(t)$ is positive semidefinite for all $t \geq t_0$, is Eq. (8.32),

$$F(t)A(t,\tau) - \frac{\partial A(t,\tau)}{\partial t} - A(t,t)H(t)A(t,\tau) = 0$$

$t_0 \leq \tau \leq t$, still a necessary and sufficient condition for Eq. (8.31),

$$\int_{t_0}^{t} [F(t)A(t,\tau) - \frac{\partial A(t,\tau)}{\partial t} - A(t,t)H(t)A(t,\tau)]E[z(\tau)z'(\sigma)]\, d\tau = 0$$

$t_0 \leq \sigma < t$, to hold? Consider the same question if $Q(t)$ is positive definite for $t \geq t_0$, but $R(t)$ is positive semidefinite.

8.9. Consider the scalar system $\dot{x} = w(t)$, $z(t) = x(t) + v(t)$ for $t \geq 0$, where $Q(t) = \frac{9}{4}$, $R(t) = 1$, $S(t) = \frac{1}{2}$, and $P(t_0) = 28$.

(a) Determine $P(t|t)$ and $K(t)$ for all $t \geq 0$.

(b) As $t \to \infty$, show that $P(t|t) \to 1$.

(c) If the filter gain is determined ignoring the fact that $S(t) = \frac{1}{2}$, that is, under the assumption that $S(t) = 0$, what is the actual filtering variance as $t \to \infty$?

8.10. In the development of the equations for optimal fixed-point smoothing, show that Eq. (8.60),

$$\frac{\partial A(t,\tau)}{\partial t} + A(t,t)H(t)A^0(t,\tau) = 0$$

$t_1 \leq \tau \leq t$, is a necessary and sufficient condition for Eq. (8.59),

$$\int_{t_1}^{t} \left[\frac{\partial A(t,\tau)}{\partial t} + A(t,t)H(t)A^0(t,\tau) \right] E[z(\tau)z'(\sigma)]\, d\tau = 0$$

to hold for $t_1 \leq \sigma < t$. Assume that $R(t)$ is positive definite for all $t \geq t_1 \geq t_0$.

8.11. A particle, constrained to move in only one dimension, leaves the origin at time $t = 0$ with a velocity which is gaussian distributed with mean zero and variance $\sigma_0^2 = $ constant > 0. A random force in the form of a zero mean gaussian white noise $\{w(t), t \geq 0\}$ with variance $\sigma_w^2 = $ constant > 0 acts to accelerate the particle. The position of the particle is tracked continuously in the presence of a scalar additive zero mean gaussian white noise $\{v(t), t \geq 0\}$, which is independent of $\{w(t), t \geq 0\}$ and has variance $\sigma_v^2 = $ constant > 0. For simplicity, assume that the particle has unit mass.

(a) Set up, but do not solve, the equations which are necessary to obtain the optimal filtered estimates of the particle's position and velocity for $t \geq 0$.

(b) Set up, but do not solve, the equations which are necessary to estimate the velocity with which the particle left the origin at $t = 0$ utilizing optimal fixed-point smoothing.

8.12. What are the equations for optimal fixed-point smoothing if $\{w(t), t \geq t_0\}$ and $\{v(t), t \geq t_0\}$ are correlated with $E[w(t)v'(\tau)] = S(t)\delta(t - \tau)$ for all $t, \tau \geq t_0$?

8.13. For the problem of optimal fixed-interval smoothing over the interval $t_0 \leq t \leq t_1$, show that:

(a) The optimal fixed-interval smoothing filter is defined by the relation

$$\dot{\hat{x}}(t|t_1) = F(t)\hat{x}(t|t_1) + A(t,t)[\hat{x}(t_1|t) - \hat{x}(t|t)]$$

for $t_0 \leq t \leq t_1$ subject to the boundary condition $\hat{x}(t_1|t_1)$ and that the optimal $n \times n$ gain matrix $A(t,t)$ is given by the relation $A(t,t) = G(t)Q(t)G'(t)P^{-1}(t|t)$.

(b) The covariance matrix of the corresponding smoothing error is governed by the matrix differential equation

$$\dot{P}(t|t_1) = [F(t) + G(t)Q(t)G'(t)P^{-1}(t|t)]P(t|t_1)$$
$$+ P(t|t_1)[F(t) + G(t)Q(t)G'(t)P^{-1}(t|t)]' - G(t)Q(t)G'(t)$$

for $t_0 \leq t \leq t_1$ subject to the boundary condition $P(t_1|t_1)$. Assume that $\{w(t), t \geq t_0\}$ and $\{v(t), t \geq t_0\}$ are independent.

8.14. Show what modifications, if any, must be made in the results in Prob. 8.13 if $\{w(t), t \geq t_0\}$ and $\{v(t), t \geq t_0\}$ are correlated with $E[w(t)v'(\tau)] = S(t)\delta(t - \tau)$ for all $t, \tau \geq t_0$.

8.15. Set up, but do not solve, the equations which are necessary to obtain the fixed-interval smoothed estimates of the position and velocity of the particle in Prob. 8.11. Consider the system over the time interval $0 \leq t \leq T$ where T is fixed.

8.16. Starting with the appropriate Wiener-Hopf integral equation, develop the equations for optimal fixed-lag smoothing for $t \geq t_0$ where the lag is $T = $ constant > 0. Assume that $\{w(t), t \geq t_0\}$ and $\{v(t), t \geq t_0\}$ are independent.

8.17. In the system model of Eqs. (7.1) and (7.2),

$$\dot{x} = F(t)x + G(t)w(t)$$

and

$$z(t) = H(t)x(t) + v(t)$$

respectively, where $t \geq t_0$, suppose that $\{v(t), t \geq t_0\}$, instead of being a zero mean gaussian white noise, is a time-correlated process which is governed by the relation

$$\dot{v} = A(t)v + \xi(t)$$

where $t \geq t_0$, $A(t)$ is a continuous $m \times m$ matrix, and $\{\xi(t), t \geq t_0\}$ is an m-dimensional zero mean gaussian white noise. Suppose further that $v(t_0)$ is independent of $x(t_0)$, $\{\xi(t), t \geq t_0\}$ and $\{w(t), t \geq t_0\}$, and has a positive definite covariance matrix $E[v(t_0)v'(t_0)] = V(t_0)$. Let $\{\xi(t), t \geq t_0\}$ be such that

$$E[x(t_0)\xi'(t)] = 0$$

$$E[\xi(t)\xi'(\tau)] = W(t)\delta(t - \tau)$$

and

$$E[w(t)\xi'(\tau)] = 0$$

for all $t, \tau \geq t_0$, where $W(t)$ is $m \times m$, positive definite and continuous.

(a) Can the augmented state vector approach along with Theorem 8.1 be applied in this case to determine the optimal filtered estimate $\hat{x}(t|t)$ for $t \geq t_0$? If so, set up the appropriate equations in partitioned form. If not, explain.

(b) Analogous to the procedure of differencing discrete-time measurements in which the measurement error is subject to time correlation to obtain a new measurement in which the measurement error is white (see Probs. 4.13 and 6.19), it is possible in the present case to use differentiation to accomplish the same result. Defining the new measurement to be

$$z^*(t) = \dot{z}(t) - A(t)z(t)$$

show that this new measurement can be written in the form

$$z^*(t) = H^*(t)x(t) + \zeta(t)$$

where

$$H^*(t) = \dot{H}(t) + H(t)F(t) - A(t)H(t)$$

and $\{\zeta(t), t \geq t_0\}$ is an m-dimensional zero mean gaussian white noise for which

$$E[\zeta(t)\zeta'(\tau)] = [H(t)G(t)Q(t)G'(t)H'(t) + W(t)]\delta(t - \tau)$$

$$E[w(t)\zeta'(\tau)] = Q(t)G'(t)H'(t)\delta(t - \tau)$$

and

$$E[x(t_0)\zeta(t)] = 0$$

for all $t, \tau \geq t_0$. Note that this formulation requires differentiability of $H(t)$ [2, 6, 7, 8].

(c) Apply Theorem 8.1 to the formulation in part b to obtain the optimal filter. Develop a block diagram for the optimal filter in which no operations requiring differentiation of $z(t)$ are present [2, 6, 7, 8].

(d) Show that the correct initial conditions for this approach are

$$\hat{x}(t_0|t_0) = P(t_0)H'(t_0)[H(t_0)P(t_0)H'(t_0) + V(t_0)]^{-1}z(t_0)$$

and

$$P(t_0|t_0) = P(t_0) - P(t_0)H'(t_0)[H(t_0)P(t_0)H'(t_0) + V(t_0)]^{-1}H(t_0)P(t_0)$$

where $P(t_0) = E[x(t_0)x'(t_0)]$ is assumed given [2, 6, 7, 8].

(e) List any additional assumptions which are necessary in part c, such as differentiability of $H(t)$, etc. Can the assumption that $W(t)$ be positive definite be weakened to requiring that $W(t)$ be positive semidefinite in part c? Explain your answer.

REFERENCES

1. Kalman, R. E., and R. S. Bucy, New Results in Linear Filtering and Prediction Theory, *J. Basic Eng.*, vol. 83, p. 95, 1961.
2. Bucy, R. S., and P. D. Joseph, Filtering for Stochastic Processes and Applications to Guidance, Interscience Division of John Wiley & Sons, Inc., New York, 1968.
3. Lee, Y. W., "Statistical Theory of Communication," John Wiley & Sons, Inc., New York, 1960.
4. Newton, G. C., Jr., L. A. Gould, and J. F. Kaiser, "Analytical Design of Linear Feedback Controls," John Wiley & Sons, Inc., New York, 1967.
5. Meditch, J. S., On Optimal Fixed-Point Linear Smoothing, *Int. J. Control*, vol. 6, p. 189, 1967. Also, The Wiener-Hopf Solution of the Optimal-Fixed Point Smoothing Problem, *Tech. Rept.* 67-111, IPAC Systems Laboratory, Northwestern University, Evanston, Illinois, June, 1967.
6. Bryson, A. E., Jr., and D. E. Johansen, Linear Filtering for Time-Varying Systems Using Measurements Containing Colored Noise, *IEEE Trans. Autom. Control*, vol. AC-10, p. 4, 1965.
7. Bucy, R. S., Optimal Filtering for Correlated Noise, *J. Math. Anal. Appl.*, vol. 20, p. 1, 1967.
8. Stear, E. B., and A. R. Stubberud, Optimal Filtering for Gauss-Markov Noise, *Int. J. Control*, vol. 8, p. 123, 1968.

9

Stochastic Optimal Control for Discrete Linear Systems

9.1 INTRODUCTION

The general problem of controlling a system which is subject to disturbances and measurement errors such that some measure of the system's behavior is optimized was discussed briefly in Chap. 1. In this chapter and the next, we address ourselves to a particular class of problems within the above general framework. Specifically, we restrict our attention to the Gauss-Markov system models of Secs. 4.3 and 4.4 in which an additive control input is present. We take as the performance measure a functional which involves the expected value of a quadratic form in the state and control variables over a fixed interval of time. Although this performance measure is nowhere near as general with respect to the control problem as was the class of performance measures which we considered in the estimation problem, it does possess features which make it useful in applications. The resulting problem, for either the discrete-time or the continuous-time case, is called the stochastic linear regulator problem. We consider the discrete-time case here and the continuous-time one in Chap. 10.

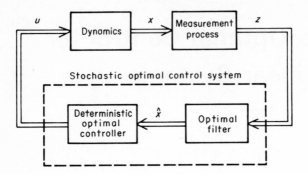

Fig. 9.1 Block diagram illustrating separation principle.

We give a detailed formulation of the discrete-time stochastic linear regulator problem in Sec. 9.2. We then introduce the principle of optimality [1] in Sec. 9.3, and use it to obtain the solution of the deterministic version of the problem in which it is assumed that the system disturbance is absent and that all of the state variables can be measured exactly. The purpose of treating the deterministic problem is twofold: first, to provide a relatively simple introduction to the technique which is used later to solve the stochastic problem, and second, to develop a result which plays a significant role in the interpretation of the solution of the latter problem.

We solve the discrete-time stochastic linear regulator problem in Sec. 9.4 and interpret it in the light of the result in Sec. 9.3. The interpretation leads to a separation principle which states that the optimal control system consists of the optimal filter in cascade with the deterministic optimal controller derived in Sec. 9.3. The result is indicated schematically in Fig. 9.1.

9.2 PROBLEM FORMULATION

SYSTEM MODEL

Our system model is defined by the relations

$$x(k + 1) = \Phi(k + 1, k)x(k) + \Gamma(k + 1, k)w(k) + \Psi(k + 1, k)u(k)$$

$$\tag{9.1}$$

and

$$z(k + 1) = H(k + 1)x(k + 1) + v(k + 1) \tag{9.2}$$

for $k = 0, 1, \ldots$. The model is the one which was formulated in Sec. 4.3 and for which the optimal estimation problem was solved in Chaps. 5 and 6 under the assumption that $u(k) = 0$ for all $k = 0, 1, \ldots$.

For convenience of reference in this chapter, let us review the definitions of the various terms in Eqs. (9.1) and (9.2):

$x = n$ vector (state)

$u = r$ vector (control)

$w = p$ vector (disturbance)

$z = m$ vector (measurement)

$v = m$ vector (measurement error)

$\Phi = n \times n$ state transition matrix

$\Gamma = n \times p$ disturbance transition matrix

$\Psi = n \times r$ control transition matrix

$H = m \times n$ measurement matrix

$x(0)$ = zero mean gaussian random n vector with positive semi-definite covariance matrix $P(0)$

$\{w(k), k = 0, 1, \ldots\}$ = zero mean gaussian white sequence which is independent of $x(0)$ and has a positive semidefinite covariance matrix $Q(k)$, $k = 0, 1, \ldots$

$\{v(k + 1), k = 0, 1, \ldots\}$ = zero mean gaussian white sequence which is independent of $x(0)$ and $\{w(k), k = 0, 1, \ldots\}$, and has a positive semidefinite covariance matrix $R(k + 1)$, $k = 0, 1, \ldots$

$\{u(k), k = 0, 1, \ldots\}$ = control sequence which is either known or can be specified as desired

From Sec. 4.3, we recall the following properties of the system model which we shall need:

1. $\{x(i), i = 0, 1, \ldots\}$ is a zero mean Gauss-Markov sequence.
2. $x(i)$ and $w(i)$ are statistically independent for all $i = 0, 1, \ldots$. [See also Eq. (5.25).]
3. $z(i)$ and $w(j)$ are statistically independent for all $j \geq i, i = 1, 2, \ldots$. [See also Eq. (5.26).]

PERFORMANCE MEASURE

We consider the problem of controlling the state x of the system of Eq. (9.1) over some fixed interval of time $[0,N]$, N = positive integer, by constructing a control sequence $\{u(k), k = 0, 1, \ldots, N - 1\}$ which minimizes the performance measure

$$J_N = E\left\{ \sum_{i=1}^{N} [x'(i)A(i)x(i) + u'(i-1)B(i-1)u(i-1)] \right\} \qquad (9.3)$$

In Eq. (9.3), $A(i)$ and $B(i-1)$ are symmetric positive semidefinite matrices, which are $n \times n$ and $r \times r$, respectively, and E denotes the expected value.

We note immediately that a number of restrictions have been imposed by the choice of Eq. (9.3) as the performance measure for control. One reason for this is that there is no result in stochastic control theory which is analogous to Sherman's result, Theorem 5.1, in estimation theory. Hence, it is necessary to consider rather specialized performance measures such as the one above.

It is clear from Eq. (9.3) that J_N is the expected value of a positive semidefinite quadratic form in the state and control. The expected value is used since we are dealing with a stochastic process. Hence, as in the estimation problem, we are assessing performance over an ensemble of systems. The summation is used to account for behavior of both the state and control over the time interval of interest. The indexing is chosen so that only those states which can be manipulated and only those controls which can affect them appear in J_N. Thus, $x(0)$ is absent because it cannot be affected by any control, while $u(0)$ is present because its choice affects $x(1)$, $x(2)$, . . . , $x(N)$. Similarly, $u(N-1)$ is the final control input, and it affects only $x(N)$ at which time the problem terminates.

The usual interpretation associated with J_N is that it is a "system error plus control effort" measure of performance. This follows from its quadratic nature, since for each $i = 1, 2, \ldots, N$, the two terms on the right-hand side of Eq. (9.3) are monotone nondecreasing functions of x and u, respectively. In the first term, then, the larger the deviation of x from zero, the greater the contribution to J_N, and similarly for u.

The first term on the right-hand side of the equation implies that the desired state is zero, i.e., the origin in n-dimensional euclidean space, since if $x(i) = 0$, $i = 1, 2, \ldots, N$, the term's contribution to J_N is zero regardless of $A(i)$. If, at each time point i, the desired state is some arbitrary $x^d(i)$, then $x(i)$ would be replaced by $x(i) - x^d(i)$ in Eq. (9.3). We choose $x^d(i) = 0$ for all i as a matter of convenience. We remark that the quadratic nature of the term implies that the measure of error here is one of error-squared, and, more generally, of weighted-error-squared because of the freedom in choosing $A(i)$.

Because the significance of terms such as $a_{jk}(i)E[x_j(i)x_k(i)]$, $j \neq k$, as measures of error is not obvious, $A(i)$ is usually chosen to be diagonal. For example, in a second-order system, the first term in Eq. (9.3) would typically be expressed as

$$E\left[\sum_{i=1}^{N} x'(i)A(i)x(i)\right] = E\left\{\sum_{i=1}^{N} [a_{11}(i)x_1{}^2(i) + a_{22}(i)x_2{}^2(i)]\right\}$$

where the choice of $a_{11}(i)$ relative to $a_{22}(i)$ determines the weighting of the error $x_1(i)$ relative to the error $x_2(i)$.

Since the second term in Eq. (9.3) is a monotone nondecreasing function of u for each i and is summed over i, it is a measure of the "intensity" of the control which is input to the system. This intensity is usually termed *control effort* and is sometimes also called *control energy*. The latter name arises as a consequence of the quadratic nature of the term. As in the case of the first term, we speak of "weighted control effort" because of the relatively arbitrary nature of $B(i-1)$. Also, as in the case of the first term, $B(i-1)$ is usually diagonal.

Viewing now the combination of the two terms, we see that J_N is a measure which provides for a trade-off between system error and control effort. The relative importance of the two terms is reflected in the choice of $A(i)$ with respect to $B(i-1)$. However, regardless of the relative choice, we note that, in general, J_N penalizes for either excessive error or excessive control. Because J_N is monotone and nondecreasing in both x and u, we wish to select the control sequence $\{u(i-1), i = 1, 2, \ldots, N\}$ to minimize J_N.

There is considerable latitude in the choice of $A(i)$ and $B(i-1)$, so that the performance measure can be specified for a variety of goals. For example, suppose that only the error at the terminal time N is of concern in a particular control problem, and we are willing to expend whatever control effort is necessary to minimize this error. Then obviously, we choose $A(i) = 0$ for $i = 1, 2, \ldots, N-1$ and $B(i-1) = 0$ for $i = 1, 2, \ldots, N$, and obtain

$$J_N = E[x'(N)A(N)x(N)] \tag{9.4}$$

A further simplification occurs if we let $A(N)$ be the $n \times n$ identity matrix, in which case J_N is the expected value of the square of the error vector at the terminal time.

Pursuing this example still further, suppose that the deviation of only the first two components of $x(N)$ from the origin is of concern; then we pick

$$A(N) = \begin{bmatrix} a_{11} & 0 & \cdot & \cdot & \cdot & \cdot & \cdot & \cdot \\ 0 & a_{22} & 0 & & \cdot & \cdot & \cdot \\ \cdot & \cdot & \cdot & 0 & 0 & & \cdot & \cdot & \cdot \\ \cdot & \cdot & \cdot & \cdot & \cdot & \cdot & \cdot & \cdot & \cdot & \cdot & \cdot \end{bmatrix}$$

with a_{11} and a_{22} positive.

Problems of this type are called *terminal error problems* for an obvious reason.

If we are concerned that the performance measure in Eq. (9.4) might place severe demands on our control resources, we introduce a penalty for this by writing

$$J_N = E\left[x'(N)x(N) + \alpha \sum_{i=1}^{N} u'(i-1)u(i-1) \right]$$

where α = positive constant. In this case, J_N is a measure of terminal error plus control effort where the relative weighting is determined by α. Intuitively, we would expect less control effort to be expended in the second example than in the first, but at the cost of a larger terminal error.

PHYSICALLY REALIZABLE CONTROLS AND PROBLEM STATEMENT

At this point, it is clear that our problem consists in specifying a control sequence $\{u(k), k = 0, 1, \ldots, N-1\}$ which minimizes J_N where the relationship between $u(k)$ and $x(k+1)$ is given by Eq. (9.1). However, the solution of the problem so posed may lead to control sequences which cannot be mechanized in practice, e.g., control sequences which require input data that is not physically available when required. Additionally, since our aim is to manipulate the system's state, it is desirable to have the control sequence depend upon information which is available about the state, namely, the measurements. The control sequence then involves feedback. In the sequel, we restrict our attention to control sequences which, at any given time point, depend only upon information about the system's state which is physically available for processing at that time.

For any given $k = 0, 1, \ldots, N-1$, it is obvious that the available data on the system's state consists of the sequence of measurements $\{z(1), z(2), \ldots, z(k)\}$ and the mean value $\bar{x}(0)$ of the initial state $x(0)$. The control vector at k can then be written in the form

$$u(k) = \mu_k[z^*(k), \bar{x}(0)] \tag{9.5}$$

where $z^*(k)$ is the mk vector

$$z^*(k) = \begin{bmatrix} z(1) \\ \cdot \\ \cdot \\ \cdot \\ z(k) \end{bmatrix}$$

and μ_k is an r-dimensional vector-valued function of the indicated variables. We remark that μ_k is to be determined so as to minimize J_N, and it is not necessarily restricted to be of the same form for all k. Also, although $\bar{x}(0) = 0$ for our system model, we shall assume it arbitrary for the present for the sake of generality.

Because a control vector of the type which is defined by Eq. (9.5) depends only on physically available data, it is called a *physically realizable control,* and μ_k, $k = 0, 1, \ldots, N - 1$, is said to specify a *physically realizable control law.* We note that for $k = 0$, $u(0)$ can only be a function of $\bar{x}(0)$. We observe also that if μ_k turns out to be independent of $\{z(1), \ldots, z(k)\}$ for all k, $u(k)$ will be an open-loop control.

Our problem can now be stated as follows.

PROBLEM STATEMENT

Determine a physically realizable control law of the form in Eq. (9.5) for the system of Eqs. (9.1) and (9.2) which minimizes the performance measure in Eq. (9.3).

Such a control law is called an *optimal control,* and the problem itself is termed the *discrete stochastic linear regulator problem.* The adjective regulator arises from the first term in J_N, which is a measure of the regulation in the resulting closed-loop system.

Discussion There are three important points which must be borne in mind about the above problem. First, it is a fixed-time problem. That is, a control law which acts over an a priori specified interval of time is to be determined. This is in contrast to free terminal-time problems in which the terminal time is explicitly involved in the optimization. A classical example here is the time-optimal control problem in which it is desired to achieve some specified goal in minimum time.

Second, no amplitude-type bounds are placed on the control vector. This would be desirable if it were known that the system of Eq. (9.1) is subject to saturation at its input. However, the second term in J_N, as we have already noted, penalizes for excessive values of the control vector. This feature can be used to limit saturation tendencies to some extent.

Third, the value of the state at the terminal time is allowed to be free, i.e., no specific constraints are imposed upon it. We can readily envision situations in which it would be desirable to have such constraints. For example, we might wish to minimize

$$J_N = E \left[\sum_{i=1}^{N} u'(i - 1)B(i)u(i - 1) \right]$$

subject to the constraint that $E[x'(N)x(N)] \leq \beta$, where β is a positive constant. Problems of this type can only be treated indirectly in our work since $x(N)$ appears in the performance measure.

9.3 DETERMINISTIC PROBLEM

PROBLEM FORMULATION

As mentioned in Sec. 9.1, it will be useful to consider first an idealized version of the stochastic linear regulator problem in which there is no system disturbance and all the state variables can be measured exactly. In this case, Eqs. (9.1), (9.3), and (9.5) become

$$x(k + 1) = \Phi(k + 1, k)x(k) + \Psi(k + 1, k)u(k) \tag{9.6}$$

$$J_N = \sum_{i=1}^{N} [x'(i)A(i)x(i) + u'(i - 1)B(i - 1)u(i - 1)] \tag{9.7}$$

and

$$u(k) = \mu_k[x^*(k)] \tag{9.8}$$

respectively, where $k = 0, 1, \ldots, N - 1$, and $x^*(k)$ is the $n(k + 1)$ vector

$$x^*(k) = \begin{bmatrix} x(0) \\ \cdot \\ \cdot \\ \cdot \\ x(k) \end{bmatrix}$$

We note that J_N no longer involves the expected value, since we are now dealing with a deterministic process, and that the measurement equation is eliminated, since $z = x$. Also, we see that since all of the state variables can be measured exactly, the uncertainty associated with the initial state $x(0)$ is removed and $\bar{x}(0)$ in Eq. (9.5) is replaced by the actual value $x(0)$ of the initial state. The deterministic linear regulator problem is, therefore, the following.

PROBLEM STATEMENT

Determine a control law of the form in Eq. (9.8) for the system of Eq. (9.6) which minimizes the performance measure in Eq. (9.7).

The resulting system will have the block diagram which is given in Fig. 9.2. The task in the problem is to specify the controller which will operate upon the state to determine the control vector which min-

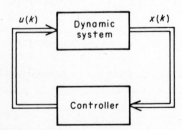

Fig. 9.2 Block diagram of closed-loop deterministic control system.

imizes the performance measure. It should be noted that the physical realizability condition, Eq. (9.8), dictates that, in general, the resulting control law involve feedback of all the preceding values of the state.

PRINCIPLE OF OPTIMALITY

The above problem and the one in Sec. 9.2 can readily be handled by utilizing the principle of optimality, which is due to Bellman [1].

Theorem 9.1 *An optimal control has the property that whatever the initial state and initial control are, the remaining control must constitute an optimal one with regard to the state which results from the initial control.*

Proof The proof is by contradiction and we carry it out in relatively general terms.

Consider a system S with control input u and state x over some time interval $t_1 \leq t \leq t_2$. Both u and x may be subject to constraints; t_1 and t_2 may be either fixed or free a priori; and t may be either a discrete- or continuous-time index on $[t_1, t_2]$. Let J denote some performance measure which is to be optimized (minimized or maximized) over $[t_1, t_2]$ by specifying some control $u(t)$. For definiteness, let us say that $u(t)$ is to be chosen to minimize J. The proof for maximization of J is essentially the same.

Assume that there exists at least one $u(t)$ which satisfies all constraints and minimizes J on $[t_1, t_2]$. Denote this optimal control by $u^o(t)$ and its corresponding performance measure by J^o.

Consider some time point $t' \in (t_1, t_2)$ and assume that over the time interval $[t', t_2]$ it is possible to find a control $u^*(t)$ for which J is less than it was for $u^o(t)$ over the same interval starting at the same state. Then it follows that the control

$$u^{**}(t) = \begin{cases} u^o(t) & t \in [t_1, t') \\ u^*(t) & t \in [t', t_2] \end{cases}$$

yields the performance measure $J^{**} < J^o$ over $[t_1, t_2]$. However, $u^o(t)$ is, by hypothesis, an optimal control for the system and cannot be improved upon. Hence, we have a contradiction, and the theorem is proved.

We note that since $u^o(t)$ is not necessarily unique, it may be possible to find a $u^*(t) \neq u^o(t)$ over $[t', t_2]$ such that

$$u^{**}(t) = \begin{cases} u^o(t) & t \in [t_1, t') \\ u^*(t) & t \in [t', t_2] \end{cases}$$

yields $J^{**} = J^o$ over $[t_1, t_2]$.

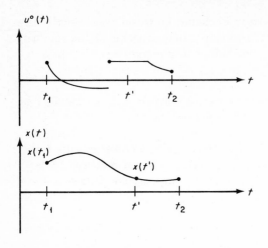

Fig. 9.3 Illustration of *principle of optimality.*

A very simple interpretation of the principle of optimality follows from a consideration of Fig. 9.3. Suppose that for some continuous-time system whose initial state is $x(t_1)$, the indicated control $\{u^o(t), t_1 \leq t \leq t_2\}$ minimizes some performance measure J over the interval $[t_1, t_2]$. Then the principle of optimality simply states that $\{u^o(t), t' \leq t \leq t_2\}$ minimizes the same J for the same system over the interval $[t', t_2]$ if the system's state at t' is the one which resulted from $u^o(t)$ acting over the interval $[t_1, t')$.

The principle of optimality has proven to be a powerful result for use in the solution of control system optimization problems. In discrete-time problems, it is used to reduce the problem from one of determining an entire control sequence at once to one of determining the elements of the sequence singly and recursively. In continuous-time problems, its application reduces a calculus of variations problem to one of solving a particular type of partial differential equation, or it permits one to develop necessary and/or sufficient conditions for optimal control which are themselves constructive. These points will be amply illustrated in the sequel.

DETERMINISTIC PROBLEM SOLUTION

We begin by defining V_N to be the minimum value of the performance measure J_N in Eq. (9.7),

$$V_N = \min_{u(0)} \cdots \min_{u(N-1)} \sum_{i=1}^{N} [x'(i)A(i)x(i) + u'(i-1)B(i-1)u(i-1)] \quad (9.9)$$

The problem is immediately seen to be one of picking rN variables, namely $u(0)$, $u(1)$, . . . , $u(N-1)$, to minimize the indicated quadratic form. If we were to utilize differential calculus for this purpose, we would obtain a system of rN algebraic equations which must be solved subject to the constraints expressed by Eq. (9.6). The computational demands would obviously be quite excessive except for some trivial cases involving small values of r and N.

The problem as defined by Eq. (9.9) can also be viewed as an N-stage decision process in which the N decisions, $u(0), \ldots, u(N-1)$, are to be made such that the given quadratic form is minimized. Rather than attempting to make the N decisions simultaneously, it would be desirable to develop a procedure for making the decisions one at a time, i.e., to reduce the N-stage problem to N one-stage problems. The principle of optimality provides the vehicle for doing just this. We proceed by induction, starting with the last stage of control in our problem.

SINGLE-STAGE

Let us suppose that our problem is simply that of picking a control which minimizes the performance measure for the last stage of control, i.e., we have only a single-stage optimization problem.

We define

$$V_1 = \min_{u(N-1)} \left[x'(N)A(N)x(N) + u'(N-1)B(N-1)u(N-1) \right]$$

$$(9.10)$$

However, from Eq. (9.6),

$$x(N) = \Phi(N, N-1)x(N-1) + \Psi(N, N-1)u(N-1)$$

Substituting this result into Eq. (9.10) and expanding, we obtain

$$
\begin{aligned}
V_1 = \min_{u(N-1)} \{ & [\Phi(N, N-1)x(N-1) \\
& + \Psi(N, N-1)u(N-1)]'A(N)[\Phi(N, N-1)x(N-1) \\
& + \Psi(N, N-1)u(N-1)] + u'(N-1)B(N-1)u(N-1) \} \\
= \min_{u(N-1)} \{ & x'(N-1)\Phi'(N, N-1)A(N)\Phi(N, N-1)x(N-1) \\
& + x'(N-1)\Phi'(N, N-1)A(N)\Psi(N, N-1)u(N-1) \\
& + u'(N-1)\Psi'(N, N-1)A(N)\Phi(N, N-1)x(N-1) \\
& + u'(N-1)[\Psi'(N, N-1)A(N)\Psi(N, N-1) \\
& \hspace{4cm} + B(N-1)]u(N-1) \}
\end{aligned}
$$

Since $A(N)$ is symmetric, we see that the second term is the transpose of the third, and since both are scalars, the two terms are equal.

Hence, we write

$$V_1 = \min_u [x'\Phi'A\Phi x + 2x'\Phi'A\Psi u + u'(\Psi'A\Psi + B)u] \qquad (9.11)$$

where we have momentarily dropped the time arguments for convenience.

We obtain the minimum in Eq. (9.11) by setting the gradient of the terms in brackets with respect to u equal to zero. We have, consequently,

$$2x'\Phi'A\Psi + 2u'(\Psi'A\Psi + B) = 0$$

Solving for u, we see that

$$u(N-1) = -[\Psi'(N, N-1)A(N)\Psi(N, N-1) + B(N-1)]^{-1}$$
$$\Psi'(N, N-1)A(N)\Phi(N, N-1)x(N-1) \qquad (9.12)$$

where we have made use of the fact that $A(N)$ and $B(N-1)$ are symmetric. We note that the resulting control law is physically realizable, and additionally is linear and involves feedback of the current state $x(N-1)$.

We define

$$S(N-1) = -[\Psi'(N, N-1)A(N)\Psi(N, N-1) + B(N-1)]^{-1}$$
$$\Psi'(N, N-1)A(N)\Phi(N, N-1) \qquad (9.13)$$

and write the control law

$$u(N-1) = S(N-1)x(N-1) \qquad (9.14)$$

The $r \times n$ matrix S is called the *feedback control gain matrix*. We note that $S(N-1)$, and moreover $u(N-1)$, exist if and only if the $n \times n$ matrix $[\Psi'(N, N-1)A(N)\Psi(N, N-1) + B(N-1)]$ is nonsingular.

Let us now evaluate V_1. Substituting Eq. (9.12) into (9.11), we have

$$\begin{aligned} V_1 &= x'\Phi'A\Phi x - 2x'\Phi'A\Psi(\Psi'A\Psi + B)^{-1}\Psi'A\Phi x \\ &\quad + x'\Phi'A\Psi(\Psi'A\Psi + B)^{-1}\Psi'A\Phi x \\ &= x'(N-1)\Phi'(N, N-1)\{A(N) - A(N)\Psi(N, N-1) \\ &\quad [\Psi'(N, N-1)A(N)\Psi(N, N-1) + B(N-1)]^{-1} \\ &\quad \Psi'(N, N-1)A(N)\}\Phi(N, N-1)x(N-1) \end{aligned}$$

Defining

$$W(N) = A(N) \qquad (9.15)$$

and

$$M(N - 1) = \Phi'(N, N - 1)\{W(N) - W(N)\Psi(N, N - 1)$$
$$[\Psi'(N, N - 1)W(N)\Psi(N, N - 1) + B(N - 1)]^{-1}$$
$$\Psi'(N, N - 1)W(N)\}\Phi(N, N - 1) \quad (9.16)$$

we have

$$V_1 = x'(N - 1)M(N - 1)x(N - 1) \quad (9.17)$$

The reason for the definitions in Eqs. (9.15) and (9.16) will become apparent later. The important point to note is that V_1, the minimum value of the performance measure for the single-stage control, is a quadratic form in $x(N - 1)$, the system's state at the beginning of our problem, or equivalently, the initial state for the single-stage problem. We remark also that, by virtue of their definitions, $W(N)$ and $M(N - 1)$ are symmetric $n \times n$ matrices.

Finally, substituting the definition in Eq. (9.15) into Eq. (9.13), we get

$$S(N - 1) = -[\Psi'(N, N - 1)W(N)\Psi(N, N - 1)$$
$$+ B(N - 1)]^{-1}\Psi'(N, N - 1)W(N)\Phi(N, N - 1) \quad (9.18)$$

DOUBLE-STAGE

We turn now to the question of optimal control for a two-stage process. We define

$$V_2 = \min_{u(N-2)} \min_{u(N-1)} \{[x'(N - 1)A(N - 1)x(N - 1)$$
$$+ u'(N - 2)B(N - 2)u(N - 2)]$$
$$+ [x'(N)A(N)x(N) + u'(N - 1)B(N - 1)u(N - 1)]\}$$

However, from the principle of optimality, this equation can be expressed as

$$V_2 = \min_{u(N-2)} [x'(N - 1)A(N - 1)x(N - 1)$$
$$+ u'(N - 2)B(N - 2)u(N - 2) + V_1] \quad (9.19)$$

since the choice of $u(N - 1)$ does not affect $x(N - 1)$.

Substituting into Eq. (9.19) from Eq. (9.17), we see that

$$V_2 = \min_{u(N-2)} [x'(N - 1)A(N - 1)x(N - 1)$$
$$+ u'(N - 2)B(N - 2)u(N - 2) + x'(N - 1)M(N - 1)x(N - 1)]$$

or

$$V_2 = \min_{u(N-2)} [x'(N - 1)W(N - 1)x(N - 1)$$
$$+ u'(N - 2)B(N - 2)u(N - 2)] \quad (9.20)$$

where we have defined

$$W(N - 1) = M(N - 1) + A(N - 1) \tag{9.21}$$

We note that $W(N - 1)$ is $n \times n$ and symmetric.

Comparison of Eqs. (9.20) and (9.10) reveals that the two equations are of the same form and the former is obtained from the latter by making the following substitutions:

$$N \rightarrow N - 1$$
$$N - 1 \rightarrow N - 2$$
$$A(N) = W(N) \rightarrow W(N - 1)$$
$$B(N - 1) \rightarrow B(N - 2)$$

Consequently, the value of $u(N - 2)$ which minimizes the right-hand side of Eq. (9.20) will be of the same form as $u(N - 1)$ in Eq. (9.12), subject to the above substitutions. By analogy to Eqs. (9.14) and (9.18), we write

$$u(N - 2) = S(N - 2)x(N - 2) \tag{9.22}$$

and

$$S(N - 2) = -[\Psi'(N - 1, N - 2)W(N - 1)\Psi(N - 1, N - 2)$$
$$+ B(N - 2)]^{-1}$$
$$\Psi'(N - 1, N - 2)W(N - 1)\Phi(N - 1, N - 2) \tag{9.23}$$

respectively, where $W(N - 1)$ is given by Eq. (9.21).

The value of V_2 is easily obtained by repeating the steps which led to Eq. (9.17), utilizing the substitutions given earlier. Then,

$$V_2 = x'(N - 2)M(N - 2)x(N - 2) \tag{9.24}$$

where

$$M(N - 2) = \Phi'(N - 1, N - 2)\{W(N - 1)$$
$$- W(N - 1)\Psi(N - 1, N - 2)$$
$$[\Psi'(N - 1, N - 2)W(N - 1)\Psi(N - 1, N - 2)$$
$$+ B(N - 2)]^{-1}\Psi'(N - 1, N - 2)W(N - 1)\}\Phi(N - 1, N - 2) \tag{9.25}$$

Again, we take note of the symmetry of $M(N - 2)$.

The computational procedure for determining the two feedback control gain matrices $S(N - 1)$ and $S(N - 2)$ for a two-stage optimal control problem is now obvious. We begin by substituting $W(N) = A(N)$ into Eq. (9.18) to obtain $S(N - 1)$ and into Eq. (9.16) to get $M(N - 1)$.

The latter and $A(N-1)$ are put into Eq. (9.21) and the result, $W(N-1)$, is substituted into Eq. (9.23) to yield $S(N-2)$. Finally, if the value of V_2 is desired, $W(N-1)$ is substituted into Eq. (9.25), giving $M(N-2)$, which, along with the knowledge of $x(N-2)$, gives V_2 according to Eq. (9.24).

It is important to note that computations proceed recursively backward in time. Also, it should be noted that repetition of some of the computations can be avoided by substituting into Eq. (9.25) from Eq. (9.23) to obtain

$$M(N-2) = \Phi'(N-1, N-2)W(N-1)\Phi(N-1, N-2)$$
$$+ \Phi'(N-1, N-2)W(N-1)\Psi(N-1, N-2)S(N-2)$$

and similarly for $M(N-1)$. Then, computations could proceed in the following sequence: $W(N) = A(N) \to S(N-1) \to M(N-1) \to W(N-1) \to S(N-2) \to M(N-2) \to V_2$.

$(j-1)$ STAGES

By finite induction, we have now for some $j \geq 3$ that the optimal control at time $N-j+1$ for a process involving $j-1$ stages is characterized by the set of relations:

$$W(N-j+2) = M(N-j+2) + A(N-j+2) \tag{9.26}$$

$$S(N-j+1) = -[\Psi'(N-j+2, N-j+1)$$
$$W(N-j+2)\Psi(N-j+2, N-j+1)$$
$$+ B(N-j+1)]^{-1}\Psi'(N-j+2, N-j+1)$$
$$W(N-j+2)\Phi(N-j+2, N-j+1) \tag{9.27}$$

$$u(N-j+1) = S(N-j+1)x(N-j+1) \tag{9.28}$$

$$M(N-j+1) = \Phi'(N-j+2, N-j+1)W(N-j+2)$$
$$\Phi(N-j+2, N-j+1)$$
$$+ \Phi'(N-j+2, N-j+1)W(N-j+2)$$
$$\Psi(N-j+2, N-j+1)S(N-j+1) \tag{9.29}$$

$$V_{j-1} = x'(N-j+1)M(N-j+1)x(N-j+1) \tag{9.30}$$

where W and M are symmetric $n \times n$ matrices.

j STAGES

For j stages of control, it follows from the principle of optimality that

$$V_j = \min_{u(N-j)} [x'(N-j+1)A(N-j+1)x(N-j+1)$$
$$+ u'(N-j)B(N-j)u(N-j) + V_{j-1}]$$

Substituting into this result from Eq. (9.30) and defining

$$W(N - j + 1) = M(N - j + 1) + A(N - j + 1) \qquad (9.31)$$

we have

$$V_j = \min_{u(N-j)} [x'(N - j + 1)W(N - j + 1)x(N - j + 1)$$
$$+ u'(N - j)B(N - j)u(N - j)] \quad (9.32)$$

However, from Eq. (9.6),

$$x(N - j + 1) = \Phi(N - j + 1, N - j)x(N - j)$$
$$+ \Psi(N - j + 1, N - j)u(N - j)$$

and Eq. (9.32) becomes

$$V_j = \min_u [(\Phi x + \Psi u)'W(\Phi x + \Psi u) + u'Bu]$$

$$= \min_u [x'\Phi'W\Phi x + x'\Phi'W\Psi u + u'\Psi'W\Phi x + u'(\Psi'W\Psi + B)u]$$

$$= \min_u [x'\Phi'W\Phi x + 2x'\Phi'W\Psi u + u'(\Psi'W\Psi + B)u] \qquad (9.33)$$

where we have made use of the fact that W is symmetric in arriving at the last line, and have momentarily dropped the time arguments. We do, however, make note of the fact that in Eq. (9.33) $x = x(N - j)$, $u = u(N - j)$, $\Phi = \Phi(N - j + 1, N - j)$, $\Psi = \Psi(N - j + 1, N - j)$, $B = B(N - j)$, and $W = W(N - j + 1)$.

Setting the gradient of the right-hand side of Eq. (9.33) with respect to u equal to zero, we have

$$2x'\Phi'W\Psi + 2u'(\Psi'W\Psi + B) = 0 \qquad (9.34)$$

which we solve to obtain

$$u(N - j) = -[\Psi'(N - j + 1, N - j)$$
$$W(N - j + 1)\Psi(N - j + 1, N - j)$$
$$+ B(N - j)]^{-1}\Psi'(N - j + 1, N - j)$$
$$W(N - j + 1)\Phi(N - j + 1, N - j)x(N - j) \quad (9.35)$$

Defining

$$S(N - j) = -[\Psi'(N - j + 1, N - j)$$
$$W(N - j + 1)\Psi(N - j + 1, N - j)$$
$$+ B(N - j)]^{-1}\Psi'(N - j + 1, N - j)$$
$$W(N - j + 1)\Phi(N - j + 1, N - j) \quad (9.36)$$

we write the control law as

$$u(N - j) = S(N - j)x(N - j) \tag{9.37}$$

Comparing Eqs. (9.31), (9.36), and (9.37) with Eqs. (9.26), (9.27), and (9.28), respectively, we observe that the former three are the same as the latter three, with the exception of the obvious reduction of all time indices by one unit.

The evaluation of V_j is now easily carried out by substituting Eq. (9.35) into Eq. (9.33) to obtain

$$\begin{aligned} V_j &= [x'\Phi'W\Phi x - 2x'\Phi'W\Psi(\Psi'W\Psi + B)^{-1}\Psi'W\Phi x \\ &\qquad\qquad + x'\Phi'W\Psi(\Psi'W\Psi + B)^{-1}\Psi'W\Phi x] \\ &= x'(N - j)M(N - j)x(N - j) \end{aligned} \tag{9.38}$$

where

$$\begin{aligned} M(N - j) &= \Phi'(N - j + 1, N - j)W(N - j + 1) \\ &\quad \Phi(N - j + 1, N - j) + \Phi'(N - j + 1, N - j) \\ &\quad W(N - j + 1)\Psi(N - j + 1, N - j)S(N - j) \end{aligned} \tag{9.39}$$

Equations (9.39) and (9.38) are clearly of the same form as Eqs. (9.29) and (9.30), respectively, with the obvious time index change.

The optimal control for the discrete-time deterministic linear regulator problem is now defined by Eqs. (9.31) to (9.39). The indexing on j is $j = 1, 2, \ldots, N$, and computations are initiated by substituting $W(N) = A(N)$ into Eq. (9.36) for $j = 1$ to obtain $S(N - 1)$.

Some simplification in the notation is possible here if we make a change in the time index. In particular, letting $k = N - j$, we have, from Eqs. (9.31), (9.36), (9.37), (9.39), and (9.38), respectively, that

$$W(k + 1) = M(k + 1) + A(k + 1) \tag{9.40}$$

$$\begin{aligned} S(k) &= -[\Psi'(k + 1, k)W(k + 1)\Psi(k + 1, k) \\ &\quad + B(k)]^{-1}\Psi'(k + 1, k)W(k + 1)\Phi(k + 1, k) \end{aligned} \tag{9.41}$$

$$u(k) = S(k)x(k) \tag{9.42}$$

$$\begin{aligned} M(k) &= \Phi'(k + 1, k)W(k + 1)\Phi(k + 1, k) \\ &\quad + \Phi'(k + 1, k)W(k + 1)\Psi(k + 1, k)S(k) \end{aligned} \tag{9.43}$$

$$V_{N-k} = x'(k)M(k)x(k) \tag{9.44}$$

where $k = N - 1, N - 2, \ldots, 0$ and $W(N) = A(N)$ is used in Eq. (9.41) to initiate computations. Equation (9.40) does not enter the computations until $k = N - 2$.

The recursive nature of the computations required to generate the optimal control sequence is quite evident. As in the estimation problem, this provides considerable savings in computer memory space. Additionally, of course, the feedback control gain matrix is determined in a concise, systematic fashion.

The optimal control is obviously physically realizable in a very simple way, namely, as a linear transformation on the state at the current time. The controller in Fig. 9.2 is seen to be simply the time-varying gain matrix $S(k)$.

In retrospect, we would have been justified in restricting our attention to control laws without memory and writing Eq. (9.8) in the form

$$u(k) = \mu_k[x(k)]$$

Indeed, we could have gone, justifiably *ex post facto*, to consideration only of linear control laws,

$$u(k) = S(k)x(k)$$

where the $r \times n$ matrix $S(k)$ is initially unknown and is to be determined such that J_N in Eq. (9.7) is minimized.

Since the computations proceed backward in time, it is clear that the time history of $S(k)$ must be determined prior to system operation, i.e., it must be precomputed and stored for use later.

It is clear that although $M(k)$ must be computed at each stage, V_{N-k} is not necessarily of interest, and, therefore, need not be computed except at $k = 0$ to obtain the minimum value of the performance measure for all N stages of optimal control. [See Eq. (9.9).]

It remains for us to convince ourselves that the control $u(k)$ which is given by Eqs. (9.40) to (9.43) actually minimizes the performance measure. We recall from differential calculus that the vanishing of the gradient with respect to u, which led in the general case to Eq. (9.34), is only a necessary condition for V_j to be a minimum. That is, $u(N - j)$ in Eq. (9.35) only guarantees that V_j attains a stationary value. Intuitively, we feel that our choice of u does minimize the performance measure since we can make J_N arbitrarily large by choosing u arbitrarily large for each time point in the control interval $[0,N]$.

Letting q denote the quadratic form on the right-hand side of Eq. (9.33), we have

$$\nabla_u q = 2x'\Phi'W\Psi + 2u'(\Psi'W\Psi + B)$$

and its transpose

$$[\nabla_u q]' = 2[\Psi'W\Phi x + (\Psi'W\Psi + B)u]$$

Taking the gradient with respect to u of this result, we obtain

$$\nabla_u[\nabla_u q]' = 2(\Psi'W\Psi + B)$$

Then a sufficient condition that we have a minimum is that the matrix

$$[\Psi'(N - j + 1, N - j)W(N - j + 1)\Psi(N - j + 1, N - j)$$
$$+ B(N - j)]$$

be positive definite for all $j = 1, 2, \ldots, N$.

We summarize the above results in the following theorem.

Theorem 9.2 *The optimal control law for the deterministic linear regulator problem is the linear feedback control law*

$$u(k) = S(k)x(k) \tag{9.45}$$

where the $r \times n$ feedback control matrix $S(k)$ is determined recursively from the set of relations

$$W(k + 1) = M(k + 1) + A(k + 1) \tag{9.46}$$

$$S(k) = -[\Psi'(k + 1, k)W(k + 1)\Psi(k + 1, k)$$
$$+ B(k)]^{-1}\Psi'(k + 1, k)W(k + 1)\Phi(k + 1, k) \tag{9.47}$$

$$M(k) = \Phi'(k + 1, k)W(k + 1)\Phi(k + 1, k)$$
$$+ \Phi'(k + 1, k)W(k + 1)\Psi(k + 1, k)S(k) \tag{9.48}$$

for $k = N - 1, N - 2, \ldots, 0$, where $W(N) = A(N)$ and the $r \times r$ matrix $[\Psi'(k + 1, k)W(k + 1)\Psi(k + 1, k) + B(k)]$ is required to be positive definite for all k. The minimum value of the performance measure for $(N - k)$ stages of control is

$$V_{N-k} = x'(k)M(k)x(k) \tag{9.49}$$

For those cases in which we are interested in the value of the performance measure only for the entire N-stage process, explicit computation of $M(k)$ except for $k = 0$ is superfluous and can be easily obviated, yielding the following result.

Corollary 9.1 *The optimal control law for the deterministic linear regulator problem is governed by the set of relations*

$$u(k) = S(k)x(k) \tag{9.45}$$

$$S(k) = -[\Psi'(k + 1, k)W(k + 1)\Psi(k + 1, k) + B(k)]^{-1}$$
$$\Psi'(k + 1, k)W(k + 1)\Phi(k + 1, k) \tag{9.47}$$

$$W(k) = \Phi'(k + 1, k)W(k + 1)\Phi(k + 1, k)$$
$$+ \Phi'(k + 1, k)W(k + 1)\Psi(k + 1, k)S(k) + A(k) \tag{9.50}$$

for $k = N - 1, N - 2, \ldots, 0$, where $W(N) = A(N)$ and the $r \times r$

matrix $[\Psi'(k + 1, k)W(k + 1)\Psi(k + 1, k) + B(k)]$ *is required to be positive definite for all* k.

Proof From Eq. (9.46),

$$M(k) = W(k) - A(k) \tag{9.51}$$

which, when substituted into Eq. (9.48), leads immediately to Eq. (9.50).

Computation of the optimal control utilizing Eqs. (9.45), (9.47), and (9.50) is obvious. Also, it is clear that the computation of the value of the performance measure for any number of stages of control can be executed by utilizing Eq. (9.51) to obtain $M(k)$.

The results which we have presented in Theorem 9.2 and Corollary 9.1 were developed by Kalman and Koepcke [2].

Example 9.1   Let us consider the particular case where

$$J_N = \sum_{i=1}^{N} x'(i)A(i)x(i)$$

with $A(i)$ positive definite for all i, and assume that the system

$$x(k + 1) = \Phi(k + 1, k)x(k) + \Psi(k + 1, k)u(k) \tag{9.52}$$

has as many control variables as it does state variables ($r = n$) with $\Psi(k + 1, k)$ nonsingular for all $k = 0, 1, \ldots, N - 1$.

We note here that the performance measure involves only weighted error squared. In other words, we are willing to expend whatever control effort is necessary to minimize the indicated performance measure.

Since $B(i - 1) = 0$ for all $i = 1, \ldots, N$, and $\Psi(k + 1, k)$ is nonsingular for all values of k which are of interest, Eq. (9.47) becomes

$$S(k) = -[\Psi'(k + 1, k)W(k + 1)\Psi(k + 1, k)]^{-1}$$
$$\Psi'(k + 1, k)W(k + 1)\Phi(k + 1, k)$$
$$= -\Psi^{-1}(k + 1, k)W^{-1}(k + 1)[\Psi'(k + 1, k)]^{-1}$$
$$\Psi'(k + 1, k)W(k + 1)\Phi(k + 1, k)$$
$$= -\Psi^{-1}(k + 1, k)\Phi(k + 1, k) \tag{9.53}$$

Then, with $u(k) = S(k)x(k)$, we see that Eq. (9.52) becomes

$$x(k + 1) = \Phi(k + 1, k)x(k) - \Psi(k + 1, k)\Psi^{-1}(k + 1, k)\Phi(k + 1, k)x(k) = 0$$

for all $k = 0, 1, \ldots, N - 1$. Obviously, the initial state is driven to zero in the first stage of control. Moreover, we note from Eq. (9.50) that

$$W(k) = \Phi'(k + 1, k)W(k + 1)\Phi(k + 1, k)$$
$$- \Phi'(k + 1, k)W(k + 1)\Psi(k + 1, k)\Psi^{-1}(k + 1, k)\Phi(k + 1, k) + A(k)$$
$$= A(k)$$

It then follows from Eqs. (9.51) and (9.49) for $k = 0$ that $M(0) = W(0) - A(0)$ $= 0$ and $V_N = 0$; that is, the control is "perfect."

These results could have been anticipated by attempting to find a control $u(k)$ in Eq. (9.52) for which $x(k + 1) = 0$. Since $\Psi(k + 1, k)$ is nonsingular for all k, the obvious choice is

$$u(k) = -\Psi^{-1}(k + 1, k)\Phi(k + 1, k)x(k)$$

However, when $\Psi(k + 1, k)$ is singular, this result is of no use. For our problem, the feedback gain matrix would be given by the first line in Eq. (9.53).

Example 9.2   As a more specific example, let us consider the scalar system

$$x(k + 1) = x(k) + 2u(k)$$

for $k = 0, 1, \ldots, N - 1$ and let the performance measure be

$$J_N = x^2(N) + 4\beta \sum_{i=1}^{N} u^2(i - 1)$$

where β is a positive constant.

For this problem, we have $\Phi(k + 1, k) = 1$, $\Psi(k + 1, k) = 2$, $A(N) = W(N) = 1$, $A(N - 1) = \cdots = A(1) = 0$, and $B(N - 1) = \cdots = B(0) = 4\alpha$.

Then, from Eqs. (9.47) and (9.50), we obtain

$$S(k) = -[4W(k + 1) + 4\alpha]^{-1}2W(k + 1)$$

$$= \frac{-W(k + 1)}{2[W(k + 1) + \beta]}$$

and

$$W(k) = W(k + 1) - \frac{W^2(k + 1)}{W(k + 1) + \beta}$$

$$= \frac{\beta W(k + 1)}{W(k + 1) + \beta}$$

$$= -2\beta S(k)$$

respectively, for $k = N - 1, \ldots, 0$ with $W(N) = 1$.

Noting that

$$S(k - 1) = \frac{-W(k)}{2[W(k) + \beta]}$$

and substituting into this result for $W(k)$, we have the recursive relation

$$S(k - 1) = \frac{2\beta S(k)}{2[-2\beta S(k) + \beta]}$$

$$= \frac{S(k)}{1 - 2S(k)}$$

for the feedback gain where $k = N - 1, \ldots, 1$. The boundary condition for this expression is seen to be

$$S(N - 1) = \frac{-W(N)}{2[W(N) + \beta]} = \frac{-1}{2(\beta + 1)}$$

We obtain immediately the following table of values from which it is clear

k	$S(k)$	$W(k)$
N	. . . .	1
$N - 1$	$-1/2(\beta + 1)$	$\beta/(\beta + 1)$
$N - 2$	$-1/2(\beta + 2)$	$\beta/(\beta + 2)$
.		
.		
1	$-1/2(\beta + N - 1)$	$\beta/(\beta + N - 1)$
0	$-1/2(\beta + N)$	$\beta/(\beta + N)$

that the general expression for the feedback gain is

$$S(k) = \frac{-1}{2(\beta + N - k)}$$

where $k = 0, 1, \ldots, N - 1$. The system equation now becomes

$$x(k + 1) = x(k) - \frac{1}{\beta + N - k} x(k)$$

$$= \left[\frac{\beta + N - k - 1}{\beta + N - k}\right] x(k)$$

If we choose $\beta \gg N$, for example, then

$$S(k) \approx \frac{-1}{2\beta}$$

and

$$x(k + 1) = \left[1 - \frac{1}{\beta}\right] x(k)$$

In this case, we are weighting control effort in the performance measure far more than terminal error. This leads to a small value for the gain, viz., a small negative constant for all k. Obviously, the control does very little to affect the terminal error and in the limit as $\beta \to \infty$ would do nothing; that is, $x(k + 1) = x(k)$, giving $x(N) = x(0)$.

From the table, we have, for $\beta \gg N$, that $W(0) \approx 1$, and, therefore, from Eq. (9.51) that $M(0) \approx 1$. The value of the performance measure is then

$$V_N = x^2(0)$$

On the other hand, if we make $\beta = 0$, i.e., the performance measure involves only terminal error,

$$S(k) = \frac{-1}{2(N - k)}$$

and the magnitude of the feedback gain increases monotonically from $1/2N$ for $k = 0$ to $\frac{1}{2}$ for $k = N - 1$. From the table, we see that $W(0) = 0$ for $\beta = 0$ and, therefore, the performance measure in this case is

$$V_N = 0$$

This means, of course, that the terminal error is zero.

Obviously, other values of β in the range $0 < \beta < \infty$ will yield performance which is somewhere between the two extremes which we have considered here.

9.4 STOCHASTIC PROBLEM

We now obtain the solution to the stochastic linear regulator problem posed in Sec. 9.2 by essentially paralleling the approach which was used in the preceding section for the deterministic problem.

In a manner analogous to our treatment of the deterministic problem, we begin by defining

$$V_N = \min_{u(0)} \cdots \min_{u(N-1)} E \left\{ \sum_{i=1}^{N} [x'(i)A(i)x(i) + u'(i-1)B(i-1)u(i-1)] \right\} \quad (9.54)$$

In Eq. (9.54), we require that the minimization be carried out over physically realizable $u(k)$, $k = 0, 1, \ldots, N - 1$, as defined by Eq. (9.5).

SINGLE-STAGE

For the single-stage optimization problem which starts at time $N - 1$ and terminates at time N, we write

$$V_1 = \min_{u(N-1)} E[x'(N)A(N)x(N) + u'(N - 1)B(N - 1)u(N - 1)]$$

$$(9.55)$$

From Eq. (9.1),

$$x(N) = \Phi(N, N - 1)x(N - 1) + \Gamma(N, N - 1)w(N - 1) + \Psi(N, N - 1)u(N - 1)$$

and Eq. (9.55) becomes

$$V_1 = \min_u E[(\Phi x + \Gamma w + \Psi u)'A(\Phi x + \Gamma w + \Psi u) + u'Bu]$$

$$= \min_u E[x'\Phi'A\Phi x + x'\Phi'A\Gamma w + x'\Phi'A\Psi u$$

$$+ w'\Gamma'A\Phi x + w'\Gamma'A\Psi u + w'\Gamma'A\Gamma w + u'\Psi'A\Phi x$$

$$+ u'\Psi'A\Gamma w + u'(\Psi'A\Psi + B)u]$$

where $x = x(N - 1)$, $w = w(N - 1)$, $u = u(N - 1)$, $A = A(N)$, $B = B(N - 1)$, $\Phi = \Phi(N, N - 1)$, $\Gamma = \Gamma(N, N - 1)$, and $\Psi = \Psi(N, N - 1)$. We remark that the indicated expected value is over x, w, and u.

We note that the terms in the brackets are each scalars and that A is symmetric. Hence, the second and fourth terms are equal to each other, as are the third and seventh terms, and the fifth and eighth terms. Consequently,

$$V_1 = \min_u E[x'\Phi'A\Phi x + 2x'\Phi'A\Gamma w + 2x'\Phi'A\Psi u$$

$$+ 2w'\Gamma'A\Psi u + w'\Gamma'A\Gamma w + u'(\Psi'A\Psi + B)u] \quad (9.56)$$

We recall from Sec. 9.2 that for our model $x(i)$ and $w(i)$ are statistically independent for all $i = 0, 1, \ldots$ Since $\{w(i), i = 0, 1, \ldots\}$ is zero mean, it follows that the second term on the right-hand side of Eq. (9.56) vanishes; i.e.,

$$E[2x'(N - 1)\Phi'(N, N - 1)A(N)\Gamma(N, N - 1)w(N - 1)]$$

$$= 2E[x'(N - 1)]\Phi'(N, N - 1)A(N)\Gamma(N, N - 1)E[w(N - 1)] = 0$$

In addition, since $u(N - 1)$ is required to be physically realizable, viz., $u(N - 1) = \mu_{N-1}[z^*(N - 1), \bar{x}(0)]$, and, for our system model, $w(j)$ and $z(i)$ are statistically independent for all $j \geq i$, $i = 1, 2, \ldots$, we see that

$$E[2w'(N - 1)\Gamma'(N, N - 1)A(N)\Psi(N, N - 1)u(N - 1)]$$

$$= 2E[w'(N - 1)]\Gamma'(N, N - 1)A(N)\Psi(N, N - 1)E[u(N - 1)]$$

It then follows that the fourth term on the right-hand side of Eq. (9.56) also vanishes, since $w(N - 1)$ has zero mean.

Consequently, we now have

$$V_1 = \min_u E[x'\Phi'A\Phi x + 2x'\Phi'A\Psi u + w'\Gamma'A\Gamma w + u'(\Psi'A\Psi + B)u]$$

$$(9.57)$$

We recall from Sec. 3.4 that one of the properties of conditional expectation is that $E(x) = E[E(x|y)]$ where the outer expected value on the right-hand side is over y.   Utilizing this result, we write Eq. (9.57) as

$$V_1 = \min_u E\{E[x'\Phi'A\Phi x + 2x'\Phi'A\Psi u + w'\Gamma'A\Gamma w$$

$$+ u'(\Psi'A\Psi + B)u|z^*(N - 1), \bar{x}(0)]\} \quad (9.58)$$

where the outer expected value is over $z^*(N - 1)$.   Although the inner expected value is conditioned on $\bar{x}(0)$ as well as $z^*(N - 1)$, the outer one is taken over only $z^*(N - 1)$, since $\bar{x}(0)$ is not a random vector.   In this connection, the probability density function of the inner expected value is of the form $f = f(x,w,u|z^*)$.

We observe now that the performance measure can be minimized by minimizing the inner expected value in Eq. (9.58) with respect to u for all $z^*(N - 1)$ and $\bar{x}(0)$.

The physical realizability condition requires that u be some deterministic function of $\bar{x}(0)$ and the random vector $z^*(N - 1)$.   Consequently, in Eq. (9.58), we have the terms

$$E[2x'\Phi'A\Psi u|z^*(N - 1), \bar{x}(0)] = 2E[x'|z^*(N - 1), \bar{x}(0)]\Phi'A\Psi u$$

and

$$E[u'(\Psi'A\Psi + B)u|z^*(N - 1), \bar{x}(0)] = u'(\Psi'A\Psi + B)u$$

Then, taking the gradient of the inner expected value in Eq. (9.58) with respect to u and setting the result equal to zero, we obtain the expression

$$2E[x'|z^*(N - 1), \bar{x}(0)]\Phi'A\Psi + 2u'(\Psi'A\Psi + B) = 0$$

Taking the transpose, solving for u, and replacing the time arguments, we see that

$$u(N - 1) = -[\Psi'(N, N - 1)A(N)\Psi(N, N - 1)$$
$$+ B(N - 1)]^{-1}\Psi'(N, N - 1)A(N)\Phi(N, N - 1)$$
$$E[x(N - 1)|z^*(N - 1), \bar{x}(0)]$$

where we have utilized the fact that $A(N)$ and $B(N - 1)$ are symmetric.

Recognizing that $E[x(N - 1)|z^*(N - 1), \bar{x}(0)]$ is simply the optimal filtered estimate of $x(N - 1)$, we have

$$u(N - 1) = -[\Psi'(N, N - 1)A(N)\Psi(N, N - 1) + B(N - 1)]^{-1}$$
$$\Psi'(N, N - 1)A(N)\Phi(N, N - 1)\hat{x}(N - 1|N - 1)$$
$$(9.59)$$

Comparing Eq. (9.59) with Eq. (9.12), we observe that the optimal control for the single-stage stochastic linear regulator problem is the same as that for the single-stage deterministic problem, with the exception that $x(N - 1)$ in the latter is replaced by $\hat{x}(N - 1|N - 1)$.   We have now the first step in the development of the separation principle—the optimal control is a gain matrix times the optimal filtered estimate, each computed independently.

As in the deterministic case, we define $W(N) = A(N)$, and

$$S(N - 1) = -[\Psi'(N, N - 1)W(N)\Psi(N, N - 1) + B(N - 1)]^{-1}$$
$$\Psi'(N, N - 1)W(N)\Phi(N, N - 1) \quad (9.60)$$

and write

$$u(N - 1) = S(N - 1)\hat{x}(N - 1|N - 1) \quad (9.61)$$

We evaluate V_1 by substituting Eq. (9.59) into Eq. (9.57) with $A(N) = W(N)$. We then have

$$V_1 = E[x'\Phi'W\Phi x - 2x'\Phi'W\Psi(\Psi'W\Psi + B)^{-1}\Psi'W\Phi\hat{x}$$
$$+ w'\Gamma'W\Gamma w + \hat{x}'\Phi'W\Psi(\Psi'W\Psi + B)^{-1}\Psi'W\Phi\hat{x}]$$

We simplify this expression by first noting that

$$\hat{x}(N - 1|N - 1) = x(N - 1) - \tilde{x}(N - 1|N - 1)$$

Then, letting Λ denote the $n \times n$ matrix $\Phi'W\Psi(\Psi'W\Psi + B)^{-1}\Psi'W\Phi$, the second and fourth terms can be combined in the following way:

$$-2x'\Lambda\hat{x} + \hat{x}'\Lambda\hat{x} = (\hat{x} - 2x)'\Lambda\hat{x}$$
$$= -(\tilde{x} + x)'\Lambda(x - \tilde{x})$$
$$= -\tilde{x}'\Lambda x + \tilde{x}'\Lambda\tilde{x} - x'\Lambda x + x'\Lambda\tilde{x}$$
$$= \tilde{x}'\Lambda\tilde{x} - x'\Lambda x$$

where we have made use of the fact that Λ is symmetric.

Hence,

$$V_1 = E[x'\Phi'W\Phi x - x'\Phi'W\Psi(\Psi'W\Psi + B)^{-1}\Psi'W\Phi x$$
$$+ \tilde{x}'\Phi'W\Psi(\Psi'W\Psi + B)^{-1}\Psi'W\Phi\tilde{x} + w'\Gamma'W\Gamma w]$$
$$= E\{x'\Phi'[W - W\Psi(\Psi'W\Psi + B)^{-1}\Psi'W]\Phi x\}$$
$$+ E[\tilde{x}'\Phi'W\Psi(\Psi'W\Psi + B)^{-1}\Psi'W\Phi\tilde{x} + w'\Gamma'W\Gamma w] \quad (9.62)$$

The first term on the right-hand side of Eq. (9.62) is of the same form as V_1 for the deterministic problem, except that now the expected value operator is present.

The second term in Eq. (9.62) is due to the uncertainty which is present, namely, the filtering error and the system disturbance. In particular, this second term can be computed explicitly in the following way. We let λ_{ij} denote the elements of Λ and η_{ij} the elements of the

$p \times p$ symmetric matrix $\Gamma'W\Gamma$. Then

$$E[\tilde{x}'\Lambda\tilde{x} + w'\Gamma'W\Gamma w] = E\left(\sum_{i=1}^{n}\sum_{j=1}^{n} \lambda_{ij}\tilde{x}_i\tilde{x}_j + \sum_{i=1}^{p}\sum_{j=1}^{p} \eta_{ij}w_iw_j\right)$$

$$= \sum_{i=1}^{n}\sum_{j=1}^{n} \lambda_{ij}p_{ij} + \sum_{i=1}^{p}\sum_{j=1}^{p} \eta_{ij}q_{ij}$$

where the p_{ij} are the elements of the $n \times n$ filtering error covariance matrix $P(N - 1|N - 1) = E[\tilde{x}(N - 1|N - 1)\tilde{x}'(N - 1|N - 1)]$ and the q_{ij} are the elements of the $p \times p$ system disturbance covariance matrix $Q(N - 1) = E[w(N - 1)w'(N - 1)]$. We denote this term by $\alpha(N - 1)$.

Alternately, we note that

$$E[\tilde{x}'\Lambda\tilde{x} + w'\Gamma'W\Gamma w] = E\{\text{tr}[\Lambda\tilde{x}\tilde{x}' + \Gamma'W\Gamma ww']\}$$

$$= \text{tr}(\Lambda P + \Gamma'W\Gamma Q)$$

In order to keep our results here as nearly parallel as possible to those for the deterministic problem, we define

$$M = \Phi'[W - W\Psi(\Psi'W\Psi + B)^{-1}\Psi'W]\Phi$$

to simplify Eq. (9.62).

Let us now replace the time arguments, and summarize our results for the single-stage stochastic optimal linear regulator problem. We have

$$u(N - 1) = S(N - 1)\hat{x}(N - 1|N - 1) \tag{9.63}$$

$$S(N - 1) = -[\Psi'(N, N - 1)W(N)\Psi(N, N - 1) + B(N - 1)]^{-1}$$
$$\Psi'(N, N - 1)W(N)\Phi(N, N - 1) \tag{9.64}$$

$$W(N) = A(N) \tag{9.65}$$

$$V_1 = E[x'(N - 1)M(N - 1)x(N - 1)] + \alpha(N - 1) \tag{9.66}$$

$$M(N - 1) = \Phi'(N, N - 1)\{W(N) - W(N)\Psi(N, N - 1)$$
$$[\Psi'(N, N - 1)W(N)\Psi(N, N - 1) + B(N - 1)]^{-1}$$
$$\Psi'(N, N - 1)W(N)\}\Phi(N, N - 1) \tag{9.67}$$

$$\alpha(N - 1) = E\{\tilde{x}'(N - 1|N - 1)\Phi'(N, N - 1)W(N)\Psi(N, N - 1)$$
$$[\Psi'(N, N - 1)W(N)\Psi(N, N - 1) + B(N - 1)]^{-1}$$
$$\Psi'(N, N - 1)W(N)\Phi(N, N - 1)\tilde{x}(N - 1|N - 1)$$
$$+ w'(N - 1)\Gamma'(N, N - 1)W(N)\Gamma(N, N - 1)$$
$$w(N - 1)\}$$

$$= E[w'(N - 1)\Gamma'(N, N - 1)W(N)\Gamma(N, N - 1)w(N - 1)$$
$$- \tilde{x}'(N - 1|N - 1)\Phi'(N, N - 1)W(N)\Psi(N, N - 1)$$
$$S(N - 1)\tilde{x}(N - 1|N - 1)] \tag{9.68}$$

DOUBLE–STAGE

Turning now to the two-stage problem, we write

$$V_2 = \min_{u(N-2)} \min_{u(N-1)} E\{[x'(N-1)A(N-1)x(N-1)$$
$$+ u'(N-2)B(N-2)u(N-2)]$$
$$+ [x'(N)A(N)x(N) + u'(N-1)B(N-1)u(N-1)]\}$$

where the expected value is over $x(N-1)$, $x(N)$, $u(N-2)$, and $u(N-1)$; and $u(N-2)$ and $u(N-1)$ are required to be physically realizable.

Utilizing the principle of optimality, we have

$$V_2 = \min_{u(N-2)} E[x'(N-1)A(N-1)x(N-1)$$
$$+ u'(N-2)B(N-2)u(N-2) + V_1] \quad (9.69)$$

From Eqs. (9.66) and (9.68), we see that

$$E(V_1) = E\{E[x'(N-1)M(N-1)x(N-1)] + \alpha(N-1)\}$$
$$= E[x'(N-1)M(N-1)x(N-1)] + \alpha(N-1)$$

Hence, Eq. (9.69) can be written

$$V_2 = \min_{u(N-2)} E[x'(N-1)A(N-1)x(N-1)$$
$$+ u'(N-2)B(N-2)u(N-2)$$
$$+ x'(N-1)M(N-1)x(N-1)] + \alpha(N-1)$$
$$= \min_{u(N-2)} E[x'(N-1)W(N-1)x(N-1)$$
$$+ u'(N-2)B(N-2)u(N-2)] + \alpha(N-1) \quad (9.70)$$

where

$$W(N-1) = M(N-1) + A(N-1)$$

It is clear that $\alpha(N-1)$ can be taken out from under the minimization, since its value is unaffected by the choice of $u(N-2)$.

With the exception of the additive constant $\alpha(N-1)$, Eq. (9.70) is of the same form as Eq. (9.55) for the single-stage problem. The time arguments here are, of course, one less than they were in Eq. (9.55), and $W(N) = A(N)$ in the latter is replaced by

$$W(N-1) = M(N-1) + A(N-1)$$

to obtain the former. Hence, by repeating the above steps, we have

$$u(N-2) = S(N-2)\hat{x}(N-2|N-2)$$
$$S(N-2) = -[\Psi'(N-1, N-2)W(N-1)\Psi(N-1, N-2)$$
$$+ B(N-2)]^{-1}\Psi'(N-1, N-2)$$
$$W(N-1)\Phi(N-1, N-2)$$

$$V_2 = E[x'(N-2)M(N-2)x(N-2)] + \alpha(N-2)$$

$$
\begin{aligned}
M(N-2) = {}& \Phi'(N-1, N-2)\{W(N-1) \\
& - W(N-1)\Psi(N-1, N-2)[\Psi'(N-1, N-2) \\
& \quad W(N-1)\Psi(N-1, N-2) + B(N-2)]^{-1} \\
& \qquad \Psi'(N-1, N-2)W(N-1)\}\Phi(N-1, N-2)
\end{aligned}
$$

and

$$
\begin{aligned}
\alpha(N-2) = {}& \alpha(N-1) + E[w'(N-2)\Gamma'(N-1, N-2) \\
& W(N-1)\Gamma(N-1, N-2)w(N-2) \\
& - \tilde{x}'(N-2|N-2)\Phi'(N-1, N-2) \\
& W(N-1)\Psi(N-1, N-2)S(N-2)\tilde{x}(N-2|N-2)]
\end{aligned}
$$

The nature of $\alpha(N-2)$ should be noted carefully. It includes the effects of estimation error and system disturbance at both $N-2$ and $N-1$. This term represents the "price" that we must pay in terms of performance because the system is subject to stochastic disturbances and because we cannot determine the system's state exactly.

Despite the more complex nature of V_2 here as contrasted with its value for the deterministic problem, we note that computation of the optimal control for the two-stage process is quite straightforward. We determine the feedback gain matrices $S(N-2)$ and $S(N-1)$ just as for the deterministic problem and use them in conjunction with optimal filtered estimates $\hat{x}(N-2|N-2)$ and $\hat{x}(N-1|N-1)$, respectively, to implement the control signals.

$(j-1)$ STAGES

As in the deterministic problem, we have for $j \geq 3$ that the optimal control at time $N-j+1$ for a process involving $j-1$ stages is characterized by the set of relations:

$$W(N-j+2) = M(N-j+2) + A(N-j+2) \tag{9.71}$$

$$
\begin{aligned}
S(N-j+1) = {}& -[\Psi'(N-j+2, N-j+1) \\
& \quad W(N-j+2)\Psi(N-j+2, N-j+1) \\
& \quad + B(N-j+1)]^{-1}\Psi'(N-j+2, N-j+1) \\
& \qquad W(N-j+2)\Phi(N-j+2, N-j+1) \quad (9.72)
\end{aligned}
$$

$$u(N-j+1) = S(N-j+1)\hat{x}(N-j+1|N-j+1) \tag{9.73}$$

$$
\begin{aligned}
V_{j-1} = {}& E[x'(N-j+1)M(N-j+1)x(N-j+1)] \\
& + \alpha(N-j+1) \quad (9.74)
\end{aligned}
$$

$$\begin{aligned}
M(N-j+1) = {}& \Phi'(N-j+2, N-j+1)\{W(N-j+2) \\
& - W(N-j+2)\Psi(N-j+2, N-j+1) \\
& [\Psi'(N-j+2, N-j+1) \\
& W(N-j+2)\Psi(N-j+2, N-j+1) \\
& + B(N-j+1)]^{-1}\Psi'(N-j+2, N-j+1) \\
& W(N-j+2)\}\Phi(N-j+2, N-j+1) \quad (9.75)
\end{aligned}$$

$$\begin{aligned}
\alpha(N-j+1) = {}& \alpha(N-j+2) \\
& + E[w'(N-j+1)\Gamma'(N-j+2, N-j+1) \\
& W(N-j+2)\Gamma(N-j+2, N-j+1)w(N-j+1) \\
& - \bar{x}'(N-j+1|N-j+1)\Phi'(N-j+2, N-j+1) \\
& W(N-j+2)\Psi(N-j+2, N-j+1) \\
& S(N-j+1)\bar{x}(N-j+1|N-j+1)] \quad (9.76)
\end{aligned}$$

where W and M are symmetric $n \times n$ matrices.

j STAGES

From the principle of optimality,

$$\begin{aligned}
V_j = \min_{u(N-j)} {}& E[x'(N-j+1)A(N-j+1)x(N-j+1) \\
& + u'(N-j)B(N-j)u(N-j) + V_{j-1}]
\end{aligned}$$

From Eq. (9.74), it is obvious that

$$\begin{aligned}
E(V_{j-1}) = {}& E[x'(N-j+1)M(N-j+1)x(N-j+1) \\
& + u'(N-j)B(N-j)u(N-j)] + \alpha(N-j+1)
\end{aligned}$$

and we have

$$\begin{aligned}
V_j = \min_{u(N-j)} {}& E[x'(N-j+1)W(N-j+1)x(N-j+1) \\
& + u'(N-j)B(N-j)u(N-j)] + \alpha(N-j+1) \quad (9.77)
\end{aligned}$$

where

$$W(N-j+1) = M(N-j+1) + A(N-j+1)$$

From Eq. (9.1),

$$\begin{aligned}
x(N-j+1) = {}& \Phi(N-j+1, N-j)x(N-j) \\
& + \Gamma(N-j+1, N-j)w(N-j) \\
& + \Psi(N-j+1, N-j)u(N-j)
\end{aligned}$$

Substituting this result into Eq. (9.77) and dropping the time arguments for simplicity, we obtain

$$V_j = \min_u E[(\Phi x + \Gamma w + \Psi u)'W(\Phi x + \Gamma w + \Psi u) + u'Bu]$$
$$+ \alpha(N - j + 1)$$

$$= \min_u E[x'\Phi'W\Phi x + x'\Phi'W\Gamma w + x'\Phi'W\Psi u + w'\Gamma'W\Phi x$$
$$+ w'\Gamma'W\Gamma w + w'\Gamma'W\Psi u + u'\Psi'W\Phi x$$
$$+ u'\Psi'W\Gamma w + u'(\Psi'W\Psi + B)u] + \alpha$$

$$= \min_u E[x'\Phi'W\Phi x + 2x'\Phi'W\Gamma w + 2x'\Phi'W\Psi u + 2w'\Gamma'W\Psi u$$
$$+ w'\Gamma'W\Gamma w + u'(\Psi'W\Psi + B)u] + \alpha \quad (9.78)$$

where we have made use of the symmetry of W.

With the exceptions of the replacement of A by W, the difference in the time arguments, and the presence of α which is a known constant unrelated to the minimization, Eq. (9.78) is the same as Eq. (9.56) for the single-stage problem. Hence, to complete our development, we merely imitate the procedure which we followed earlier.

Since $x(N - j)$ and $w(N - j)$ are statistically independent and the latter has zero mean, the second term on the right-hand side of Eq. (9.78) vanishes.

Similarly, $u(N - j) = \mu_{N-j}[z^*(N - j), \bar{x}(0)]$, and since $w(N - j)$ is statistically independent of $z(N - j)$, $z(N - j - 1)$, . . . , $z(1)$, the fourth term also vanishes. As a result,

$$V_j = \min_u E[x'\Phi'W\Phi x + 2x'\Phi'W\Psi u + w'\Gamma'W\Gamma w$$
$$+ u'(\Psi'W\Psi + B)u] + \alpha \quad (9.79)$$

Utilizing the fact that $E(x) = E[E(x|y)]$, we have

$$V_j = \min_u E\{E[x'\Phi'W\Phi x + 2x'\Phi'W\Psi u + w'\Gamma'W\Gamma w$$
$$+ u'(\Psi'W\Psi + B)u|z^*(N - j), \bar{x}(0)]\} + \alpha$$

and, as before, we minimize the performance measure by determining the value of u which minimizes the inner expected value for all $z^*(N - j)$ and $\bar{x}(0)$.

For the inner expectation, we note that

$$E[2x'\Phi'W\Psi u|z^*(N - 1), \bar{x}(0)] = 2E[x'|z^*(N - 1), \bar{x}(0)]\Phi'W\Psi u$$

and

$$E[u'(\Psi'W\Psi + B)u|z^*(N - 1), \bar{x}(0)] = u'(\Psi'W\Psi + B)u$$

Consequently, setting the gradient of the inner expected value with respect to u equal to zero, we obtain

$$2E[x' | z^*(N - j), \bar{x}(0)]\Phi'W\Psi + 2u'(\Psi'W\Psi + B) = 0$$

Taking the transpose, solving for u, and replacing the time arguments, we have

$$
\begin{aligned}
u(N - j) &= -[\Psi'(N - j + 1, N - j) \\
&\quad W(N - j + 1)\Psi(N - j + 1, N - j) \\
&\quad + B(N - j)]^{-1}\Psi'(N - j + 1, N - j)W(N - j + 1) \\
&\qquad \Phi(N - j + 1, N - j)E[x(N - j) | z^*(N - j), \bar{x}(0)] \\
&= -[\Psi'(N - j + 1, N - j) \\
&\quad W(N - j + 1)\Psi(N - j + 1, N - j) \\
&\quad + B(N - j)]^{-1}\Psi'(N - j + 1, N - j) \\
&\qquad W(N - j + 1)\Phi(N - j + 1, N - j)\hat{x}(N - j | N - j) \\
&= S(N - j)\hat{x}(N - j | N - j) \qquad\qquad (9.80)
\end{aligned}
$$

where the definition of $S(N - j)$ is obvious; that is, it is the same as in the deterministic linear regulator problem. We have now established the separation principle in general.

The evaluation of V_{N-j} proceeds in the same fashion as that of V_1. Substituting the second line of Eq. (9.80) into Eq. (9.79), repeating the steps which led to Eq. (9.62) in the single-stage problem, and utilizing the definition

$$M = \Phi'[W - W\Psi(\Psi'W\Psi + B)^{-1}\Psi'W]\Phi$$

we have

$$
\begin{aligned}
V_j &= E[x'(N - j)M(N - j)x(N - j)] \\
&\quad + E\{\bar{x}'(N - j | N - j)\Phi'(N - j + 1, N - j)W(N - j + 1) \\
&\quad \Psi(N - j + 1, N - j)[\Psi'(N - j + 1, N - j) \\
&\quad W(N - j + 1)\Psi(N - j + 1, N - j) \\
&\quad + B(N - j)]^{-1}\Psi'(N - j + 1, N - j) \\
&\quad W(N - j + 1)\Phi(N - j + 1, N - j)\bar{x}(N - j | N - j) \\
&\quad + w'(N - j)\Gamma'(N - j + 1, N - j) \\
&\qquad W(N - j + 1)\Gamma(N - j + 1, N - j)w(N - j)\} \\
&\qquad\qquad\qquad\qquad\qquad + \alpha(N - j + 1)
\end{aligned}
$$

Defining

$$
\begin{aligned}
\alpha(N - j) = \ & \alpha(N - j + 1) + E\{\tilde{x}'(N - j|N - j) \\
& \Phi'(N - j + 1, N - j)W(N - j)\Psi(N - j + 1, N - j) \\
& [\Psi'(N - j + 1, N - j)W(N - j + 1) \\
& \Psi(N - j + 1, N - j) + B(N - j)]^{-1} \\
& \Psi'(N - j + 1, N - j)W(N - j + 1) \\
& \Phi(N - j + 1, N - j)\tilde{x}(N - j|N - j) \\
& + w'(N - j)\Gamma'(N - j + 1, N - j)W(N - j + 1) \\
& \quad \Gamma(N - j + 1, N - j)w(N - j)\} \quad (9.81)
\end{aligned}
$$

we can write

$$
V_j = E[x'(N - j)M(N - j)x(N - j)] + \alpha(N - j) \qquad (9.82)
$$

Utilizing the relation for $S(N - j)$, we can put Eq. (9.81) in the slightly simpler form

$$
\begin{aligned}
\alpha(N - j) = \ & \alpha(N - j + 1) + E[w'(N - j)\Gamma'(N - j + 1, N - j) \\
& W(N - j + 1)\Gamma(N - j + 1, N - j)w(N - j) \\
& - \tilde{x}'(N - j|N - j)\Phi'(N - j + 1, N - j) \\
& W(N - j + 1)\Psi(N - j + 1, N - j)S(N - j) \\
& \quad \tilde{x}(N - j|N - j)] \quad (9.83)
\end{aligned}
$$

Making the same change in the time index here as we did in Sec. 9.3 for the deterministic problem, viz., $k = N - j$, we express Eq. (9.80) as

$$
u(k) = S(k)\hat{x}(k|k)
$$

for $k = 0, 1, \ldots, N - 1$, where $S(k)$ is determined using the algorithm in either Theorem 9.2 or Corollary 9.1. Similarly, we have

$$
V_{N-k} = E[x'(k)M(k)x(k)] + \alpha(k) \qquad (9.84)
$$

and

$$
\begin{aligned}
\alpha(k) = \ & \alpha(k + 1) + E[w'(k)\Gamma'(k + 1, k)W(k + 1)\Gamma(k + 1, k)w(k) \\
& - \tilde{x}'(k|k)\Phi'(k + 1, k)W(k + 1)\Psi(k + 1, k)S(k)\tilde{x}(k|k)] \quad (9.85)
\end{aligned}
$$

for Eqs. (9.82) and (9.83), respectively, where $k = N - 1, N - 2, \ldots,$ 0, and $M(k)$ and $W(k)$ are determined using the results in Theorem 9.2 or Corollary 9.1.

The computation of $\alpha(k)$ can be carried out recursively backward in time using Eq. (9.85). Setting $k = N - 1$ in this equation and comparing the result with Eq. (9.68) for the single-stage problem, we see

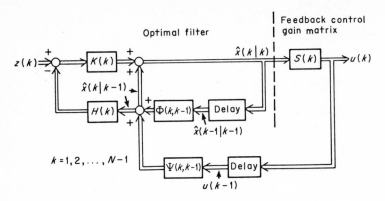

Fig. 9.4 Optimal control system.

that the correct boundary condition for the computations is $\alpha(N) = 0$. It is, of course, quite clear that computation of $\alpha(k)$ is necessary only to evaluate the performance measure, and need not be carried out in the computation of the optimal control.

At each stage, the optimal estimate of $x(k)$ must reflect the fact that the control signal $S(k - 1)\hat{x}(k - 1|k - 1)$ was input to the system. Hence, we have

$$\hat{x}(k|k) = \hat{x}(k|k - 1) + K(k)[z(k) - H(k)\hat{x}(k|k - 1)]$$

where

$$\hat{x}(k|k - 1) = \Phi(k, k - 1)\hat{x}(k - 1|k - 1) + \Psi(k, k - 1)u(k - 1)$$
$$(9.86)$$

for the optimal estimate.

We note that $\hat{x}(N|N)$ need not be determined, since the last control acts at $k = N - 1$. Also, $\hat{x}(0|0) = \bar{x}(0)$ implies that $u(0) = S(0)\hat{x}(0|0)$ is zero if $x(0)$ has zero mean.

The complete system now assumes the form given in Fig. 9.1. The detailed block diagram for the control system is shown in Fig. 9.4. Let us now summarize the result which we have obtained.

Theorem 9.3 *The optimal control system for the stochastic linear regulator problem consists of the optimal linear filter cascaded with the optimal feedback gain matrix of the deterministic linear regulator. The parameters for the two parts of the control system are determined separately. The performance measure for the complete control system is governed by Eqs. (9.84) and (9.85), where the boundary condition for the latter equation is $\alpha(N) = 0$.*

This important result, which reduces a given optimization problem to two separate optimization problems whose solutions are known, was first conjectured by Kalman and Koepcke [2], and subsequently proved independently by Joseph [3] and Gunckel [4]. It is commonly referred to as the *separation principle*. An extension of this result to more general problems has been given by Striebel [5].

The most striking feature of the separation principle is that the feedback control gain matrix is independent of all the statistical parameters in the problem, whereas the optimal filter is independent of the matrices in the performance measure.

We have only shown here that the control law $u(k) = S(k)\hat{x}(k|k)$ is a necessary condition for the performance measure in Eq. (9.3) to be a minimum. The sufficiency, however, follows easily from a consideration of the set of partial derivatives of

$$2E[x'|z^*(N - j), \bar{x}(0)]\Phi'W\Psi + 2u'(\Psi'W\Psi + B)$$

with respect to the components of u. We see that this leads to the same requirement as for the deterministic problem, viz., that the matrix

$$[\Psi'(k + 1, k)W(k + 1)\Psi(k + 1, k) + B(k)]$$

be positive definite for all $k = 0, 1, \ldots, N - 1$.

Example 9.3 Let us consider the stochastic version of the problem in Example 9.1 in which the initial state is a zero mean gaussian random n vector, there is no system disturbance, and all the state variables can be measured in the presence of an additive zero mean gaussian white sequence $\{v(k + 1), k = 0, 1, \ldots, N - 2\}$ whose covariance matrix $R(k + 1)$ is positive definite for all k. We thus have

$$J_N = E\Big[\sum_{i=1}^{N} x'(i)A(i)x(i) \Big]$$

$$x(k + 1) = \Phi(k + 1, k)x(k) + \Psi(k + 1, k)u(k)$$

$$z(k + 1) = x(k + 1) + v(k + 1)$$

From Theorem 9.3 and Example 9.1, it follows that

$$u(k) = S(k)\hat{x}(k|k) = -\Psi^{-1}(k + 1, k)\Phi(k + 1, k)\hat{x}(k|k)$$

Substituting this result into the system equation, we see that

$$x(k + 1) = \Phi(k + 1, k)x(k) - \Psi(k + 1, k)\Psi^{-1}(k + 1, k)\Phi(k + 1, k)\hat{x}(k|k)$$
$$= \Phi(k + 1, k)\bar{x}(k|k)$$

Since there is no system disturbance, it follows that

$$x(k + 1) = \bar{x}(k + 1|k)$$

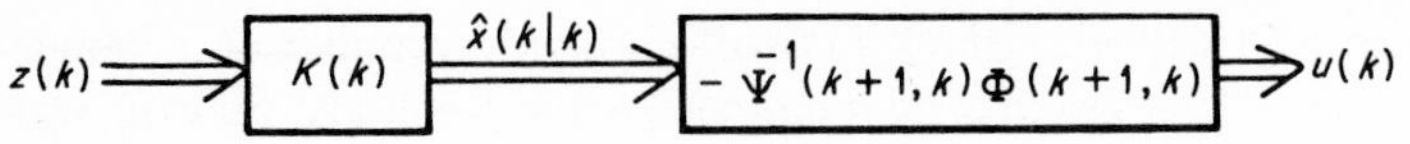

Fig. 9.5 Optimal control system for Example 9.3.

Hence, the system error at each time point is the error in prediction at that time point. This is in contrast to the deterministic case where the control was perfect and $x(k + 1) = 0$ for $k = 0, 1, \ldots, N - 1$. For the stochastic problem, we see that the control is only as "good" as the prediction accuracy.

The optimal filter for the control system is governed by the relation

$$\hat{x}(k|k) = \Phi(k, k - 1)\hat{x}(k - 1|k - 1) + \Psi(k, k - 1)u(k - 1)$$
$$+ K(k)[z(k) - \Phi(k, k - 1)\hat{x}(k - 1|k - 1) - \Psi(k, k - 1)u(k - 1)]$$

Since $u(k - 1) = -\Psi^{-1}(k, k - 1)\Phi(k, k - 1)\hat{x}(k - 1|k - 1)$, this reduces to

$$\hat{x}(k|k) = K(k)z(k)$$

The block diagram for the complete control system is given in Fig. 9.5.

From Theorem 5.5, we have here that

$$P(k|k - 1) = \Phi(k, k - 1)P(k - 1|k - 1)\Phi'(k, k - 1)$$
$$K(k) = P(k|k - 1)[P(k|k - 1) + R(k)]^{-1}$$
$$P(k|k) = [I - K(k)]P(k|k - 1)$$

for $k = 1, 2, \ldots, N - 1$, with $P(0|0) = E[x(0)x'(0)]$ assumed given.

Finally, let us consider evaluation of the performance measure. From Example 9.1 and Eq. (9.51), we have $W(k) = A(k)$ and

$$M(k) = W(k) - A(k) = 0$$

respectively. Hence, Eq. (9.83) reduces to

$$V_{N-k} = \alpha(k)$$

Substituting into Eq. (9.84) for $W(k + 1)$ and $S(k)$, and recalling that there is no system disturbance, we have

$$\alpha(k) = \alpha(k + 1) + E[\tilde{x}'(k|k)\Phi'(k + 1, k)A(k + 1)\Psi(k + 1, k)$$
$$\Psi^{-1}(k + 1, k)\Phi(k + 1, k)\tilde{x}(k|k)]$$
$$= \alpha(k + 1) + E[\tilde{x}'(k + 1|k)A(k + 1)\tilde{x}(k + 1|k)]$$
$$= \alpha(k + 1) + \operatorname{tr}[A(k + 1)P(k + 1|k)]$$

for $k = N - 1, N - 2, \ldots, 0$, where $\tilde{x}(k + 1|k) = \Phi(k + 1, k)\tilde{x}(k|k)$ is the prediction error, and $\alpha(N) = 0$ is the boundary condition. Successive evaluation of this last relation then leads to the result

$$\alpha(0) = \sum_{i=1}^{N} \operatorname{tr}[A(i)P(i|i - 1)]$$

and we have

$$V_N = \alpha(0)$$

for the optimal value of the performance measure for N stages of control.

Example 9.4   As a specific numerical example, let us consider the stochastic version of Example 9.2 in which

$$x(k + 1) = x(k) + w(k) + 2u(k)$$
$$z(k + 1) = x(k + 1) + v(k + 1)$$

and

$$J_3 = E\left[x^2(3) + \sum_{i=1}^{3} u^2(i - 1)\right] \quad .$$

We thus have a three-stage, terminal error plus control effort, stochastic linear regulator problem.

We assume that $\{w(k), k = 0, 1, 2\}$ is zero mean, gaussian, and white with constant variance $Q(k) = E[w^2(k)] = 25$, and that $\{v(k + 1), k = 0, 1, 2\}$ is zero mean, gaussian, white, and independent of $\{w(k), k = 0, 1, 2\}$ with constant variance $R(k + 1) = E[v^2(k + 1)] = 15$. We note that although the measurement at $k = 3$ permits us to obtain the optimal estimate of the terminal state, it is of no use in controlling the system, since the process terminates at that time.

We assume that $x(0)$ is a zero mean gaussian random variable which is independent of the two gaussian white sequences and has a variance $P(0) = E[x^2(0)] = 100$.

Invoking Theorem 9.3, we determine the optimal filter and the feedback control gain separately. The filter is the same as in Example 5.4, except that we now have a control signal present. Hence, the filter equation here is

$$\hat{x}(k|k) = \hat{x}(k - 1|k - 1) + 2u(k - 1) + K(k)[z(k) - \hat{x}(k - 1|k - 1) - 2u(k - 1)]$$

where

$$u(k - 1) = S(k - 1)\hat{x}(k - 1|k - 1)$$

for $k = 1, 2, 3$, where $\hat{x}(0|0) = 0$ and $S(k - 1)$ remains to be determined.

From the table in Example 5.4, we have the following data for the optimal filter:

k	$P(k|k - 1)$	$K(k)$	$P(k|k)$
0	. . . . .	. . . . .	100
1	125	0.893	13.40
2	38.4	0.720	10.80
3	35.8	0.704	10.57

It is clear that the filter need be operative only to determine $\hat{x}(1|1)$ and $\hat{x}(2|2)$, since $\hat{x}(0|0) = 0$ and $\hat{x}(3|3)$ is of no use in control.

Turning now to the question of the feedback control gain, we have precisely the problem of Example 9.2 with $\beta = \frac{1}{4}$ and $N = 3$. From the table in that example, we have

k	$S(k)$	$W(k)$
3	....	1
2	-0.400	0.200
1	-0.222	0.111
0	-0.154	0.077

The optimal control at each stage is seen to be

$$u(0) = -0.154\hat{x}(0|0) = 0 \qquad \text{since } \hat{x}(0|0) = 0$$
$$u(1) = -0.222\hat{x}(1|1)$$

and

$$u(2) = -0.400\hat{x}(2|2)$$

The block diagram for the control system is given in Fig. 9.6.

Let us now evaluate the performance measure for our problem. From Eq. (9.83) for $N = 3$ and $k = 0$, we obtain

$$V_3 = E[M(0)x^2(0)] + \alpha(0) \qquad\qquad (9.87)$$

From Eq. (9.51), $M(0) = W(0) - A(0)$. In our example, $A(0) = 0$, and we see from the second table above that $W(0) = 0.077$. Recalling that $E[x^2(0)] = P(0) = 100$, we have, therefore, that

$$V_3 = 7.7 + \alpha(0)$$

We remark that for the deterministic three-stage problem, we would have $V_3 = 0.077x^2(0)$.

Here, we note that the first term in Eq. (9.87), whose value we have just shown is 7.7, depends upon $M(0)$ which also appears in the performance meas-

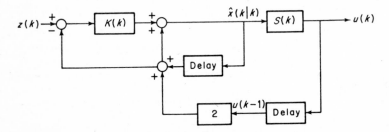

Fig. 9.6 Optimal control system for Example 9.4.

ure for the deterministic problem, and upon the variance of the initial state. Hence, we interpret this first term as the component of V_3 which is due to the uncertainty associated with the initial state. The second term, $\alpha(0)$, we know is due to the presence of the system disturbance and the fact that the filtered estimate is subject to error.

We now evaluate $\alpha(0)$ by successive application of Eq. (9.85). For our example, we have $\alpha(N) = \alpha(3) = 0$, $\Gamma(k + 1, k) = 1$, $\Phi(k + 1, k) = 1$, and $\Psi(k + 1, k) = 2$. Since we are dealing only with scalars, Eq. (9.85) reduces to

$$\alpha(k) = \alpha(k + 1) + W(k + 1)E[w^2(k)] - 2W(k + 1)S(k)E[\tilde{x}^2(k|k)]$$
$$= \alpha(k + 1) + W(k + 1)[25 - 2S(k)P(k|k)] \tag{9.88}$$

for $k = 2, 1, 0$, where we have made use of the facts that $E[w^2(k)] = Q(k) = 25$ and $E[\tilde{x}^2(k|k)] = P(k|k)$. Substituting into this relation from the above tables, we obtain the sequence of values: $\alpha(2) = 33.64$, $\alpha(1) = 39.83$, and $\alpha(0) = 46.02$. We then have

$$V_3 = 7.7 + 46.02 = 53.72$$

Of itself, this number tells us very little about the system's performance. We do know, of course, that 53.72 is the minimum value of the particular performance measure which we have selected, but so what? Our reply is in two parts.

First, we see from Eqs. (9.87) and (9.88) that V_3 is composed of three components: one due to uncertainty associated with the initial state, viz., $E[M(0)x^2(0)]$; another due to the system disturbance; and the third one due to filtering error. Therefore, we can decompose V_3 to determine a performance "budget." Carrying this out, we obtain the following results.

Component of V_3 due to:

    1. Uncertainty in $x(0)$ =   7.7
    2. System disturbance = 32.77
    3. Filtering error   = 13.25
               Total   53.72

Such a budget can be of considerable importance in systems analysis. For example, we note here that the "culprit" so far as performance is concerned is the system disturbance. If this disturbance is, say, due to internal noise, then a redesign of the system electronics may be desirable to improve performance. On the other hand, if this disturbance arises from external phenomena over which there is no control in the design, it must be tolerated. Still, some improvement in performance can be attained by utilizing a more accurate measurement scheme than the present one.

The second part of our response to the question of "so what?" is that V_3 provides a standard of reference against which we can compare the performance of other control schemes. For example, in order to economize in the design of the control system, we might decide to omit implementation of the filter and simply feed back the measurement $z(k)$ rather than the optimal estimate $\hat{x}(k|k)$ through the optimal feedback gain. Our nonoptimal control law would then be

$$u^*(k) = S(k)z(k) \tag{9.89}$$

where the asterisk denotes that u is no longer optimal in the sense of the given performance measure. For simplicity, we might arbitrarily set $z(0) = 0$ since there is no measurement at $k = 0$. This also makes $u^*(0)$ equal to the optimal $u(0)$. The question now is, "How does the nonoptimal, but relatively simple, control law which is defined by Eq. (9.89) compare with the optimal control law in terms of the performance measure

$$J_3 = E\left[x^2(3) + \sum_{i=1}^{3} u^2(i-1) \right] ?"$$

We leave the computational details as an exercise.

Finally, we remark that there are other possible nonoptimal schemes which we might consider. We could, for example, go to the extreme of using only a constant gain in the feedback path in the above scheme rather than the time-varying optimal one. This, obviously, is the simplest possible feedback control for the problem. Alternately, we could use the filter with a constant gain, that is, $K(k) = $ constant, and a constant feedback gain, $S(k) = $ constant.

We conclude this chapter by remarking that it is possible to obtain the optimal control for the stochastic linear regulator problem for cases where the elements of Φ, Γ, and Ψ are random variables which are not necessarily independent of the system disturbances and measurement errors. Although the separation principle as we have it here does not carry over to such problems, the results are nevertheless useful. A complete treatment of these problems is found in Aoki [6].

PROBLEMS

9.1. Show how the results of Theorem 9.2 must be modified if the desired state at each time point is some arbitrary $x^d(k)$ instead of the origin.

9.2. Apply the principle of optimality to obtain the optimal control law for the problem where

$$x(k+1) = \Phi(k+1, k)x(k) + \Psi(k+1, k)u(k)$$

$$J_N = \sum_{i=1}^{N} x'(i)A(i)x(i)$$

and

$$|u_j(k)| \le 1 \qquad j = 1, 2, \ldots, r$$

Assume that N is fixed, $x(0)$ is given, $A(i)$ is positive definite for all i, and that J_N is to be minimized. Note that this is the deterministic linear regulator problem with a magnitude (saturation) constraint on the elements of the control vector.

9.3. Assume that Eq. (9.8) in the formulation of the deterministic linear regulator problem is replaced by the relation

$$u(k) = S(k)x(k)$$

where $k = 0, 1, \ldots, N - 1$ and $S(k)$ is an arbitrary $r \times n$ matrix. Utilizing the principle of optimality, derive the equations for computing the $S(k)$ for which the performance measure in Eq. (9.7) is minimized. Note here that the more general physical realizability condition of Eq. (9.8) has been replaced by the requirement that the control law be a linear transformation of the state at each time point.

9.4. Consider the second-order linear system

$$\dot{x}_1 = x_2$$
$$\dot{x}_2 = u(t)$$

for $t \geq 0$ and assume that $x(0)$ is given.

(a) For a sampling interval of one second, determine the difference equation for the system in the form

$$x(k + 1) = \Phi(k + 1, k)x(k) + \Psi(k + 1, k)u(k)$$

where $k = 0, 1, \ldots$, corresponds to the time instants $t_0 = 0, t_1 = 1, t_2 = 2, \ldots$. Assume that the control signal $u(t)$ is a constant over each sampling interval, i.e., $u(t) = u(k) = $ constant for $t_k \leq t < t_{k+1}, k = 0, 1, \ldots$.

(b) Determine the optimal control for the two-stage process where the performance measure is

$$J_2 = \sum_{i=1}^{2} [x_1{}^2(i) + x_2{}^2(i)]$$

(c) What is V_2 for this problem?

9.5. Does Theorem 9.3 still hold if the first measurement is $z(0) = H(0)x(0) + v(0)$ instead of $z(1)$? Verify your answer. Assume here that $\{v(k), k = 0, 1, \ldots\}$ is a zero mean gaussian white sequence which is independent of both $x(0)$ and $\{w(k), k = 0, 1, \ldots\}$ and has a positive semidefinite covariance matrix.

9.6. Is Theorem 9.3 valid if $\{v(k + 1), k = 0, 1, \ldots\}$ and $\{w(k), k = 0, 1, \ldots\}$ are correlated with $E[w(j + 1)v'(k + 1)] = U(k)\delta_{jk}$, where $U(k)$ is a real $p \times m$ matrix? If your answer is yes, show what modifications, if any, are necessary in the development in Sec. 9.4. If your answer is no, show why.

9.7. Consider Eq. (9.58). Since the performance measure can be minimized by minimizing the inner expected value with respect to u for all $z^*(N - 1)$ and $\bar{x}(0)$, an alternate approach to the stochastic linear regulator problem can be taken by defining, in general for a j-stage process, the quantity

$$\nu_j = \min_{u(N-j)} \cdots \min_{u(N-1)} E \left\{ \sum_{i=N-j+1}^{N} [x'(i)A(i)x(i) \right.$$
$$\left. + u'(i - 1)B(i - 1)u(i - 1) | z^*(N - j), \bar{x}(0)] \right\}$$

for $j = 1, 2, \ldots, N$ and carrying out the development in terms of ν_j instead of V_j. Utilizing the principle of optimality, determine the solution of the stochastic linear regulator problem for the above approach and develop an explicit expression for ν_j.

9.8. Determine the optimal filter, the optimal feedback gain, and the optimal value of the performance measure for the two-stage scalar problem where

$$x(k + 1) = 2x(k) + u(k)$$
$$z(k) = x(k) + v(k)$$

and

$$J_2 = E[x'(2)x(2)]$$

Assume that $\{v(k), \ k = 0, 1, \ldots\}$ is a zero mean gaussian white sequence which is independent of $x(0)$ and has a constant variance of 5. Assume also that $x(0)$ is gaussian with zero mean and a variance of 5. Determine a performance "budget" for this problem analogous to the one in Example 9.4.

9.9. Following the suggestion at the end of Example 9.4, suppose that the control law in Eq. (9.89) is used instead of the optimal one. Taking $u^*(0) = 0$, evaluate the performance measure

$$J_3 = E\left[x^2(3) + \sum_{i=1}^{3} u^{*2}(i - 1) \right]$$

and compare its value with V_3.

9.10. Do the results stated in Theorem 9.2, Corollary 9.1, and Theorem 9.3 refer to a local or global minimum for the relevant performance measure? Explain your answer.

9.11. When optimal control is used, the system of Eq. (9.1) is governed by the relation

$$x(k + 1) = \Phi(k + 1, k)x(k) + \Gamma(k + 1, k)w(k) + \Psi(k + 1, k)S(k)\hat{x}(k|k)$$

for $k = 0, 1, \ldots, N - 1$. Is the stochastic process $\{x(k), k = 0, 1, \ldots, N - 1\}$ Gauss-Markov? Explain your answer.

REFERENCES

1. Bellman, R. E., "Dynamic Programming," Princeton University Press, Princeton, N.J., 1957.
2. Kalman, R. E., and R. W. Koepcke, Optimal Synthesis of Linear Sampling Control Systems Using Generalized Performance Indexes, *Trans. ASME*, vol. 80 p. 1820, 1958.
3. Joseph, P. D., and J. T. Tou, On Linear Control Theory, *Trans. AIEE*, pt. II, vol. 80, p. 193, 1961.
4. Gunckel, T. L., II, and G. F. Franklin, A General Solution for Linear Sampled Data Control, *J. Basic Eng.*, vol. 85, p. 197, 1963.
5. Striebel, C., Sufficient Statistics in the Optimum Control of Stochastic Systems, *J. Math. Anal. Appl.*, vol. 12, p. 576, 1965.
6. Aoki, M., "Optimization of Stochastic Systems—Topics in Discrete-Time Systems," Academic Press Inc., New York, 1967.

10
Stochastic Optimal Control for Continuous Linear Systems

10.1 INTRODUCTION

We proceed now to a study of the continuous stochastic linear regulator problem. The problem can be viewed as the limiting case of the discrete-time problem in Chap. 9 in which the time interval between measurements and control inputs is made arbitrarily small. We adopt this particular viewpoint here, but remark that it is by no means the only one available. In fact, the literature on the continuous stochastic linear regulator problem and certain variations of it, including problems involving nonlinear dynamic systems, is quite extensive.†

In our presentation here, we begin with a formulation of the continuous stochastic linear regulator problem in Sec. 10.2. In Sec. 10.3, we formulate an equivalent discrete-time problem, and in Sec. 10.4, we apply the results of Chap. 9 to this latter problem. We then formally obtain

† The number of relevant references is far too excessive for a complete and accurate listing to be given here. Instead, a representative listing is given at the end of the chapter to provide the interested reader with a starting point for additional study.

the limit of the resulting control algorithm as the time interval between measurements and control inputs goes to zero. The result, as we might expect, is a separation principle.

10.2 PROBLEM FORMULATION

SYSTEM MODEL

The system model is the same as the one used in the study of the continuous-time estimation problem (see Sec. 7.2) except that there is now present a control input. The model was introduced in Sec. 4.4. In particular, we have

$$\dot{x} = F(t)x + G(t)w(t) + C(t)u(t) \tag{10.1}$$

and

$$z(t) = H(t)x(t) + v(t) \tag{10.2}$$

for $t \geq t_0$. The vectors x, w, u, z, and v play the same roles, respectively, as they did in Chap. 9, except that here they are functions of the continuous-time variable t. The matrices $F(t)$, $G(t)$, $C(t)$ and $H(t)$ are, respectively, $n \times n$, $n \times p$, $n \times r$, and $m \times n$, and each is continuous in t. The initial time t_0 is fixed, and the dot denotes the time derivative.

The stochastic processes $\{w(t),\ t \geq t_0\}$ and $\{v(t),\ t \geq t_0\}$ are zero mean gaussian white noises with

$$E[w(t)w'(\tau)] = Q(t)\delta(t - \tau)$$

$$E[v(t)v'(\tau)] = R(t)\delta(t - \tau)$$

and

$$E[w(t)v'(\tau)] = 0$$

for all $t, \tau \geq t_0$, where all of the terms have been defined previously.

The initial state $x(t_0)$ is taken as a zero mean gaussian random n vector, which is independent of both of the above noise processes, and for which $E[x(t_0)x'(t_0)] = P(t_0)$, as in previous work.

PERFORMANCE MEASURE

Our problem is to determine a control input $u(t)$, $t \geq t_0$, subject to certain restrictions, such that the system of Eq. (10.1) behaves in some desired fashion. For this purpose, we choose to determine the control input such that the performance measure

$$J = E\left\{x'(t_1)\Lambda x(t_1) + \int_{t_0}^{t_1} [x'(t)A(t)x(t) + u'(t)B(t)u(t)]\,dt\right\} \tag{10.3}$$

is minimized.

In Eq. (10.3), $t_1 > t_0$ is the fixed terminal time. Hence, as in the discrete-time case, we have a fixed-time optimization problem. Also, in Eq. (10.3), Λ is a symmetric $n \times n$ positive semidefinite matrix; $A(t)$ is a symmetric $n \times n$ positive semidefinite matrix; $B(t)$ is a symmetric $r \times r$ positive definite matrix; and E denotes the expected value. Both $A(t)$ and $B(t)$ are assumed to be continuous for all $t \in [t_0, t_1]$.

Equation (10.3) is analogous to Eq. (9.3) for the discrete stochastic linear regulator problem. It is quadratic in the state and control, but here the integral replaces the sum in Eq. (9.3). The term $x'(t_1)\Lambda x(t_1)$ is used to include terminal error explicitly in the performance measure, while the integral involves system error and control effort over the entire time interval of operation. The analogy between Eqs. (10.3) and (9.3) will become more apparent in Section 10.3. One important difference should be noted, however. In Eq. (9.3), $B(i-1)$, $i = 1, 2, \ldots, N$, is required to be positive semidefinite, while $B(t)$, $t_0 \leq t \leq t_1$, in Eq. (10.3) is taken to be positive definite. The reason for this is that the resulting optimal control algorithm for the continuous-time problem requires that $B(t)$ be nonsingular for all $t \in [t_0, t_1]$. This means that the continuous-time problem does not admit performance measures such as

$$J = E[x'(t_1)\Lambda x(t_1)]$$

$$J = E\left[\int_{t_0}^{t_1} x'(t)A(t)x(t)\ dt\right]$$

or some linear combination of the two, whereas the analogous discrete-time problem can in some cases.

PHYSICALLY REALIZABLE CONTROLS AND PROBLEM STATEMENT

As in the discrete stochastic linear regulator problem, we require here that the control input be realizable in terms of data about the system's state which is physically available. We say, therefore, that a control law is physically realizable if it is of the form

$$u(t) = \mu[z(\tau), t_0 \leq \tau \leq t; \bar{x}(t_0); t] \tag{10.4}$$

where μ is an r-dimensional, vector-valued function of the set of measurements $\{z(\tau), t_0 \leq \tau \leq t\}$, the mean value of the initial state, and the current time t.

The continuous stochastic linear regulator problem statement now follows.

PROBLEM STATEMENT

Determine a physically realizable control law of the form in Eq. (10.4) for the system of Eqs. (10.1) and (10.2) which minimizes the performance measure in Eq. (10.3).

Such a control is, of course, termed an *optimal control*.

10.3 EQUIVALENT DISCRETE–TIME PROBLEM

SYSTEM MODEL

By including an additive control input in the discrete-time model which we gave in Sec. 7.3 for the continuous-time estimation problem, we obtain the model required here (see also Sec. 4.4). In particular, we have

$$x(t + \Delta t) = \Phi(t + \Delta t, t)x(t) + \Gamma(t + \Delta t, t)w(t) + \Psi(t + \Delta t, t)u(t) \tag{10.5}$$

where

$$\Phi(t + \Delta t, t) = I + F(t)\,\Delta t + 0(\Delta t^2) \tag{10.6}$$

$$\Gamma(t + \Delta t, t) = G(t)\,\Delta t + 0(\Delta t^2) \tag{10.7}$$

$$\Psi(t + \Delta t, t) = C(t)\,\Delta t + 0(\Delta t^2) \tag{10.8}$$

and t is the discrete-time index $\{t = t_0 + j\,\Delta t,\ j = 0,\ 1,\ \ldots\}$ with $\Delta t > 0$. Here, as in Chap. 7, t will be used to denote both discrete and continuous time, its usage being clear from context.

The control input $u(t)$ is required to be piecewise constant, and the interval $[t_0, t_1]$ is assumed to be divided in N increments of length Δt, where N is a positive integer and

$$\Delta t = \frac{t_1 - t_0}{N}$$

In the limit as $\Delta t \to 0$, we require that N be such that

$$N\,\Delta t = t_1 - t_0 = \text{constant}$$

A sample control input and the partitioning of the interval $[t_0, t_1]$ are depicted in Fig. 10.1. We note that the indexing on j terminates at N so far as the control problem is concerned.

In Eq. (10.5), we recall from Sec. 7.3 that $\{w(t),\ t = t_0 + j\,\Delta t,\ j = 0,\ 1,\ \ldots\}$ is a zero mean gaussian white sequence whose covariance

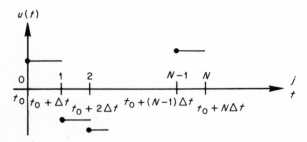

Fig. 10.1 Piecewise constant control and partitioning of control interval.

matrix is

$$E[w(t)w'(\tau)] = \frac{Q(t)}{\Delta t} \delta_{jk}$$

where $Q(t)$ is positive semidefinite for all $t \geq t_0$, τ is the discrete-time index $\{\tau = t_0 + k\,\Delta t, k = 0, 1, \ldots\}$, δ_{jk} is the Kronecker delta, and $x(t_0)$ is a zero mean gaussian random n vector which is independent of $\{w(t), t = t_0 + j\,\Delta t, j = 0, 1, \ldots\}$ and has the covariance matrix $P(t_0)$.

The discrete-time measurement equation is

$$z(t + \Delta t) = H(t + \Delta t)x(t + \Delta t) + v(t + \Delta t) \tag{10.9}$$

where $\{v(t + \Delta t), t = t_0 + j\,\Delta t, j = 0, 1, \ldots\}$ is a zero mean gaussian white sequence which is independent of $\{w(t), t = t_0 + j\,\Delta t, j = 0, 1, \ldots\}$ and $x(t_0)$, and whose covariance matrix is

$$E[v(t + \Delta t)v'(\tau + \Delta t)] = \frac{R(t + \Delta t)}{\Delta t} \delta_{jk}$$

with $R(t + \Delta t)$ positive definite for all $t \geq t_0$.

This completes our specification of the equivalent discrete-time model, and we turn now to a consideration of the performance measure.

PERFORMANCE MEASURE

Let us suppose that some arbitrary, well-behaved control input $u(t)$, which is not necessarily optimal, is applied to the system of Eq. (10.1) and let us consider the term

$$\beta = x'(t_1)\Lambda x(t_1) + \int_{t_0}^{t_1} x'(t)A(t)x(t)\,dt \tag{10.10}$$

of the performance measure in Eq. (10.3). Let us subdivide the interval $[t_0, t_1]$ into N subintervals each of length $\Delta t = t_1 - t_0/N$ and use i to index these subintervals. A representative situation is given in Fig. 10.2. As before, we require that as $\Delta t \to 0$, N be such that $N\,\Delta t = t_1 - t_0$. As a consequence, we see that $t_0 + N\,\Delta t = t_1$ for all N.

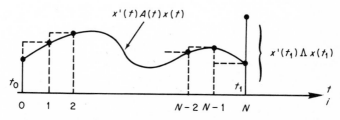

Fig. 10.2 Performance measure representation involving terminal error and integral of system error.

Utilizing the partition which is indicated in Fig. 10.2, we can express Eq. (10.10) as

$$\beta = \lim_{\Delta t \to 0} \left[x'(t_0 + N\,\Delta t)\Lambda x(t_0 + N\,\Delta t) \right.$$
$$\left. + \sum_{i=1}^{N} x'(t_0 + i\,\Delta t)A(t_0 + i\,\Delta t)x(t_0 + i\,\Delta t)\,\Delta t \right] \quad (10.11)$$

Next we consider the term

$$\gamma = \int_{t_0}^{t_1} u'(t)B(t)u(t)\,dt \quad (10.12)$$

Utilizing the same partition as above and defining the values of $u(t)$ at the left instead of the right endpoints of the subintervals [cf. Figs. (10.1) and (10.2)], we write Eq. (10.12) as

$$\gamma = \lim_{\Delta t \to 0} \sum_{i=1}^{N} u'[t_0 + (i-1)\,\Delta t]B[t_0 + (i-1)\,\Delta t]u[t_0 + (i-1)\,\Delta t]\,\Delta t$$
$$(10.13)$$

It now follows from Eqs. (10.3), and (10.10) to (10.13) that

$$J = E(\beta + \gamma)$$
$$= E\left(\lim_{\Delta t \to 0} \left\{ x'(t_0 + N\,\Delta t)\Lambda x(t_0 + N\,\Delta t) \right.\right.$$
$$+ \sum_{i=1}^{N} x'(t_0 + i\,\Delta t)A(t_0 + i\,\Delta t)x(t_0 + i\,\Delta t)\,\Delta t$$
$$\left.\left. + \sum_{i=1}^{N} u'[t_0 + (i-1)\,\Delta t]B[t_0 + (i-1)\,\Delta t]u[t_0 + (i-1)\,\Delta t]\,\Delta t \right\} \right)$$
$$= \lim_{\Delta t \to 0} E\left\{ x'(t_0 + N\,\Delta t)\Lambda x(t_0 + N\,\Delta t) \right.$$
$$+ \sum_{i=1}^{N} x'(t_0 + i\,\Delta t)A(t_0 + i\,\Delta t)x(t_0 + i\,\Delta t)\,\Delta t$$
$$\left. + \sum_{i=1}^{N} u'[t_0 + (i-1)\,\Delta t]B[t_0 + (i-1)\,\Delta t]u[t_0 + (i-1)\,\Delta t]\,\Delta t \right\}$$
$$(10.14)$$

For a given N, we define

$$J_N = E\left\{ x'(t_0 + N\,\Delta t)\Lambda x(t_0 + N\,\Delta t) \right.$$
$$+ \sum_{i=1}^{N} x'(t_0 + i\,\Delta t)A(t_0 + i\,\Delta t)x(t_0 + i\,\Delta t)\,\Delta t$$
$$\left. + \sum_{i=1}^{N} u'[t_0 + (i-1)\,\Delta t]B[t_0 + (i-1)\,\Delta t]u[t_0 + (i-1)\,\Delta t]\,\Delta t \right\}$$
$$(10.15)$$

and note that

$$J = \lim_{N \to \infty} J_N$$

if $\Delta t \to 0$ such that $N \, \Delta t = t_1 - t_0$.

Equation (10.15) defines the performance measure for our equivalent discrete-time problem. Comparing this equation with Eq. (9.3) for the discrete stochastic linear regulator problem, we obtain the following relationships:

$$A(i) = \begin{cases} A(t_0 + i \, \Delta t) \, \Delta t & \text{for } i = 1, 2, \ldots, N - 1 \\ \Lambda + A(t_1) \, \Delta t & \text{for } i = N \end{cases} \tag{10.16}$$

and

$$B(i - 1) = B[t_0 + (i - 1) \, \Delta t] \, \Delta t \quad \text{for } i = 1, 2, \ldots, N \tag{10.17}$$

PHYSICALLY REALIZABLE CONTROLS

As already remarked, and also indicated in Fig. 10.1, the discrete-time control input $u(t)$ is required to be piecewise constant. We now impose the additional requirement that it be a function only of physically available data on the system's state. To this end, we express Eq. (10.4) in the discrete-time form

$$u(t) = \mu_t[z(t_0 + i \, \Delta t), i = 0, 1, \ldots, j; \bar{x}(t_0)] \tag{10.18}$$

where $t = t_0 + j \, \Delta t, j = 0, 1, \ldots, N - 1$.

EQUIVALENT PROBLEM

For a given N, the problem defined by Eqs. (10.5), (10.9), (10.15), and (10.18), where it is desired to determine the control sequence $\{u(t), t = t_0 + j \, \Delta t, j = 0, 1, \ldots, N - 1\}$ such that J_N is minimized is precisely the discrete stochastic linear regulator problem of Chap. 9.

On the other hand, in the limit as $\Delta t \to 0$ with N such that

$$N \, \Delta t = t_1 - t_0$$

it is the continuous stochastic linear regulator problem which was formulated in Sec. 10.2.

Consequently, we are now ready to apply the results of Chap. 9, and to consider the limiting behavior of the control algorithm which we so obtain.

10.4 OPTIMAL CONTROL

CONTROL ALGORITHM

For a given N, and for any $t = t_0 + j\,\Delta t, j = 0, 1, \ldots, N - 1$, application of the results in Chap. 9 to the above discrete-time problem leads to the following set of relations for the optimal control:

$$u(t) = S(t)\hat{x}(t|t) \tag{10.19}$$

$$W(t + \Delta t) = M(t + \Delta t) + A(t + \Delta t)\,\Delta t \tag{10.20}$$

$$S(t) = -[\Psi'(t + \Delta t, t)W(t + \Delta t)\Psi(t + \Delta t, t) + B(t)\,\Delta t]^{-1}$$
$$\Psi'(t + \Delta t, t)W(t + \Delta t)\Phi(t + \Delta t, t) \tag{10.21}$$

$$M(t) = \Phi'(t + \Delta t, t)W(t + \Delta t)\Phi(t + \Delta t, t) - \Phi'(t + \Delta t, t)W(t + \Delta t)$$
$$\Psi(t + \Delta t, t)[\Psi'(t + \Delta t, t)W(t + \Delta t)\Psi(t + \Delta t, t) + B(t)\Delta t]^{-1}$$
$$\Psi'(t + \Delta t, t)W(t + \Delta t)\Phi(t + \Delta t, t) \tag{10.22}$$

In arriving at this set of relations, we have utilized Eq. (9.80); Eqs. (9.46) to (9.48) of Theorem 9.2; and Eqs. (10.16) and (10.17).

The indexing on t in the computations for $S(t)$, as defined by Eqs. (10.20) to (10.22), is $t = t_0 + (N - 1)\,\Delta t, t_0 + (N - 2)\,\Delta t, \ldots, t_0 + \Delta t, t_0$. Computation is initiated with the boundary condition

$$W(t_0 + N\,\Delta t) = W(t_1)$$
$$= \Lambda + A(t_1)\,\Delta t \tag{10.23}$$

as seen from Eq. (10.16). The sequence of computations is, of course, the same as in Chap. 9.

From Eq. (10.20),

$$M(t) = W(t) - A(t)\,\Delta t \tag{10.24}$$

Substituting this result into the left-hand side of Eq. (10.22) and solving for $W(t)$, we get

$$W(t) = \Phi'(t + \Delta t, t)W(t + \Delta t)\Phi(t + \Delta t, t)$$
$$- \Phi'(t + \Delta t, t)W(t + \Delta t)\Psi(t + \Delta t, t)$$
$$[\Psi'(t + \Delta t, t)W(t + \Delta t)\Psi(t + \Delta t, t)$$
$$+ B(t)\,\Delta t]^{-1}\Psi'(t + \Delta t, t)W(t + \Delta t)\Phi(t + \Delta t, t)$$
$$+ A(t)\,\Delta t \tag{10.25}$$

Substituting into Eq. (10.25) from Eqs. (10.6) to (10.8), and expanding the result, we obtain

$$
\begin{aligned}
W(t) =\ & [I + F(t)\,\Delta t + 0(\Delta t^2)]' W(t + \Delta t)[I + F(t)\,\Delta t + 0(\Delta t^2)] \\
& - [I + F(t)\,\Delta t + 0(\Delta t^2)]' W(t + \Delta t)[C(t)\,\Delta t + 0(\Delta t^2)] \\
& \{[C(t)\,\Delta t + 0(\Delta t^2)]' W(t + \Delta t)[C(t)\,\Delta t + 0(\Delta t^2)] + B(t)\,\Delta t\}^{-1} \\
& [C(t)\,\Delta t + 0(\Delta t^2)]' W(t + \Delta t)[I + F(t)\,\Delta t + 0(\Delta t^2)] + A(t)\,\Delta t \\
=\ & W(t + \Delta t) + F'(t)W(t + \Delta t)\,\Delta t + W(t + \Delta t)F(t)\,\Delta t \\
& - W(t + \Delta t)C(t)[C'(t)W(t + \Delta t)C(t)(\Delta t)^2 \\
& + B(t)\,\Delta t + 0(\Delta t^3)]^{-1}C'(t)(\Delta t)^2 W(t + \Delta t) + A(t)\,\Delta t + 0(\Delta t^2) \\
=\ & W(t + \Delta t) + F'(t)W(t + \Delta t)\,\Delta t + W(t + \Delta t)F(t)\,\Delta t \\
& - W(t + \Delta t)C(t)[C'(t)W(t + \Delta t)C(t)\,\Delta t + B(t)]^{-1} \\
& \qquad\qquad C'(t)W(t + \Delta t)\,\Delta t + A(t)\,\Delta t + 0(\Delta t^2) \quad (10.26)
\end{aligned}
$$

Assuming that $\lim W(t + \Delta t)$ as $\Delta t \to 0$ exists and is finite, we see from Eq. (10.26) that

$$
\lim_{\Delta t \to 0} W(t + \Delta t) = W(t)
$$

Then, rewriting Eq. (10.26) in the form

$$
\begin{aligned}
W(t) - W(t + \Delta t) =\ & F'(t)W(t + \Delta t) + W(t + \Delta t)F(t) \\
& - W(t + \Delta t)C(t)[C'(t)W(t + \Delta t)C(t) \\
& \Delta t + B(t)]^{-1}C'(t)W(t + \Delta t)\,\Delta t \\
& \qquad\qquad\qquad + A(t)\,\Delta t + 0(\Delta t^2)
\end{aligned}
$$

dividing through by Δt, and taking the limit as $\Delta t \to 0$, we obtain the matrix differential equation

$$
\begin{aligned}
-\dot{W}(t) =\ & F'(t)W(t) + W(t)F(t) \\
& - W(t)C(t)B^{-1}(t)C'(t)W(t) + A(t) \quad (10.27)
\end{aligned}
$$

where t is now the continuous-time variable with $t_0 \le t \le t_1$.

From Eq. (10.23), we see that the appropriate boundary condition for Eq. (10.27) is

$$
W(t_1) = \Lambda
$$

which is obtained by considering the limit of the expression as $\Delta t \to 0$.

We consider next the limiting behavior of Eq. (10.21). Substituting into this equation from Eqs. (10.6) and (10.8), we have

$$S(t) = -\{[C(t)\,\Delta t + 0(\Delta t^2)]'W(t + \Delta t)[C(t)\,\Delta t + 0(\Delta t^2)] + B(t)\,\Delta t\}^{-1}$$
$$[C(t)\,\Delta t + 0(\Delta t^2)]'W(t + \Delta t)[I + F(t)\,\Delta t + 0(\Delta t^2)]$$
$$= -[C'(t)W(t + \Delta t)C(t)(\Delta t)^2 + B(t)\,\Delta t + 0(\Delta t^3)]^{-1}$$
$$[C'(t)W(t + \Delta t)\,\Delta t + 0(\Delta t^2)]$$
$$= -[C'(t)W(t + \Delta t)C(t)\,\Delta t + B(t) + 0(\Delta t^2)]^{-1}$$
$$[C'(t)W(t + \Delta t) + 0(\Delta t)]$$

In the limit as $\Delta t \to 0$, it then follows that

$$S(t) = -B^{-1}(t)C'(t)W(t) \qquad (10.28)$$

for $t_0 \leq t \leq t_1$.

Also, in the limit as $\Delta t \to 0$, Eq. (10.19) becomes

$$u(t) = S(t)\hat{x}(t|t) \qquad (10.29)$$

for $t_0 \leq t \leq t_1$. Hence, the optimal control for the continuous stochastic linear regulator problem is specified by Eqs. (10.27) to (10.29).

We see that determination of the feedback control matrix $S(t)$ requires solution of the system of first-order second-degree ordinary differential equations in Eq. (10.27) subject to the boundary condition $W(t_1) = \Lambda$. It is clear that the computation of $W(t)$ is carried out backward in time from t_1. Equation (10.27) is a matrix Riccati differential equation just like the one we encountered in the continuous optimal filtering problem.

It is obvious that $W(t)$ is symmetric, and, therefore, as in the filtering problem, specified by $n(n + 1)/2$ elements.

The reason for requiring that $B(t)$ in the performance measure in Eq. (10.3) be positive definite for $t_0 \leq t \leq t_1$ is now obvious from both Eqs. (10.27) and (10.28).

As we expected, we have a separation principle here. We see from Eqs. (10.27) and (10.28) that $S(t)$ is independent of the statistical parameters in the problem, namely, $P(t_0)$, $Q(t)$, and $R(t)$. If all the state variables can be measured exactly, we note that Eq. (10.29) becomes $u(t) = S(t)x(t)$. Clearly, determination of $S(t)$ is independent of the presence or absence of uncertainty.

In the stochastic case, of course, the effect of the control input must be included in the optimal filter. Hence, we have

$$\dot{\hat{x}} = F(t)\hat{x} + K(t)[z(t) - H(t)\hat{x}] + C(t)u(t)$$

for $t_0 \leq t \leq t_1$, where all of the terms have been defined previously.

10.5 PERFORMANCE MEASURE

Returning now to the equivalent discrete-time problem, and letting t denote the discrete-time index $t = t_0 + k\,\Delta t$, $k = 0, 1, \ldots, N - 1$, we have from Eqs. (9.84) and (9.85) that

$$V(t_1 - t) = E[x'(t)M(t)x(t)] + \alpha(t) \tag{10.30}$$

and

$$\alpha(t) = \alpha(t + \Delta t) + E[w'(t)\Gamma'(t + \Delta t, t)W(t + \Delta t)\Gamma(t + \Delta t, t)w(t)]$$
$$- E[\tilde{x}'(t|t)\Phi'(t + \Delta t, t)$$
$$W(t + \Delta t)\Psi(t + \Delta t, t)S(t)\tilde{x}(t|t)] \tag{10.31}$$

respectively. In Eq. (10.30), $V(t_1 - t)$ corresponds to V_{N-k} of Eq. (9.84) and is the value of the performance measure for optimal control over the interval $[t,t_1]$, where t is the discrete-time index defined above. We recall from Sec. 9.4 that Eq. (10.31) is subject to the boundary condition $\alpha(t_1) = 0$.

Substituting into Eq. (10.30) from Eq. (10.24), we obtain

$$V(t_1 - t) = E\{x'(t)[W(t) - A(t)\,\Delta t]x(t)\} + \alpha(t)$$

which in the limit as $\Delta t \to 0$ becomes

$$V(t_1 - t) = E[x'(t)W(t)x(t)] + \alpha(t) \tag{10.32}$$

under the assumption that

$$\lim_{\Delta t \to 0} \alpha(t)$$

exists. In Eq. (10.32), t is the continuous-time variable defined over the interval $[t_0,t_1]$.

Since we are primarily interested in the value of the performance measure for the entire interval of control, we set $t = t_0$ in Eq. (10.32) and have

$$V(t_1 - t_0) = E[x'(t_0)W(t_0)x(t_0)] + \alpha(t_0)$$

It now follows that

$$V(t_1 - t_0) = \text{tr } E[W(t_0)x(t_0)x'(t_0)] + \alpha(t_0)$$
$$= \text{tr } [W(t_0)P(t_0)] + \alpha(t_0) \tag{10.33}$$

where tr is the trace, $W(t_0)$ is obtained from the solution of Eq. (10.27), and $P(t_0)$, it is recalled, is the covariance matrix of the initial state.

Let us now examine the limiting behavior of Eq. (10.31) as $\Delta t \to 0$. We first rewrite the equation in the form

$$\alpha(t) = \alpha(t + \Delta t) + E\{\text{tr}\,[\Gamma'(t + \Delta t, t)W(t + \Delta t)\Gamma(t$$
$$+ \Delta t, t)w(t)w'(t)]\} - E\{\text{tr}\,[\Phi'(t + \Delta t, t)W(t + \Delta t)\Psi(t$$
$$+ \Delta t, t)S(t)\tilde{x}(t|t)\tilde{x}'(t|t)]\}$$
$$= \alpha(t + \Delta t) + \text{tr}\left[\Gamma'(t + \Delta t, t)W(t + \Delta t)\Gamma(t + \Delta t, t)\frac{Q(t)}{\Delta t}\right]$$
$$- \text{tr}\,[\Phi'(t + \Delta t, t)W(t + \Delta t)\Psi(t + \Delta t, t)S(t)P(t|t)] \quad (10.34)$$

where we have made the substitutions $E[w(t)w'(t)] = Q(t)/\Delta t$ and

$$E[\tilde{x}(t|t)\tilde{x}'(t|t)] = P(t|t)$$

The latter, we recall, is the filtering error covariance matrix.

Substituting into Eq. (10.34) from Eqs. (10.6) to (10.8) and regrouping terms, we obtain the result

$$\alpha(t) = \alpha(t + \Delta t) + \text{tr}\,\Big\{[G'(t)\,\Delta t + 0(\Delta t^2)]W(t + \Delta t)[G(t)\,\Delta t$$
$$+ 0(\Delta t^2)]\frac{Q(t)}{\Delta t}\Big\}$$
$$- \text{tr}\,\{[I + F(t)\,\Delta t + 0(\Delta t^2)]W(t + \Delta t)[C(t)\,\Delta t + 0(\Delta t^2)]$$
$$S(t)P(t|t)\}$$
$$= \alpha(t + \Delta t) + \text{tr}\,[G'(t)W(t + \Delta t)G(t)Q(t) + 0(\Delta t)]\,\Delta t$$
$$- \text{tr}\,[W(t + \Delta t)C(t)S(t)P(t|t) + 0(\Delta t)]\,\Delta t \quad (10.35)$$

From this equation, it is clear that

$$\lim_{\Delta t \to 0} \alpha(t + \Delta t) = \alpha(t)$$

Transposing $\alpha(t + \Delta t)$ to the left-hand side of the equation and dividing through by Δt, we have, in the limit as $\Delta t \to 0$, that

$$-\dot{\alpha} = \text{tr}\,[G'(t)W(t)G(t)Q(t)] - \text{tr}\,[W(t)C(t)S(t)P(t|t)] \quad (10.36)$$

for $t_0 \leq t \leq t_1$ where we have made use of the fact that

$$\lim_{\Delta t \to 0} S(t) \qquad \text{and} \qquad \lim_{\Delta t \to 0} P(t|t)$$

exist, with the former given by Eq. (10.28) and the latter being the filtering error covariance matrix for continuous optimal linear filtering. From Eq. (10.28),

$$C'(t)W(t) = -B(t)S(t)$$

Taking the transpose and recalling that both $W(t)$ and $B(t)$ are symmetric, we have

$$W(t)C(t) = -S'(t)B(t)$$

Utilizing this result, we express Eq. (10.36) as

$$\dot{\alpha} = -\operatorname{tr}\left[G'(t)W(t)G(t)Q(t)\right] - \operatorname{tr}\left[S'(t)B(t)S(t)P(t|t)\right] \tag{10.37}$$

for $t_0 \leq t \leq t_1$. We recall that the appropriate boundary condition for this equation is $\alpha(t_1) = 0$.

Solution of Eq. (10.37) gives $\alpha(t_0)$, which is required in Eq. (10.33) to evaluate $V(t_1 - t_0)$. It is seen that $\alpha(t)$ will consist of two terms analogous to those in $\alpha(k)$ in the discrete-time problem. The first is due to the system disturbance, the second to filtering error. Hence, the interpretation of Eq. (10.33) in terms of its components is the same as for the performance measure in the discrete stochastic linear regulator problem (see Example 9.4).

10.6 SUMMARY AND EXAMPLES

We summarize our results in the following theorem.

Theorem 10.1 *The optimal control for the continuous stochastic linear regulator problem is characterized by the set of relations*

$$u(t) = S(t)\hat{x}(t|t) \tag{10.38}$$

$$S(t) = -B^{-1}(t)C'(t)W(t) \tag{10.39}$$

$$\dot{W} = -F'(t)W - WF(t) + WC(t)B^{-1}(t)C'(t)W - A(t) \tag{10.40}$$

$$\dot{\hat{x}} = F(t)\hat{x} + K(t)[z(t) - H(t)\hat{x}] + C(t)u(t) \tag{10.41}$$

for $t_0 \leq t \leq t_1$, where $W(t_1) = \Lambda$ and $\hat{x} = \hat{x}(t|t)$ is the optimal filtered estimate of the system's state.

The value of the performance measure for the optimal control is

$$V(t_1 - t_0) = \operatorname{tr}\left[W(t_0)P(t_0)\right] + \alpha(t_0) \tag{10.42}$$

where $\alpha(t_0)$ is obtained from the solution of the differential equation

$$\dot{\alpha} = -\operatorname{tr}\left[G'(t)W(t)G(t)Q(t)\right] - \operatorname{tr}\left[S'(t)B(t)S(t)P(t|t)\right] \tag{10.43}$$

for $t_0 \leq t \leq t_1$, where $\alpha(t_1) = 0$ and $P(t|t)$ is the filtering error covariance matrix.

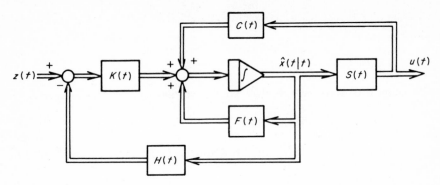

Fig. 10.3 Optimal control system for continuous stochastic linear regulator problem.

The block diagram for the optimal control system is given in Fig. 10.3.

The separation principle for the stochastic linear regulator problem was derived by Potter [18] in 1964. The work of Wonham [22] in 1968 yielded a separation principle for a much more general class of problems. In particular, the performance measure need not be quadratic, and the optimal feedback control need not be linear in the state or its optimal estimate.

We conclude this chapter with two example applications of Theorem 10.1.

Example 10.1 Let us consider the class of problems in which all the elements of the state vector of the system $\dot{x} = F(t)x + G(t)w(t) + C(t)u(t)$ can be measured exactly for $t_0 \leq t \leq t_1$, and in which the performance measure is given by Equation (10.3).

Since the state variables can be measured exactly, we have $\hat{x}(t|t) = x(t)$ with zero error, so that $P(t|t) = 0$ for $t_0 \leq t \leq t_1$. The optimal control is $u(t) = S(t)x(t)$ with $S(t)$ determined from Eqs. (10.39) and (10.40).

In order to evaluate the performance measure, we consider the first line in Eq. (10.33). Since $x(t_0)$ can be measured exactly, we see that

$$V(t_1 - t_0) = \text{tr } E[W(t_0)x(t_0)x'(t_0)] + \alpha(t_0)$$
$$= \text{tr } [W(t_0)x(t_0)x'(t_0)] + \alpha(t_0)$$

or, equivalently,

$$V(t_0 - t_0) = x'(t_0)W(t_0)x(t_0) + \alpha(t_0) \qquad (10.44)$$

Since $P(t|t) = 0$ for $t_0 \leq t \leq t_1$, Eq. (10.43) becomes

$$\dot{\alpha} = -\text{tr } [G'(t)W(t)G(t)Q(t)]$$

for $t_0 \leq t \leq t_1$, where $\alpha(t_1) = 0$ and $W(t)$ is the solution of Eq. (10.40). Integrating, we obtain

$$\alpha(t_0) = \text{tr} \int_{t_0}^{t_1} G'(t) W(t) G(t) Q(t) \, dt \tag{10.45}$$

which gives the component of the performance measure which is due to the system disturbance. It is obvious that if there were no system disturbance, we would have a purely deterministic problem with $\alpha(t_0) = 0$ and $V(t_1 - t_0) = x'(t_0) W(t_0) x(t_0)$.

As a specific case here, let us consider the scalar system $\dot{x} = w(t) + u(t)$ for $0 \leq t \leq T$, where $T = \text{constant} > 0$. We assume that $Q(t) = \sigma_w^2 > 0$ and take

$$J = E \int_0^T [\gamma^2 x^2(t) + u^2(t)] \, dt$$

as the performance measure where $\gamma = $ positive constant. The performance measure is seen to be a combination of weighted mean square error and control effort.

For this case, we have $F(t) = 0$, $G(t) = C(t) = 1$, $A(t) = \gamma^2$, $B(t) = 1$, and $\Lambda = 0$. Hence, Eq. (10.40) becomes

$$\dot{W} = W^2 - \gamma^2$$

for $0 \leq t \leq T$ with $W(T) = 0$. This equation can be solved by separation of variables to obtain

$$W(t) = \gamma \tanh \gamma(T - t) \tag{10.46}$$

Since $B(t) = C(t) = 1$, it follows from Eq. (10.39) that $S(t) = -W(t)$. The general behavior of the feedback gain as a function of time is indicated in the sketch in Fig. 10.4. We note that the peak magnitude of the gain occurs at $t = 0$ where $S(0) = -\gamma \tanh \gamma T$ and then monotonically decreases to zero. Since $\tanh \gamma T \leq 1$ for all $\gamma T > 0$, it follows that $|S(t)| \leq \gamma$ in all cases. Obviously, the greater the weighting of error relative to control effort in the performance measure, the greater the peak magnitude of the feedback gain.

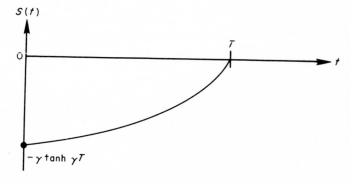

Fig. 10.4 Feedback gain as a function of time.

From Eq. (10.45), we have

$$\alpha(0) = \sigma_w{}^2 \int_0^T \gamma \tanh \gamma(T - t)\, dt$$

which, when evaluated, gives

$$\alpha(0) = \sigma_w{}^2 \log (\cosh \gamma T)$$

where the logarithm is to the base e. Substituting this result along with $W(0) = \gamma \tanh \gamma T$ into Eq. (10.44), we obtain

$$V(T) = \gamma x^2(0) \tanh \gamma T + \sigma_w{}^2 \log (\cosh \gamma T)$$

If γ is chosen large in order to place a heavy weighting on system error relative to control effort such that $\gamma T \gg 1$, we have

$$V(T) \approx \gamma x^2(0) + \sigma_w{}^2 \gamma T = \gamma[x^2(0) + \sigma_w{}^2 T]$$

In this situation, we see that the effect of the system disturbance on the performance measure is directly proportional to the time interval of control.

Example 10.2 We consider the scalar system $\dot{x} = -x + u(t)$ over the interval $[0,T]$ where the measurement is $z(t) = x(t) + v(t)$. We assume that $P(t_0) = \sigma_0{}^2 = 3$ and $R(t) = \sigma_v{}^2 = \frac{1}{2}$. Utilizing the results in Example 7.1 with $Q(t) = \sigma_w{}^2 = 0$, we have that the filtering error variance is

$$P(t|t) = \frac{3e^{-2t}}{4 - 3e^{-2t}}$$

and that the filter gain is

$$K(t) = 2P(t|t) = \frac{6e^{-2t}}{4 - 3e^{-2t}}$$

for $0 \le t \le T$.

For the control problem, we choose

$$J = E\left[\lambda x^2(T) + \int_0^T u^2(t)\, dt \right]$$

where $\lambda =$ constant > 0 as the performance measure. Equation (10.40) becomes

$$\dot{W} = 2W + W^2$$

in this example with $W(T) = \lambda$. Solving this equation by separation of variables, we obtain the result

$$W(t) = \frac{2\lambda}{\lambda + (2 - \lambda)e^{2(t-T)}}$$

Since $B(t) = C(t) = 1$, we see from Eq. (10.39) that the optimal feedback control gain is $S(t) = -W(t)$.

The optimal filter equation is

$$\dot{\hat{x}} = -\hat{x} + K(t)[z(t) - \hat{x}] + S(t)\hat{x}$$
$$= -[1 + K(t)]\hat{x} + K(t)z(t) + S(t)\hat{x}$$

for $0 \le t \le T$ with $\hat{x}(0|0) = 0$. Defining $L(t) = -[1 + K(t)]$, we have

$$\dot{\hat{x}} = L(t)\hat{x} + K(t)z(t) + S(t)\hat{x}$$

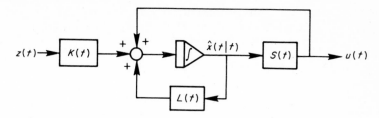

Fig. 10.5 Optimal control system for Example 10.2.

where

$$L(t) = -\left[1 + \frac{6e^{-2t}}{4 - 3e^{-2t}}\right]$$

$$= -\frac{4 + 3e^{-2t}}{4 - 3e^{-2t}}$$

The optimal control is $u(t) = S(t)\hat{x}(t|t)$, and the block diagram of the optimal control system assumes the form given in Fig. 10.5.

We evaluate $V(T)$ for the case where $\lambda = 2$; that is, terminal error is weighted twice as much as control effort in the performance measure.

In this instance, we have $W(t) = 2$ and $S(t) = -2$ for $0 \leq t \leq T$. Equations (10.42) and (10.43) become

$$V(T) = 2\sigma_0{}^2 + \alpha(0)$$
$$= 6 + \alpha(0)$$

and

$$\dot{\alpha} = -4P(t|t) = \frac{-12e^{-2t}}{4 - 3e^{-2t}} \tag{10.47}$$

respectively, with $\alpha(T) = 0$. Since there is no system disturbance, $\alpha(0)$ depends only upon the filtering error.

Integrating in Eq. (10.47) between the limits 0 and T, subject to the given boundary condition, we obtain

$$\alpha(0) = 2 \log (4 - 3e^{-2T})$$

Hence,

$$V(T) = 6 + 2 \log (4 - 3e^{-2T})$$

As a function of T, we see that the performance measure is monotone and increasing, varying between 6 and $6 + 2 \log 4 \approx 8.77$.

PROBLEMS

10.1. Consider the continuous deterministic linear regulator problem defined by the relations

$$\dot{x} = F(t)x + C(t)u(t)$$

$$J = x'(t_1)\Lambda x(t_1) + \int_{t_0}^{t_1} [x'(t)A(t)x(t) + u'(t)B(t)u(t)]\, dt$$

where $t_0 \leq t \leq t_1$. Assume that $x(t_0)$ is given, that $x(t)$ can be measured exactly, and that it is desired to determine $u(t)$ for $t_0 \leq t \leq t_1$ such that J is minimized.

(a) For a given $t \in [t_0, t_1]$ and $x(t) = x$, define

$$V(x,t) = \min_{u(\tau)} \left\{ x'(t_1)\Lambda x(t_1) + \int_t^{t_1} [x'(\tau)A(\tau)x(\tau) + u'(\tau)B(\tau)u(\tau)]\, d\tau \right\}$$

and utilize the principle of optimality to show that $V(x,t)$ must satisfy the partial differential equation

$$-\frac{\partial V}{\partial t} = \min_u [x'Ax + u'Bu + (\nabla_x V)\dot{x}]$$

for all $t \in [t_0, t_1]$, where $\nabla_x V$ is the gradient of V with respect to x, that is, the row vector $[\partial V/\partial x_1 \ \ldots \ \partial V/\partial x_n]$, and the boundary condition is $V(x,t_1) = x'(t_1)\Lambda x(t_1)$. In approaching this problem, divide the interval $[t,t_1]$ into the two intervals $[t, t + \Delta t]$ and $[t + \Delta t, t_1]$, and then utilize the principle of optimality. Assume that $V(x,t)$ is continuous and continuously differentiable for all x and t, $t_0 \leq t \leq t_1$. (This partial differential equation is a special case of the functional equation of dynamic programming for optimal control. The resulting equation after the minimization is performed is called the Hamilton-Jacobi-Bellman equation.)

(b) Performing the indicated minimization, show that the optimal $u(t)$ is

$$u(t) = -\tfrac{1}{2}B^{-1}(t)C'(t)[\nabla_x V(x,t)]'$$

and determine the Hamilton-Jacobi equation which $V(x,t)$ must satisfy.

(c) As a trial solution for the Hamilton-Jacobi equation in part b, let $V(x,t) = x'W(t)x$, where $W(t)$ is symmetric and show that $W(t)$ must satisfy Eq. (10.40),

$$\dot{W} = -F'(t)W - WF(t) + WC(t)B^{-1}(t)C'(t)W - A(t)$$

where $t_0 \leq t \leq t_1$ and $W(t_1) = \Lambda$. It then follows that $[\nabla_x V(x,t)]' = 2W(t)x$ and $u(t) = -B^{-1}(t)C'(t)W(t)x$.

(d) Show that a trial solution of the form $V(x,t) = \alpha(t) + x'\beta(t) + x'W(t)x$, where $\alpha(t)$ is a scalar function of time and $\beta(t)$ is an n-dimensional vector-valued function of time, leads to the results that $\alpha(t) = 0$ and $\beta(t) = 0$ for $t_0 \leq t \leq t_1$, and that $W(t)$ is the same as in part c above.

10.2. For a given arbitrary $u(t)$, over what variables is the expected value operation in Eq. (10.3) to be carried out?

10.3. Develop a flow chart for a digital computer solution of the continuous stochastic linear regulator problem including evaluation of the performance. Assume a stepsize of Δt in the computations.

10.4. Assuming that an arbitrary feedback gain matrix $S^*(t)$ rather then the optimal one is used in the control system in Figure 10.3, determine the expression or expressions which are required in order to evaluate the performance measure in Eq. (10.3).

10.5. In what way must the results in Theorem 10.1 be modified if $\{w(t), t \geq t_0\}$, the system disturbance, is a Gauss-Markov process instead of a gaussian white noise? Assume that $\{w(t), t \geq t_0\}$ is defined by the relation

$$\dot{w} = \Phi(t)w + \xi(t)$$

for $t \geq t_0$, where $\{\xi(t), t \geq t_0\}$ is a zero mean gaussian white noise which is independent of $w(t_0)$, $x(t_0)$, and $\{v(t), t \geq t_0\}$, and has covariance matrix $E[\xi(t)\xi'(\tau)] = Z(t)\delta(t - \tau)$, where $Z(t)$ is $p \times p$ and positive semidefinite. Assume also that $\Phi(t)$ is $p \times p$ and continuous for all t; that $w(t_0)$ is a zero mean gaussian random p vector with positive semidefinite covariance matrix $E[w(t_0)w'(t_0)] = W(t_0)$; that $w(t_0)$ is independent of $\{v(t), t \geq t_0\}$; and that $E[x(t_0)w'(t_0)] = Y(t_0)$, an $n \times p$ matrix.

10.6. Is the stochastic process $\{x(t), t_0 \leq t \leq t_1\}$ which is defined by the differential equation

$$\dot{x} = F(t)x + G(t)w(t) + C(t)S(t)\hat{x}(t|t)$$

where $S(t)$ is the optimal feedback gain matrix and $\hat{x}(t|t)$, the optimal filtered estimate of $x(t)$, is given, a Gauss-Markov process? Explain your answer.

10.7. Consider a first-order chemical reaction such as in Example 7.5, Eq. (7.70), in which it is desired to obtain a certain concentration c_0 at some specified time T by controlling the process over the interval $[0,T]$. Assume that the reaction can be characterized by the scalar differential equation

$$\dot{x} = -ax + u(t)$$

for $0 \leq t \leq T$, where $u(t)$ is the control variable, e.g., raw material flow rate. Assume that $x(0)$ is a gaussian random variable with mean $\bar{x}(0) \neq 0$ and variance σ_0^2.

Suppose that the concentration is monitored during the reaction and that the measurement can be modeled by the relation

$$z(t) = x(t) + v(t)$$

where $\{v(t), t \geq 0\}$ is a scalar zero mean gaussian white noise which is independent of $x(0)$ and has a variance $\sigma_v^2 = \text{constant} > 0$.

Taking

$$J = E\left\{\lambda[c_0 - x(T)]^2 + \int_0^T u^2(t)\,dt\right\}$$

where $\lambda = \text{constant} > 0$ as the performance to be minimized, determine the optimal control system for the reaction.

10.8. It is desired to maintain the angular velocity ω of a centrifuge, having a moment of inertia I about its spin axis, at a fixed value ω_0 over some specified time interval $[0,T]$. The angular velocity is measured by a rate sensor to within an error which can be taken as a zero mean gaussian white noise, and it is to be controlled by varying the applied torque. Assume that $\omega(0)$ is a gaussian random variable which is independent of the measurement error and has a mean value ω_0 and variance σ_0^2. Take

$$J = E\int_0^T \{a[\omega_0 - \omega(t)]^2 + u^2(t)\}\,dt$$

where $a = \text{constant} > 0$ and $u(t)$ is the applied torque as the performance measure and determine the control system for which J is minimized. List any additional assumptions which you make.

REFERENCES

1. Adorno, D. S., Optimum Control of Certain Linear Systems with Quadratic Loss, *Inform. Control*, vol. 5, p. 1, 1962.
2. Bellman, R. E., Dynamic Programming and Stochastic Control Processes, *Inform. Control*, vol. 1, p. 228, 1958.

3. ——, "Adaptive Control Processes—A Guided Tour," Princeton University Press, Princeton, New Jersey, 1961.

4. ——, On the Foundations of a Theory of Stochastic Variational Problems, *Proc. Symp. Appl. Math.*, vol. 13, Am. Math. Soc., Providence R.I., 1962.

5. Fel'dbaum, A. A., Dual Control Theory, *Automation Remote Control*, vol. 21, p. 1240, p. 1453, 1960; vol. 22, p. 3, p. 129, 1961.

6. ——, On Optimal Control of Markov Objects, *Automation Remote Control*, vol. 23, p. 993, 1962.

7. Fleming, W. H., Some Markovian Optimization Problems, *J. Math. Mech.*, vol. 12, p. 131, 1963.

8. Florentin, J. J., Optimal Control of Continuous-Time, Markov, Stochastic Systems, *J. Elec. Control*, vol. 10, p. 473, 1961.

9. ——, Partial Observability and Optimal Control, *J. Elec. Control*, vol. 13, p. 263, 1962.

10. Ho, Y. C., and R. C. K. Lee, A Bayesian Approach to Problems in Stochastic Estimation and Control, *Proc. 1964 Joint Automatic Control Conf.*, Stanford University, Stanford, California, p. 382, 1964.

11. Kalman, R. E., Control of Randomly Varying Linear Dynamical Systems, *Proc. Symp. Appl. Math.*, vol. 12, Am. Math. Soc., Providence, Rhode Island, 1962.

12. Krasovskii, N. N., and Z. A. Lidskii, Analytical Design of Controllers in Systems with Random Attributes, *Automation Remote Control*, vol. 22, p. 1021, p. 1141, p. 1289, 1961.

13. Kushner, H. J., Optimal Stochastic Control, *IRE Trans. Autom. Control*, vol. AC-7, p. 120, 1962.

14. ——, On the Dynamical Equations of Conditional Probability Density Functions with Applications to Optimum Stochastic Control Theory, *J. Math. Anal. Appl.*, vol. 8, p. 332, 1964.

15. ——, Some Problems and Some Recent Results in Stochastic Control, *1965 IEEE Intern. Convention Rec.*, pt. 6, p. 108, 1965.

16. McLane, P. J., and P. W. U. Graefe, Optimal Regulation of a Class of Linear Stochastic Systems Relative to Quadratic Criteria, *Int. J. Control*, vol. 5, p. 135, 1967.

17. Meditch, J. S., Near-Optimal Stochastic Linear Controls, Boeing document D1-82-0647, Boeing Scientific Research Laboratories, Seattle, Washington, September, 1967.

18. Potter, J. E., "A Guidance-Navigation Separation Theorem," AIAA/ION Astrodynamics, Guidance, and Control Conference, AIAA Paper No. 64-653, Los Angeles, California, August 24–26, 1964.

19. Schultz, P. R., An Optimal Control Problem with State Vector Measurement Errors, in C. T. Leondes, ed., "Advances in Control Systems—Theory and Applications," vol. 1, Academic Press Inc., New York, 1964.

20. Wonham, W. M., Stochastic Problems in Optimal Control, *Tech. Rept.* TR 63-14, Research Institute for Advanced Studies (RIAS), Martin Company, Baltimore, May, 1963.

21. ——, Lecture Notes on Stochastic Control, Lecture Notes 67-2, Center for Dynamical Systems, Division of Applied Mathematics, Brown University, Providence, Rhode Island, February, 1967.

22. ——, On the Separation Theorem of Stochastic Control, PM-38, NASA Electronics Research Center, Cambridge, Massachusetts, January 22, 1968. Also, *SIAM J. Control*, vol. 6 p. 312, 1968.

Name Index

Subject Index